Indian Mathematics

Engaging with the World
from Ancient to Modern Times

Indian Mathematics

Engaging with the World
from Ancient to Modern Times

George Gheverghese Joseph

University of Manchester, UK
National University of Singapore, Singapore
McMaster University, Canada

World Scientific

NEW JERSEY · LONDON · SINGAPORE · BEIJING · SHANGHAI · HONG KONG · TAIPEI · CHENNAI · TOKYO

Published by

World Scientific Publishing Europe Ltd.

57 Shelton Street, Covent Garden, London WC2H 9HE

Head office: 5 Toh Tuck Link, Singapore 596224

USA office: 27 Warren Street, Suite 401-402, Hackensack, NJ 07601

Library of Congress Cataloging-in-Publication Data

Names: Joseph, George Gheverghese.

Title: Indian mathematics : engaging with the world from ancient to modern times /
George Gheverghese Joseph (NUS, Singapore & University of Manchester, UK &
McMaster University, Canada).

Description: New Jersey : World Scientific, 2016. | Includes bibliographical references.

Identifiers: LCCN 2016000001| ISBN 9781786340603 (hc : alk. paper) |
ISBN 9781786340610 (pbk : alk. paper)

Subjects: LCSH: Mathematics--India--HIstory. | Mathematics--India. | Mathematics, Medieval.

Classification: LCC QA27.I4 J664 2016 | DDC 510.954--dc23

LC record available at http://lccn.loc.gov/2016000001

British Library Cataloguing-in-Publication Data

A catalogue record for this book is available from the British Library.

Desk Editors: V. Vishnu Mohan/Mary Simpson

Typeset by Stallion Press
Email: enquiries@stallionpress.com

Printed in Singapore

To Leela

Without whom this book would have just been a pipe dream

Contents

Acknowledgements

Like any book of this nature, this could not have been written without the help of many people, both past and present. The extensive references to their works in the text and bibliography would hopefully serve as a grateful acknowledgement of the debt owed. It was particularly fortuitous that while I was preparing this book there appeared a number of publications which were helpful in constructing the general narrative. They have been acknowledged as footnotes for their specific contributions in different chapters and their work referred to in various parts of this book. In particular, I would like to acknowledge the work of Madhukar Mallayya whose expertise and insight were useful in writing the chapters in the latter part of the book relating to Kerala mathematics and in the survey of the historical development of trigonometry that he carried out as a member of a research project of which the author was the Principal Investigator.

Dennis Almeida's contributions have been wide ranging as a member of the Research Project alluded to and an inspiration on issues relating transmissions and circulation of mathematical ideas across cultures. While researching this book, I came across an interesting initiative. The NPTEL (National Programme on Technology Enhanced Learning) has constructed a number of online courses in science and technology, including one entitled "Mathematics in India — From Vedic Period to Modern Times", by three stalwarts,

M. D. Srinivas, M. S. Sriram and K. Ramsubramaniam. Their focus is on the development of mathematical ideas and techniques with history and comparative analysis taking a secondary role. I found it a useful source of reference and illumination and wish to acknowledge my debt to the authors of this project. Finally, during the time I have been working on the book several people have given me advice, constructive criticism and encouragement. They have included Burjor Avari, Julian William, Dennis Almeida, Eddie D'Sa, Mary Searle-Chatterjee, Bill Farebrother, and the list goes on. It is appropriate given my insufficient response to some of the advice given that I exclude those mentioned above for any errors of fact and interpretation.

It would be very remiss of me not to acknowledge the help and patience of V. Vishnu Mohan and Mary Simpson who have been responsible for the efficient project management of this book.

George Gheverghese Joseph
University of Manchester, UK

Prelude

The genesis of this book can be traced to a growing recognition of three major global changes that occurred during the closing decades of the twentieth century. They are: (i) the rising economic power of the non-Western world; (ii) the growing crisis of planetary sustainability and (iii) the loss of authoritative sources of transcendence such as Marxism and religion These changes require us to revisit the paradigm of the 19th century that had sought to explain if not celebrate the unique rise of the West.

Two important concepts helped to modify the old paradigm: circulation and transcendence. 'Circulatory histories'* are now gaining acceptance following the analysis of historical data of transnational and trans-local flows. Emphasising circulatory histories over linear and bounded national or civilisational histories are now recognised as valid and need little justification. However, the acceptance of the idea of transcendence is another matter. Although it is not difficult to show that important changes in world history have in one way or another been influenced by transcendental sources of imagination, inspiration, commitment and resolve, the subject is handled with wariness in academic circles, who see them as

---

*Knowledge is not something to be discovered; it circulates through revelations from knowledgeable source to less knowledgeable source. Discoveries are therefore particular moments in a history of knowledge in circulation.

insufficient in themselves as explanations unless linked to societal structures and environmental conditions. It was the study of the origins, evolution and transmission of the zero that led me initially to see clearly how transcendence and circulation impinge and interact on one another.

A few years ago on a radio programme I was asked: "Where did zero originate?" Fortunately, I was allowed time to develop an answer without presuming, as many programmes are prone to do today, that listeners have the attention span of a grasshopper. Trying to gather my thoughts, I resorted to the familiar ploy of taking refuge in definitions. Now, if zero merely signifies a magnitude or a direction separator (i.e. separating 'above' the zero level from 'below' the zero level), the Egyptian zero (*nfr*), goes back at least four thousand years. If zero serves merely as a place-holder symbol, indicating the absence of a magnitude at a specified place position (as for example, the '0' in 101 shows the absence of any 'tens' in 101), then such a zero was already present in the Mesopotamian number system long before the first recorded occurrence of the Indian zero. If zero is shown by just an empty space within a well-defined positional number system, such a zero was present in Chinese mathematics some centuries before the Indian zero. The Chinese case showed that the absence of a symbol for zero did not prevent it from becoming an efficient computational tool that could handle even solutions of higher degree order equations involving fractions. However, the zero alluded to in the question asked was a multi-faceted mathematical object: a symbol, a number, a magnitude, a direction separator and a place-holder, all in one operating within a fully established positional number system. Such a zero has occurred only twice in history — the Indian zero and the Mayan zero which appeared in splendid isolation in Central America around the beginning of the Common Era.

To explain the appearance of the Indian or the Mayan zero, it is important to examine the context in which the two independent inventions occurred. From existing evidence, much of it fragmentary in the Mayan case, we surmise that both were numerate cultures with a passion for astronomy and its other half, astrology. As we will see later, the Indian culture from an early time showed interest in and even fascination for very large numbers for their own sake accompanied by a delight in calculations with these numbers. This may have been true of the Mayan culture as well. Both cultures were obsessed with the passage of time but in very different ways. The Indian interest was tied up with a widespread belief of a never-ending cycle of births and rebirths where the primary objective for individual salvation was to break this cycle. It was believed from very early time that this could be achieved in different ways, including offering sacrifices on specially constructed altars that conformed to specific shapes and sizes and carried out on auspicious days. And it was the problems of constructing these altars that gave birth to both practical and theoretical geometry in India. This is a theme explored in Chapter 3 of this book.

In the Mayan case, preoccupation with passage of time took the form of a great fear that the world would suddenly come to an end unless the gods (and especially the Sun God) were propitiated by sacrifice undertaken at auspicious times of the year dictated by specific astronomical occurrences. Human sacrifices were conducted on altars laid out for this specific purpose. The victims were chosen carefully for the possession of certain prized characteristics: beautiful virgins, high-born war captives, etc.

In both the Mayan and the Indian cases there was an essential need for accurate measurement of the passage of time, and hence the keeping of detailed calendars and an elaborate division into eras. The need for such precise calculations may have stimulated the development of efficient number systems with a fully developed zero. And it was but an accident of history and geography that the Indian zero prevailed while the Mayan zero eventually disappeared into oblivion.

The dissemination westwards (and then worldwide) of the Indian zero is an integral and well-known part of the history of Indian and later Indo-Arabic numerals. What is often emphasised is the *manner* in which the zero moved west from India to the Islamic world and then to various parts of Europe. And, given that such a movement did take place, the question arises as to whether the transmission of such a novel concept was constrained by cultural and linguistic filters operating in the recipient cultures. And did these filters subsequently inhibit our understanding of the concept of the Indian zero and the arithmetic of operations with that zero?[†] What we are aware of today is the widespread 'discomfort' of both adults and school children with the concept of zero itself and any mathematical operations involving zero. Consider, for example, what the usual response would be to the following questions:

- Is zero a positive or negative number?
- Is zero an odd or even number?
- Divide 2 by zero.

Even among university students of mathematics a discussion of these questions tends to be vague. There seems to be a singular absence of any attempt to inform students about 'calculating with zero'. This was a standard topic found in Indian texts on mathematics from the time of Brahmagupta (b. 598 CE).

In any systematic study of the history of mathematics both the methodological aspects of the transmission process as well as the pedagogical implications of an imperfect filtering process should be looked at carefully, especially wherever a claim is made that a mathematical object or method is 'borrowed' from one culture by another. It is particularly important in studying the history of Indian mathematics that any claim to Indian originality should be scrutinised both from a local and global perspective. The Indian zero provides a good example of how the concepts of 'circulation' and 'transcendence' serve as useful guides to a cross-cultural study of the history of mathematics.

This book is broadly divided into three parts for the benefit of readers with different backgrounds and interests. The first five chapters contain historical and cross-cultural emphasis. A short

---

[†]Cultural transmission of zero is not merely a *copying* process, but also a *reconstructive* process in which cognitive biases play an important role. A major bias that inhibits accurate transmission is a tendency for people to make different inferences regarding operations with zero and other mathematical objects (De Cruz and De Smedt, 2013).

survey of the history of Indian mathematics and its contacts with the outside world is contained in Chapter 2. The next six chapters are more mathematical, beginning with Aryabhata I (b. 476 CE). The last two chapters culminate with the arrival of modern mathematics. More advanced mathematics is contained in the Appendices.

Chapter 1 of the book, entitled 'Rewriting the History of Indian Mathematics: Some Outstanding Issues', will hopefully set the scene. It begins with an examination of the conflicting perspectives offered on the history of Indian mathematics[‡] and how these have bedevilled a proper study of the subject. This chapter proceeds to identify some real problems in studying the history of Indian mathematics, principally the lack of a consensus on the chronology of Indian mathematics before the fifth century CE. For example, one of the earliest astronomical texts of India, the *Vedanga Jyotisa*, has been variously dated as being composed from about 28,000 years ago to about 2,500 years ago! At the centre of this incredible divergence on dating is the vexed question of the validity of finding the dates of certain texts on the basis of astronomical, literary or other even more ephemeral evidence. The other issues highlighted in this chapter include those regarding the sources and interpretations of texts and of differing characterisation of Indian mathematical traditions.

Chapter 2 provides a short survey of the history of Indian mathematics and the nature and extent of *global* interactions of different mathematical traditions, and more specifically the impact of Indian mathematics on other mathematical traditions and vice versa. Chapters 1 and 2 constitute a non-technical introductory framework within which the more substantive content of the later chapters are presented.

Chapter 3 assesses some early evidence on Indian mathematics, notably from the Harappan and the Vedic periods (*c.* 2000–500 BCE), for the suggested links between these cultures (now that this has now become a matter of public debate with politicians joining in!), and

---

[‡]The term 'Indian mathematics' is used throughout this book to describe mathematics of the Indian sub-continent, including countries situated geographically in that area.

their possible links with the outside world. Chapter 4 contains a detailed look at the Jain and Buddhist inputs into the mathematical and scientific traditions of ancient India. This is not only interesting because of their contribution to early number theory and pure mathematics, but also in the case of the latter it was a vehicle through which certain mathematical and astronomical ideas were taken outside India. The resulting influence of Indian mathematics not only on the West but also on China, Korea, Tibet and other cultures south and east of India are touched upon in this book.

From the earliest times, mathematics in India was closely linked to a mixed bag of astronomy, 'calendrics' (i.e. constructing calendars) and astrology. Contemporary to or slightly after the Jain and Buddhist periods of Indian history were the beginnings of the astronomical tradition, preoccupied with the charting of the solar year, including solstices, equinoxes, lunar periods, solar and lunar eclipses and planetary movements. This was a period during which transmitted Hellenistic influences were believed to have gone into the shaping of Indian astronomy. Chapter 5 will examine only briefly evidence of this transmission since early Indian astronomy is of peripheral interest in this book.

Chapters 6–9 cover the period of Indian mathematical resurgence, beginning with Aryabhata I in the second half of the fifth century CE and ending with Bhaskara II (or Bhaskaracharya) and Narayana Pandita in the early centuries of the second millennium of the Common Era. Although the mathematics of this period remains the best explored area of Indian mathematics, this book provides not only a fuller account of the mathematics of that thousand years (500–1500 CE) but also a discussion of two issues often neglected: (i) the nature and the mechanics of the transmission of ideas from the sub-continent of India to the rest of the world and vice versa; (ii) the agents and the processes of dissemination of mathematical and astronomical ideas within the sub-continent, as evidenced by the increasing production of texts and translations into regional languages. The mathematical activities uncovered will hopefully alter the perception that mathematics in India after Bhaskara II made only "spotty progress until modern times".

A commonly held view of Indian mathematics is that if the Indians made any contribution to world mathematics, it was basically in what is often termed as 'elementary mathematics'. This perception needs further scrutiny. It is well known that two powerful tools contributed to the creation of modern mathematics in the seventeenth century: the discovery of the general algorithms of calculus and the development and application of infinite series techniques. These two streams of discovery reinforced each other in their simultaneous development since each served to extend the range of application of the other. But what is less known until recently is that the origin of the analysis and derivations of certain infinite series, notably those relating to the arctangent, sine and cosine, was not in Europe, but in an area in South India that now falls within the State of Kerala. From a region covering about 500 square kilometres north of Cochin and during the period between the 14th and 16th centuries, there emerged discoveries that predate similar work of Gregory, Newton and Leibniz by about 300 years.

There are several questions worth exploring about the activities of this group of mathematicians and astronomers from Kerala, (hereafter referred to as the 'Kerala School'), apart from the mathematical content of their work. In one substantial chapter and three Appendices (Chapter 10) we examine the social and historical context in which the Kerala School developed and the mathematical motivation underlying their interest in infinite series. We conclude our discussion of the Kerala School by identifying some circumstantial and other evidence to suggest that the mathematics underlying 'the calculus' in Europe could have travelled from Kerala to Europe through different agents, including priests, navigators, craftsmen and traders.

Traditional treatment of the history of Indian mathematics is significant for what it leaves out. What is often omitted is the flourishing mathematical tradition existing or introduced into India during the medieval times where the sources were Persian and Arabic texts or texts in regional vernaculars. A brief examination of these sources of mathematics will identify major differences between these streams of mathematical activity of those working within the Sanskrit tradition and the others. By comparing the research preoccupations of different groups roughly contemporaneous with one another — for example, the Kerala mathematicians who occupied themselves with work on infinite series inspired by Aryabhata and his School and mathematicians working within the Graeco–Arab–Persian tradition whose interests were primarily in Greek geometry and Ptolemaic astronomy — it becomes possible to identify the main epistemological differences between the two mathematical traditions. The two parallel traditions did meet in a few cases involving astronomy but hardly ever on matters relating to pure mathematics. And this lack of contact was a missed opportunity that had considerable repercussions for the development of Indian mathematics and astronomy. This will be the subject matter of Chapter 12.

There has been a tendency on the part of some historians of Indian mathematics to ignore the importance of astronomy in providing both the motivation and the instruments in the historical development of Indian mathematics. To list three examples of astronomy being the prime movers in the creation of new mathematical techniques: solution of indeterminate equations, emergence of trigonometric functions (notably sine function) and the growth of spherical trigonometry, and finally the discovery of infinite series. After all, practically all the Indian mathematicians that we know of today were also were astronomers: Aryabhata, Bhaskara I, Varahamihira, Brahmagupta, Bhaskara II, Madhava, Paramesvara, Nilakantha, Jyesthadeva, ... the list is many. To represent the relevance and the creative force of astronomy, a whole chapter (Chapter 11) is devoted to the handmaiden of astronomy, the subject of trigonometry.

The last chapter (Chapter 13) of the book will examine the various phases in the introduction of the modern Western mathematics into India. During the first phase, an attempt was made to incorporate ideas and concepts of modern mathematics into the indigenous mathematical tradition. However, especially after Macaulay's damning indictment of indigenous knowledge in 1835, the last vestiges of an indigenous education programme that tried to blend traditional learning with western ideas had virtually been wiped out by 1850. Western mathematics became the dominant strain acceptable to the rulers and to the ruled. But this did not prevent the emergence of exceptional individuals such as Master Ramchandra (b. 1860) who offered an approach to advanced algebra based on traditional methods, and Srinivas Ramanujan, about some 50 years later whose early interest in and approach to mathematics were probably influenced by indigenous tradition. What remains today are the remnants of traditional mathematics, mainly found today in India among practitioners of occupations such as traditional architecture and astrology.

In writing this book I am hoping to interest a wider audience, a readership hopefully as broad as that of the '*The Crest of the Peacock*'. And at the same time, the book contains a comprehensive survey of Indian mathematics exploring subjects and individuals who are often neglected. A bold historian reveals usually just as much about herself, and her driving motivations, as any actual disclosure of historical pattern. In the case of the *Crest*, it arose from a sense of outrage that people were being written out of history by a dominant culture that had imposed economic, military and psychological dominance over a vast area of the globe. In this book, which I see in some ways as a sequel, the outrage is replaced by a sense of amity to all. It is hoped that in locating Indian mathematics in a global context, drawing out parallel developments in other cultures and carefully examining the question of circulation of mathematical ideas across boundaries, the book will appeal to a wider public.

Chapter 1

Rewriting the History of Indian Mathematics: Some Outstanding Issues

1.1 What is 'Indian' in the History of Indian Mathematics?

A few years ago at the end of a talk, the author was asked by a member of the audience: 'You talk about Harappa culture being the beginnings of Indian civilisation. But the ancient cities of Harappa and Mohenjo-daro are in Pakistan. Shouldn't the culture be more accurately described as the beginnings of Pakistani civilisation?' It was a similar type of logic that may have led to a tongue-in-cheek remark by the Nobel Laureate, Amartya Sen, that Panini, the great Sanskrit grammarian, was an Afghani and not an Indian! The serious issue raised here is what does the 'Indian' in Indian mathematics signify?

As Palat (2002) has pointed out, to many outsiders until recently and especially in Europe, the word 'India' was virtually synonymous with Asia. To the ancient Greeks who gave us the terms 'Asia' and 'India', the latter signified the end of the inhabited world, a view that persisted as late as 1523, when Maximilian of Transylvania wrote, "the natives of all unknown countries are commonly called Indians"!

The term 'India' defined an amorphous entity, a geographical abstraction and a storehouse of dreams — dreams of untold wealth of precious metals, rare woods and of spices, fantasies regarding places peopled with mythical men and animals and monsters, dreams of abundance and fulfilled desires for a penniless and repressed Europe working off their inhibitions, and dreams that would fire the enthusiasm of explorers such as Vasco Da Gama and Columbus to sail over unknown seas in search of India and at each port of call leave a legacy of their misunderstanding by labelling the inhabitants that they came across as Indians.

India became a concrete socio-political reality and jurisdictional entity only after the British incursion when they named their colony as Indian Empire. The emergence of Pakistan as a political entity in the middle of the 20th century was in turn a result of a tidying up operation before the hasty retreat of the British. The splintering of Pakistan led to an independent Bangladesh. Therefore, in talking about 'Indian' mathematics, we are less concerned with a socio-political or jurisdictional entity and more with the geographical abstraction and a possible cultural continuity and unity over many millennia uniting many diverse people of the Indian sub-continent through religion, art, literature or customs. Often this is more apparent to an outsider than a native.

Indian mathematics, as shown in Figure 1.1, is best represented as a *mosaic* where influences over the ages have come from different sources, including Mesopotamia, Greece, China and the Islamic world. There have been home-grown influences, of which "dissenting" movements such as Jainism and Buddhism have been important in

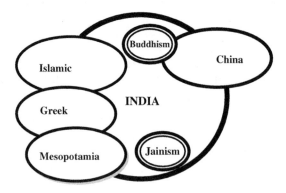

Figure 1.1. Indian mathematics: a mosaic.

shaping the nature and scope of Indian mathematics. In the course of this book we will examine in varying details both these 'external' and 'internal' influences.[1]

The characterisation of the mathematics of the Indian sub-continent raises other problems apart from one of definition. It may well be asked: Would it not be appropriate following the examples of labelling as 'Greek' or 'Islamic' mathematics to describe the mathematics of the Indian sub-continent as 'Sanskritic' mathematics or 'Hindu' mathematics since a number of the texts were written in some form of Sanskrit and most people involved in the creation or dissemination of the subject were 'Hindus' according to the present usage of that term?

The term 'Hindu' has never been one that the inhabitants of the Indian sub-continent have ever applied to themselves until recently. Historically, it derives from the Sindh River, now more widely known by the name 'Indus'. The label 'Hindu' as a religious category is of relatively recent origin. It emerged for a number of different reasons: a search for a religious identity to oppose Christian evangelism during the British rule, the British imperial agenda of 'divide and rule' that pitted the 'Hindu' against the 'Muslim', and the development of romantic notions regarding the spirituality and the other-worldliness of the 'Hindu'. It only became a political category after the partition of the Indian sub-continent in 1947. And as a social category, the term could well be used as an all embracing description of the sects, lifestyles, the agreements and disagreements found in the geographical boundaries between the Hindu Kush Mountains in the North and Kanyakumari (Cape Comorin) in the South. In that sense, the term encompasses everything and at the same time nothing, as do certain schools of 'Hindu' philosophy. And in the same sense all inhabitants of India, sharing common life styles and values, may be labelled 'Hindus'.

It is therefore tempting to use the term 'Hindu' mathematics to describe the mathematical tradition of the sub-continent. And all too often, outside India, some historians of mathematics have continued to do so, as for example in their continuing references to the 'Hindu-Arabic' rather than 'Indo-Arabic' numerals. It was decided not to use the label 'Hindu' in this book mainly because of the religious and political dimension to its usage today. As it is evident from later chapters, the mathematics of the Indian sub-continent came from different sources, including Buddhist, Jain and later Islamic groups. And while most seminal texts were written in Sanskrit, there were important works in regional languages, notably those from the states of South India and in Persian and Arabic in North India. For these reasons, we chose the label 'Indian' to highlight the mathematical activities of the sub-continent.

1.2   Other Problems in Studying the History of Indian Mathematics

Figure 1.2 pinpoints problems in studying the history of Indian mathematics. They may be broadly categorised as the four missing 'C's': 'Completeness', 'Continuity', 'Consensus' and 'Chronology'.

Briefly, the lack of 'Completeness' may be traced to: (i) the neglect of commentaries on the main texts[2]; (ii) the neglect of supplementary evidence found in religious and mythological sources, and (iii) the neglect of the 'material' basis in the creation and use of mathematical knowledge. The lack of 'Continuity' may have arisen as a result of: (a) 'biased scholarship', (b) an overemphasis on the value of written evidence over oral evidence, and (c) the discontinuities

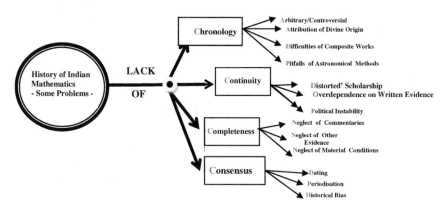

Figure 1.2. Problems of writing histories of Indian mathematics.

introduced by periods of political instability. And finally, the lack of 'Consensus' arose in the form of disagreements on dating and the division of history arbitrarily into distinct and identifiable periods, and which may be attributed to the overall neglect of inter-cultural and inter-temporal comparisons in constructing a time-line for Indian mathematics. We hope in the course of this book to explore and rectify where possible these resulting lacunae in historical scholarship.

An important objective of this book is to restore continuity in the history of Indian mathematics wherever possible. The first three missing 'C's' would impact on the critical problem of the fourth missing 'C', i.e. Chronology. They will be referred to at different points in this book; and the purpose of introducing them here is to give an early signal to the readers of the difficulties that they pose.

It is generally accepted that a well-conceived and consistent chronology constitute the basic frame around which a coherent history should be written. Unfortunately, in the case of India, both with regard to its pre-historic and ancient periods, the dates of some of the seminal events and of written sources are matters of serious contestations and diverse opinions.

Figure 1.3 below provides a broad chronology for Indian math-ematics consistent with the recommendations of the Chronology Committee set up in 1950.[3] It recognises five periods of Indian math-ematics, with the name of the mathematicians/astronomers and their works listed beside one another. The only omissions are the Persian-Arabic School of mathematicians and astronomers, who were more or less contemporaneous with their Kerala counterparts, and the mathematics of the different regions of India. We devote Chapter 12 to this fascinating but neglected areas of Indian mathematics.

1.3 Differing Perspectives on Indian Mathematics and Astronomy

In CE 1068, a Moorish historian from Spain, Said al-Andalusi (1029–1070), wrote a book entitled *Tabaqat al-Uman* (*Book of the Categories of Nation*).[4] In it he listed the accomplishments of

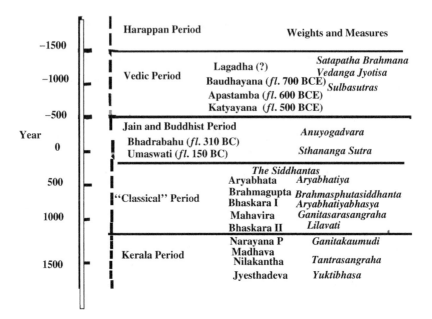

Figure 1.3. Chronology of Indian mathematics.

different 'nations'. Attributing the source of his information to some unnamed 'king of China' he wrote:

> The kings of the world are five in number and all the people of the world are their subjects. They are the king of China, the king of India, the king of the Turks, the king of the Furs [Persians] and the king of the Romans. The king of China is the king of humans because the people of China are more obedient to authority and are stronger followers of government diktats than all the other peoples of the world. The king of India is the 'king of wisdom because of the Indian's careful study of '*ulam*' [sciences] and their [contribution to the] advancement of all the branches of knowledge. The king of the Turks is the 'king of lions' because of the courage and the ferocity of the Turks. The king of Persia is the 'king of kings' because of the richness, glory and importance of his kingdom, and since Persia had subdued the kings of the populated centres of the world which were at the same time the most fertile of the climatic regions. And finally the king of the Romans is 'king of men' because the Romans, of all the peoples, have the most beautiful faces, the best-built bodies, and the most robust physiques.

Referring specifically to the Indians,[5] al-Andalusi added:

"The Indians, as known to all nations for many centuries, are the [embodiment] of wisdom, the source of fairness and objectivity. They are people of sublime pensiveness.... and (possess) useful and rare inventions...(To) their credit, the Indians have made great strides in the study of numbers and of geometry. They have acquired immense information and reached the zenith in their knowledge of the movements of the stars, and in unravelling the secrets of the skies as well as in their pursuit of other mathematical studies. In addition, they have surpassed all others in their knowledge of medical science and the efficacy of various drugs, the characteristics of compounds and the peculiarities of substances."

A more critical assessment of Indian mathematics and astronomy came from a slightly older contemporary of al-Andalusi, living in the eastern part of the Islamic world, the notable scholar Abu Rayhan al-Biruni (973–1048 CE). Taken to India as a captive by Sultan Muhamed Ghazni during his destructive invasion of that country, al-Biruni used his stay there to study Sanskrit and acquaint himself with the sciences, geography and culture of the Indians. His book, entitled *Kitab ta'rikh al-Hind* was translated in 1888 by Sachau as *Alberuni's India*. Al-Biruni's role in bringing Indian science to the notice of the wider world has been widely recognised, although not his imperfect knowledge of Sanskrit and his frequent confusion about the chronology and the content of standard texts and acquaintance with the controversies of the day.[6] Hence, the following statement by him[7]:

They (the Indians)...believe that there is no country but theirs,...no science like theirs. They are haughty, foolishly vain,...and stolid...If they travelled and mixed with other nations, they would soon change their mind, for their ancestors were not as narrow-minded as the present generation

In his final assessment of Indian science after making a number of positive comments on Indian astronomy and mathematics, al-Biruni compared Indian science to "a mixture of pearls and dung",[8]

a statement quoted extensively ever since by historians of science, particularly by those who have their own reservations about ancient Indian science. Nevertheless, a common element in these early evaluations is the insistence on the uniqueness of Indian mathematics. By the 19th century, however, and contemporaneous with the rise of European colonisation in the East, the views of European scholars developed ethnocentric overtones.

In 1875, Sedillot asserted that not only was Indian science indebted to Europe (p. 461) but also that the Indian numbers were an 'abbreviated form' of Roman numbers (p. 462) and that Sanskrit was merely 'muddled' Greek (p. 467). Although Sedillot's assertions may be judged today as being based on imperfect knowledge, they implied a clear assumption of the superiority of European science, the intellectual heir of Greek learning.[9]

In a similar vein Bentley (1823) also cast doubt on the chronology of India by locating Aryabhata and other Indian mathematicians several centuries later than was actually the case. He was of the opinion that 'Brahmins' had actively fabricated evidence to locate Indian mathematicians earlier than they existed.[10]

Inadequate understanding of Indian mathematics was not confined to earlier scholars. More recently, Smith (1923/25), an eminent historian of mathematics, claimed that, without the introduction of Western civilisation in the 18th and 19th centuries, India would have stagnated mathematically. He went on to add that: "Not since Bhaskara (i.e. Bhāskara II, b. 1114) has she produced a single native genius in this field" (Vol. 1, p. 435).

It is clear that Smith was either unaware of or ignored the works of scholars such as Whish (1835) and Warren (1825) who were among the first Westerners to acknowledge the achievements of the Kerala School.[11] Furthermore, the alleged hiatus in astronomical and mathematical activity after Bhāskara II (b. 1114) ignores some of the influential work produced subsequently, some of which we will be examining in later chapters.[12]

During this period astronomy and mathematics attained new heights in Kerala as we will see in Chapter 10. These advances have continued to be ignored: Edwards' text on calculus (1979) made no mention of the work of the Kerala School in his history of the calculus, or in articles on the history of infinite series by historians of mathematics such as Abeles (1993) and Fiegenbaum (1986). This neglect may be attributed to invincible ignorance or the deeply entrenched Eurocentrism that accompanied European colonisation over a wide range of activities, including the writing of the history of mathematics. The rise of nationalism in 19th century Europe and the consequent search for the roots of European civilisation, led to a preoccupation with Greece and the myth of Greek culture as the cradle of all knowledge and values. Europe then became heir to Greek learning and values. Rare exceptions to this skewed version of history were provided by Burgess (1860) and Peacock (1849).[13]

However, by the latter half of the 20th century European scholars had started to analyse the mathematics of the Kerala School using largely secondary sources of Rajagopal and his associates (1944, 1949, 1951, 1952, 1978, 1986) and Sarasvati Amma (1963, 1979). The achievements of the Kerala School and their chronological priority over similar developments in Europe were now being aired in several publications. (Baron, 1983; Katz, 1995; Calinger, 1999; Plofker, 2009). However, these evaluations seem often accompanied by a strong defence of the European claim for the invention of the generalised calculus.

For example, Baron (1969) states:

> The fact that the Leibniz–Newton controversy hinged as much on priority in the development of certain infinite series as on the generalisation of the operational processes of integration and differentiation and their expression in terms of a specialised notation does not justify the belief that the [Kerala] development and use for numerical integration establishes a claim to the invention of the infinitesimal calculus. (p. 65)

These comparisons appear to support Leibniz and Newton as inventors of the generalised infinitesimal calculus. But how justified is a qualitative comparison between two developments founded on different epistemological bases? It is worthwhile reiterating the point made later in the book that the initial development of the calculus in

17th century Europe followed the paradigm of Euclidean geometry in which generalisation was important and in which the concept of infinite remained a difficult issue (Katz, 1992; Scott, 1981).

On the other hand, from the 15th century onwards the Kerala mathematicians, following the different epistemology espoused by Aryabhata I in the late 5th century CE and employing computational methods, were able to grasp the notion of the infinitesimal and thus provide a rationale for explaining the infinite series and even the infinitesimal concepts in seminal texts such as the *Bijaganita* of Bhaskaracharya (12th century) and Jyesthadeva's *Yuktibhasa* (16th century).

A similar ambivalence regarding Indian science continued to more recent times, complicated by the emergence of the British Raj in India. The mathematical heritage of the Indian sub-continent in the minds of a number of historians contained a few highlights: the introduction of a system of decimal place value numerals incorporating zero, the use of negative numbers, the invention of the sine function and, more recently, the derivation of power series in Kerala. The rest showed a lacunae owing mainly to difficulties in grappling with primary sources, much of it presented in the unfamiliar garb of medieval Indian astronomical and astrological texts in highly compressed Sanskrit, and in the alienating context of the socio-economic background of Indian history. This problem arose quite early around the time of the first Indo–European contacts.

At present, a yawning gap is clearly discernible between the embarrassed silence and neglect of Indian mathematics on the part of Western and Indian scholars and the emerging triumphalism of certain Indians and their foreign camp-followers; between a view of Western scholarship as custodians and guardians of historical veracity and proportionate judgement of all things Indian and a politically-driven clarion call for 'retrospective privileging'[14] on the part of some Indians.

In the case of Indian scholarship, there has been a further problem. The historians of Indian mathematics working within India have generally tended to overlook the social and global context in which their subject arose. Instead they concentrate on advances made in 'real' mathematics. Often this has taken the form of writing about specific results in mathematics using modern terminology and notation. They argue that such results were known to Indians long before they were known in the West. At the same time social historians and anthropologists studying India seem more pre-occupied with issues of caste and religious divisions between communities and tend to overlook the social cohesiveness implied in the common numerate practices in the whole community. And between these two non-communicating groups, the mathematical activities of the common people have fallen through the cracks and therefore remain unacknowledged.[15]

Endnotes

[1] The 'Indian' in Indian mathematics is not merely defined by geography and history or what labels foreigners used. It could be argued that from the Vedic times onwards, there was a unity of ideas and methods that involved specific ways of oral and symbolic expression involving a form of recursive reasoning and avoidance of modes of argument such as indirect methods of proof, so favoured by the ancient Greeks. This will be discussed briefly in later chapters.

[2] The word *commentary* oversimplifies the complex strategy of knowledge production in traditional India. An original text was normally improved upon by discussion (dialectics or *tarka*) and then subjected to interpretation (*vyakhya*), commentary (*bhasya*), compilation (*samhita*) and analysis (*sangraha*). These terms, appended to the names of specific texts, indicate the refinements and critical scrutiny that the whole process of knowledge production went through.

[3] For further details on the Chronology Committee and their deliberations, see Gupta (1990).

[4] The quotations are from a translation of al-Andalusi's book by Salem and Kumar (1996), found on pp. 11–14.

[5] It is interesting that a thousand years later, a historian of science, Cohen (1994) changed this order of ranking civilisations from Rome and South-East Asia (those *without* science), India (one *with* some science), China and the Islamic world (those with advanced science) and at the top, Western civilisation, the only one that produced modern science.

[6] As evidence of al-Biruni's imperfect acquaintance with the controversies in India, consider the following quotation at the end of Chapter 26 of *India* where he refers to Brahmagupta's *Brahmasphutasiddhanta*.

Followers of Aryabhata maintain that the earth is moving and the heaven resting. People have tried to refute them by saying that, if such were the case, stones and trees would fall from the earth.

Al-Biruni then adds intriguingly: "There are, however, other reasons which make this impossible. This question is most difficult to solve." Al-Biruni's difficulty arose from his misunderstanding of what Brahmagupta had said, which was:

> If the Earth moves through one minute of arc in one respiration, from where does it start its motion and where does it go? And, if it rotates (at the same place), why do tall and lofty objects not fall down?

One can only infer that either al-Biruni's informant did not explain the passage correctly or that al-Biruni did not understand it himself, or that the passage was not transmitted properly in Arabic.

[7] It is likely that scores of Indians met al-Biruni, taught him Sanskrit, gave him manuscript copies, explained their contents and probably translated some passages. He rarely mentions names of these informants in India and hardly had any appreciative word for them although he is critical of the carelessness and the obscurantism of scribes that he had met. One can only assume al-Biruni, brought up on a diet of a methodology of Greco-Islamic Euclidean tradition, found it difficult to fathom the alien Sanskrit science and reacted negatively which set the trend for future. As we will see in a later chapter in a text, entitled "Treatise on the proportion rules of the Indians (or 'Rule of Three')," he shows a good understanding the different approaches to the subject of ratio between the Indian and Euclidean tradition.

[8] "I can only compare their mathematical and astronomical literature, as far as I know it, to a mixture of pearl shells and sour dates, or of pearls and dung or of costly crystals and common pebbles. Both kinds of things are equal in their eyes, since they cannot raise themselves to the methods of a strictly scientific deduction". It could be assumed that al-Biruni had difficulties with scientific Sanskrit expressed in verses (or *slokas*).

[9] "On one side, there is a perfect language, the language of Homer, approved by many centuries, by all branches of human cultural knowledge, by arts brought to high levels of perfection. On the other side, there is [in India] Tamil with innumerable dialects and that Brahmanic filth which survives to our day in the environment of the most crude superstitions" (p. 460).

[10] "We come now to notice another forgery, the *Brahma Siddhanta Sphuta*, the author of which I know (?). The object of this forgery was to throw Varaha Mihira, who lived about the time of Akber, back into

antiquity... Thus we see how Brahma Gupta, a person who lived long before Aryabhata and Varaha Mihira, is made to quote them, for the purpose of throwing them back into antiquity.... It proves most certainly that the *Braham Siddhanta* cited, or at least a part of it, is a complete forgery, probably framed, among many other books, during the last century by a junta of Brahmins, for the purpose of carrying on a regular systematic imposition". (p. 151)

For the record, Aryabhata was born in 476 CE and the dates when Varhamihira, Brahmagupta and Emperor Akbar lived were 505, 598 and 1550 respectively. So it is safe to suggest that Bentley's hypothesis was a product of either ignorance or Eurocentric fabrication. Nevertheless Bentley's altered chronology and attitude had the effect not only of lessening the achievements of the Indian mathematics but also of making redundant any conjecture of possible transmission to Europe.

[11] There are three East India employees involved in this interesting story of the discovery of Kerala mathematics. Charles Whish who became a Criminal Judge at Cuddaph in South Malabar and died in 1833 at the early age of 38, had reported as early as 1825 to the Madras Literary Society of the remarkable Kerala work, although the report was published by the Royal Asiatic Society in 1834, based on a paper he read in 1832. John Warren was the acting Director of the Madras Observatory between 1805–1811 and published *Kalasankalita* in 1825, a compendium of the different methods employed by the calendar-makers for reckoning time. In it he wrote that several infinite series were to be found in different parts of India. George Hynes was in the East India Medical Service before he was appointed as Secretary of the Committee of Public Instruction and was then closely associated with the Madras Literary Society. Hynes, one would suspect from the correspondence between the three people involved, acted as a brake in publicising the Indian origins of the infinite series, and expressed his doubts in somewhat intemperate terms. Presumably because of his relative seniority, both Whish and Warren were reluctant to oppose his view that the work on infinite series was communicated "by Europeans to some learned Natives in modern times." It took Whish over 10 years before stating in his paper that: "I have ascertained beyond a doubt that the invention of infinite series of these forms originated in Malabar." For further details, see U.K.V. Sarma *et al.* (2010, pp. 116–118).

[12] The following list is by no means exhaustive:

—Thakkura Pheru who wrote the *Ganitasarakaumudi* in the first quarter of the fourteen century;

—Narayana Pandita who composed a major mathematical treatise entitled the *Ganitakaumudi* in 1356;

—Mahendra Suri the composer of *Yantraraja* in 1370 containing a table of ninety Rsines (R = 3600);

—Jnanaraja who wrote the *Siddhantasundara* in 1503;

—Nityananda who wrote the *Siddhantaraja* in 1639, a Sanskrit translation of a Indo-Persian *zij*;

—Munisvara who authored the *Siddhantasarvabhauma* in 1646;

—Kamalakara who wrote the *Siddhantatatva viveka* in 1658;

—Jagannatha Samrat who wrote the *Rekhaganita* in 1718 (a translation of the Arabic version of 'Elements') and the *Siddhantasamrat* in 1732 (a translation of Ptolemy's *Almagest*).

Some of these texts will be referred to in later chapters

[13] They wrote:

> Professor Whitney seems to hold the opinion, that the Hindus derived their astronomy and astrology almost wholly from the Greeks... I think he does not give the Hindus the credit due to them, and awards to the Greeks more credit than they are justly entitled to (Burgess, 1860, p. 387).
>
> (I)t is unnecessary to quote more examples of the names even of distinguished men who have written in favour of a hypothesis [of the Greek origin of numbers and of their transmission to India] so entirely unsupported by facts (Peacock, 1849, p. 420).

[14] 'Retrospective Privileging' implies 'looking back' and granting a special place or privilege to a particular group or culture. The most powerful and sustained form of retrospective privileging has been the Eurocentric Vision, the centre-piece of a 'Greek Miracle' and the construction of the 'Dark Ages'. For details, see Joseph (2011).

[15] Attempts at trying to understand Indian mathematics in its own right have been bedevilled by a problem. Apart from the inadequacy of trying to fit it into the 'grand narrative' mode favoured by certain histories of mathematics, there has been a tendency to promote untested claims for the autonomy and antiquity of the Indian scientific tradition, in response to the earlier colonial disparagement of Indian intellectual traditions. Debates about history, religion and cultures have therefore become highly politicised as a result.

Chapter 2

Indian Mathematics in World Mathematics: An Overview

2.1 Indian Mathematics: Unravelling its Roots*

In a celebration of the extraordinary versatility of mathematics, the Indian mathematician Mahavira (*c.* 950 CE) wrote:

> In all activities, worldly or spiritual, *ganita* (i.e. calculation) is of the essence. In the art of love, in the performance of music, in dancing and drama, in the art of cooking, in the practice of medicine and architecture, in poetic composition, in the rules of logic and grammar, in this and many other pursuits, *ganita* has its place. In relation to the movement of the sun and other heavenly bodies, in calculating eclipses and the conjunction of planets, in estimating the *triprasna* (i.e. the direction, position and time) of the moon — indeed in all these [*ganita*] is used. The number, the diameter and perimeter of islands, oceans and mountains, of the inhabitants who reside there, the lengths of their life, their eight attributes, all these are found by the application of *ganita* ... Why keep talking at length? In all three worlds involving moving and non-moving entities, there is nothing that can be (done) without the science of calculation (*ganita*). [*Ganitasarasangraha*: Chapter 1, verses 9–16]

Indian mathematics from its very beginning saw its primary function as generating rules for systematic and efficient methods of

---

*The reader is encouraged to look at Figure 1.3 in the last chapter for a chronological framework for this and subsequent chapters.

calculation. This was so from the time of the first recorded evidence of mathematics going back to the Vedic period.

2.1.1   Geometry of vedic altars

A widely held view is that Indian mathematics originated in the service of religion. Support for this view is sought in the complexity of motives behind the recording of the *Sulbasutras*, the earliest written mathematical source available today, dated conservatively around 800 to 500 BCE, dealing with the measurement and construction of sacrificial altars. This view ignores the skills in mensuration and practical arithmetic that must have existed in the Harappan (or Indus Valley) culture dating to at least 2500 BCE.[1] Even a superficial study of the Harappan cities shows its builders were extremely capable town planners and engineers possessing a fair knowledge of practical geometry and applied mathematics in general.[2]

The *Sulbsutras* contain instructions for the construction of sacrificial altars (*vedi*) and for the location of sacred fire (*agni*) that had to conform to clearly laid down requirements regarding their shapes and areas if they were to be effective instruments of sacrifice. There were two main types of ritual, one for worship at home and other for communal worship. Square and circular altars were usually sufficient for household rituals, while more elaborate altars of shapes consisting of combinations of rectangles, triangles, and trapeze were required for public worship. A more elaborate public altar was shaped like a giant falcon about to take flight (*vakrapaksa-syenacit*), as shown in Figure 3.7 in Chapter 3. It was believed that offering a sacrifice on such an altar would enable the soul of the supplicant to be conveyed straight to heaven.

In the construction of larger altars conforming to certain shapes and prescribed areas or perimeters, two geometrical problems arose. One was the problem of finding a square equal in area to two or more given squares; the other was the problem of converting other shapes (e.g., a circle or a trapezium or a rectangle) into a square of equal area or vice versa. These constructions, discussed in Chapter 3, were achieved through a judicious combination of concrete geometry (in particular, the principle of "dissection and re-assembly"), ingenious algorithms, and the application of the so-called Pythagorean theorem. The existence of early Indian geometry found in the *Sulbasutras*, provides a telling argument against the idea of a "Greek miracle" which has been a dominant theme in Western historiography for so long. Seidenberg (1962, 1983) and Van der Waerden (1983) have pointed out that the geometries of early Greece and ancient India have a common source given certain inherent similarities. In both cases, altar constructions had ritual significance in their insistence on altars conforming to specific shapes and magnitudes, and in their preoccupation with the problem of constructing a square equal in area to a given rectangle (or two other squares). Both cultures arrived at similar solutions using the Pythagorean theorem. However, there was a crucial difference from the beginning between the two geometries: in India, unlike Greece, it was the area of the altar and not its volume that was important from a ritual perspective.[3]

The progress of pure geometry, after its promising beginning during the Vedic period, appears patchy. But there was one area in which the Indian contribution was significant; and that was in the study of the properties of a cyclical quadrilateral (a quadrilateral inscribed in a circle). In the *Brahmasphutasiddantha*, Brahmagupta (b. 598 CE) gave correct results for the area and diagonals of cyclical quadrilaterals. This interest continued for about a thousand years with notable refinements provided by Narayana Pandita in his *Ganitakaumudi* in the 14th century, in Paramesvara's *Lilavatibhasya* in the 15th century and Jysthadeva's *Yuktibhasa* in the 16th century.

2.1.2   The origins of Indian numerals

If you ask anyone to name the single-most contribution that India made to world mathematics, the answer would be our number system and particularly the zero (usually dated around the first few centuries of the Common Era). In the perceptive words of the French mathematician, Pierre-Simon Laplace (1749–1827):

> The ingenious method of expressing every possible number using a set of 10 symbols (each symbol having a place value and an absolute value) emerged in India. The idea would seem so simple nowadays that its significance and profound importance is no longer appreciated. Its simplicity lies in the way it facilitated calculation and placed arithmetic foremost amongst useful inventions. The importance of this invention is more readily appreciated when one considers that it was beyond the two greatest men of Antiquity, Archimedes and Apollonius.

Fascination with large numbers has remained an abiding characteristic of Indian civilisation. As early as 800 BCE, there appeared names for powers of 10 up to 62. The importance of number-names in the evolution of the decimal place-value notation in India cannot be overstated. The word-numeral system was the logical outcome of proceeding by multiples of 10.[4]

Such a system presupposed a scientifically based vocabulary of names in which the principles of addition, subtraction and multiplication could also be incorporated. The system required: (i) the naming of the first nine digits (*eka, dvi, tri, catur, pancha, sat, sapta, asta, nava*); (ii) obtaining a second group of nine numbers by multiplying each of the nine digits by ten (*dasa, vimsat, trimsat, catvarimsat, panchasat, sasti, saptati, astiti, navati*); and (iii) constructing a third group of numbers of increasing integral powers of 10, starting with hundred (10^2) and going up to trillion (10^{12}) [*sata, sahasara, ayut, niyuta, prayuta, arbuda, nyarbuda, samudra, madhya, anta, parardha*].

In this, an early system of word-numerals, the place-value principle is clearly present. The number, one thousand nine hundred and ninety four can be expressed by words or word combination showing $[(1 \times 1000) + (9 \times 100) + (9 \times 10) + 4]$. Compare this with the Roman representation, MCMLXXXXIV, in which only the principles of addition and subtraction are incorporated in the representation of numbers [i.e. $(2000 - 100) + (50 + 10 + 10 + 10 + 10) + (5 - 1)$].

To understand the persistence of word-numerals in Indian mathematics even after the spread of the place-value system of

Indian numerals,[5] it is necessary to recognise the importance of the oral mode of preserving and disseminating knowledge. A common characteristic of written texts in India from time immemorial was the *sutra* style of writing which presented information in a cryptic form, leaving out details and rationale to be filled in by teachers and commentators. In short pithy statements, often expressed in verse, a *sutra* enabled the reader to memorise the content easily.

As a replacement for the older word-numeral system consisting of names of numbers, a new system was devised to help versification. In an early system, known as *bhuta-samkhya*, numbers were indicated by well-known objects or ideas commonly associated with the numbers. Thus, *zero* was represented by *sunya* (void) or *ambara akasa* ('heavenly space') or . . . ; *one* by *rupa* (moon) or *bhumi* (earth) or . . . ; *two* by *netra* (eyes) or *paksha* (waxing and waning of the moon) or . . . ; *three* by *kala* (time: past, present and future) or *loka* (heaven, earth and hell) or . . . , and so on. With multiple words available for each number, the choice of a particular word for a number was dictated by literary considerations. This form of notation continued for many years in both secular and religious writings because it was aesthetically pleasing and offered a useful way of memorising numbers and rules.[6]

There were two major problems with the *bhuta samkhya* system. First, there is an "exclusionist" element in that to decode the words for their numerical values required familiarity with philosophical and religious texts from which the correspondences were often established in the first place. Second, at times the same word stood for two or more different numbers, particularly where some writers had their own preferences when it came to choosing words to correspond to numbers, as for example, when *paksa* has been used for 2 as well as 15 and *dik* for 8, 10 as well as 4.

A third system of numerical notation originated in the work of Aryabhata (b. 476 CE). In his *Aryabhatiya*, there is an alphabetic scheme for representing numerals, based on distinguishing between classified (*varga*) and unclassified (*avarga*) consonants and vowels in Sanskrit.[7] The *vargas* fall into five phonetic groups: *ka-varga* (guttural), *ca-varga* (palatal), *ta-varga* (lingual), *ta-varga* (dental) and *pa-varga* (labial). Each group has five letters associated to it, where the letters run from k to m in the Sanskrit alphabet, representing numbers from one to 25. There are seven *avargas* consisting of semi-vowels and sibilants representing numerical values $30, 40, 50, \ldots, 90$. An eighth *avarga* was used to extend the number to the next place value. The ten vowels denoted successive integral powers of 10 from 100 onwards.

This form of representation, compared to *bhuta-samkhya*, had the advantage of brevity and clarity but the disadvantage of having limited potential for formations of words that are pronounceable and meaningful, both necessary requirements for easy memorisation. For example, according to the Aryabhata's astronomical system the representation of the number of revolutions of the moon in a *yuga* (57 753 336) is the unpronounceable and meaningless word *cayagiyinusuchlr*![8]

From a refinement of this alphabet–numeral system of notation emerged the *katapayadi* system which the founder of the Kerala school of astronomy, Varurici, popularised around the seventh century CE. In this system, every number in the decimal place-value system is represented by a word, each letter of the word represents a digit. The Sanskrit consonants क (k) to झ (jh) indicate one to nine, and so does त (t) to ध (dh); प (p) to म (m) stand for one to five and य (y) to ह (h) for one to eight.[9] A vowel not preceded by a consonant stands for zero but vowels following consonants have no special value. In the case of conjunct consonants, only the last consonant has a numerical value. Number-words are conventionally read from right to left so that the letter denoting the "units" is given first and so on.[10]

2.1.3   Literacy and numeracy: the resulting synergy

The close relationship between literacy and numeracy, implied by such varied systems of numerical notation, may find its roots in the way that Sanskrit developed in its formative period after its separation from other languages of the Indo-European family. A long tradition of oral communication of knowledge left a singular mark on the nature and transmission of knowledge, whether religious or scientific, in Indian culture. After many years, as Sanskrit became a written language, three kinds of Sanskrit developed with descending degrees of artificiality: grammatical, logical and mathematical Sanskrit.[11]

Mathematical Sanskrit remained the least artificial of the three kinds of written Sanskrit, with the greatest artificiality found in the development of *grammatical* Sanskrit by Panini and Patanjali, followed 500 years later by the *logical* Sanskrit of *Nyaya* that culminated a 1000 years later in *Navya-Nyaya*.[12] This has some important implications for a comparative study of the historical development of Indian and Western mathematics which can only be touched upon here.[13] The chronological order of the development of artificial scientific languages in the West is a reversal of the Indian experience with logic following mathematics and linguistics being a late developer. In India, mathematical Sanskrit never quite became an artificial language, although it employed abbreviations and artificial notations as shorthand for practical procedures. And logical Sanskrit never became, like its Western counterpart, an important adjunct to 'mathematical philosophy'. Recognising differences between the two traditions help to solve the puzzle of why mathematics developed so differently.

2.1.4 The appearance of zero

We have already noted in the Prelude that possibly the greatest contribution of India to world mathematics is the place value system of written numbers (numerals) incorporating a zero, the ancestor of our world-wide system of number representation. There are historical records of only three other numbers systems which were based on the positional principle. Predating all other system was the Babylonian system that evolved around the third millennium BCE. A sexagesimal (i.e. base 60) scale was employed with a simple collection of an appropriate number of symbols to write numbers less than 60. But it was imperfectly developed being partly additive and partly 'place-valued', for within the base of 60, a decimal system was used. Indeed, the absence of a symbol for zero until the early Hellenistic period limited the usefulness of the system for computational and representational purposes.

The Chinese rod numeral system was essentially a decimal base system. The numbers $1, 2, \ldots, 9$ are represented by rods whose orientation and location determine the place value of the number represented and whose colour show whether the quantity was positive (red) or negative (black). In terms of computation, the representation of zero by a blank space posed no problem for, unlike the Babylonian system, the blank was itself a numeral.

The third positional system was the Mayan, essentially a vigesimal (base 20) system incorporating a symbol for zero which was recognised as a numeral in its own right. It had, however, a serious irregularity since its 'place' units were $1, 20, 18 \times 20, 18 \times 20^2, 18 \times 20^3$, and so on. This anomaly reduces its efficiency in arithmetical calculation. For example, one of the most useful facilities of our number system is the ability to multiply a given number by 10 by adding a zero to it. An addition of a Mayan zero to the end of a number would not in general multiply the number by 20 because of the mixed base system employed.[14]

The discussion so far of the different positional systems highlights two important points. First, a place-value system could and did exist without any symbol for zero. But the zero symbol as part of the numerical system never existed and could not have come into being without place value. Second, the relative strength of the Indian number system is inextricably tied up with the Indian concept of zero. A further examination of the origins and use of zero in Indian mathematics will be found in Chapter 4.

2.1.5 *Ganita* (i.e. Mathematics) comes of age

The Vedic religion with its sacrificial rites began to wane and other religions began to replace it. One of these was Jainism, a dissenting religion and philosophy which was founded in India around the sixth century BCE. Although the period after the decline of the Vedic religion up to the time of Aryabhata I around 500 CE used to be considered as a 'Dark Age' in Indian mathematics, recently it has been recognised as a time when many mathematical ideas were maturing. In fact Aryabhata is now thought of as summarising earlier mathematical developments as well as beginning the next phase in the development of Indian mathematics. The main topics of Jain mathematics in around 150 BCE were: the theory of numbers, arithmetical operations, geometry, operations with fractions, simple equations, cubic equations, bi-quadric equations and permutations and combinations. More surprisingly, the Jains developed a theory of the infinite containing different levels of infinity, showed a primitive understanding of the theory of indices and had some notion of logarithms to base 2. These are discussed in Chapter 4.

One of the difficult problems facing historians of mathematics is deciding on the date of the Bakhshali manuscript. If this is a work which is indeed from 400 CE, or at any rate a copy of a work which was originally written at that time, then our understanding of the achievements of earlier mathematics (pure and applied) will be greatly enhanced. Where there is uncertainty over the date of the Bakhshali manuscript, we should avoid rewriting the history of the Jain period in light of the mathematics contained in this remarkable document.

If the Vedic religion gave rise to a study of mathematics for constructing sacrificial altars, then it was Jain cosmology which led to ideas of the infinite in Jain mathematics. Later mathematical advances were often driven by the study of astronomy. Perhaps it would be more accurate to say that astrology was the driving force since it was that which required accurate information about the planets and other heavenly bodies. Religion too played a major role in astronomical investigations in India for accurate calendars had to be prepared to allow religious observances to occur at the correct times. Mathematics became an applied science in India for many centuries with practitioners developing methods to solve practical problems.[15]

By about 500 CE the 'classical' era of Indian mathematics began with the work of Aryabhata I. His work was both a summary of the mathematics known earlier and the beginning of a new era for astronomy and mathematics. His ideas of astronomy were truly revolutionary. He replaced the two demons, the *Dhruva Rahu* believed to have engineered the phases of the Moon and the *Parva*

Rahu who caused an eclipse by 'covering' the Moon or Sun, with a modern theory of eclipses He introduced trigonometry in order to carry out his astronomical calculations, based on the Greek epicycle theory; and he obtained integer solutions for indeterminate equations that arose from investigation of astronomical theories.[16]

2.1.6   Early Indian algebra

The beginnings of Indian algebra may be traced to the *Sulbasutras* and later the *Bakhshali Manuscript,* for both contain simple examples involving the solutions of what we would now label as simple and simultaneous equations.[17] But it was only from the time of Aryabhata that algebra grew into a distinct branch of mathematics. Brahmagupta (b. 598 CE) called it *kuttaka ganita*, or simply *kuttaka,* which later came to refer to a particular branch of algebra dealing with methods of solving indeterminate equations to which the Indians made early and substantial contribution.

An important feature of early Indian algebra which distinguishes it from other mathematical traditions was the use of symbols such as the letters of the alphabet to denote unknown quantities. It is this very feature of algebra that one immediately associates with the subject today The Indians were probably among the first to make systematic use of this method of representing unknown quantities. A general term for the unknown was *yavat tavat,* shortened to the algebraic symbol *ya*. In Brahmaguptas's work appears Sanskrit letters and abbreviations of names of different colours used to represent several unknown quantities. For example, the letter *ka* stood for *kalaka*, meaning 'black', and the letter *ni* for *nilaka* meaning 'blue'. With an efficient and well-established numeral system and the beginnings of syncopated algebra, the Indians were able to solve determinate and indeterminate equations of first and second degrees and involving in certain cases more than one unknown. It is possible that a number of these methods reached the Islamic world before being transmitted further westwards by a similar process and often involving the same actors as the ones that were alluded to earlier.

2.1.7   The Indian contribution to trigonometry

The beginnings of a systematic study of trigonometry are found in the works of the Alexandrians, Hipparchus (*c.* 150 BCE), Menelaus (*c.* 100 CE) and Ptolemy (*c.* 150 CE). However, from about the time of Aryabhata I, the character of the subject changed to resemble its modern form. Later, it was transmitted to the Islamic world who introduced further refinements. The knowledge then spread to Europe, where a detailed account of trigonometry is found in a book entitled *De triangulis omni modis*, by Regiomontanus (1464)

In early Indian mathematics, trigonometry formed an integral part of astronomy. References to trigonometric concepts and relations are found in astronomical texts such as *Suryasiddhanta* (*c.* 400 CE), Varahmihira's *Panchasiddhanta* (*c.* 500 CE), Brahmagupta's *Brahmasphutasiddhanta* (628 CE) and the influential work of Bhaskara II entitled *Siddhantasiromani* (1150 CE). Infinite expansion of trigonometric functions, building on earlier work, formed the basis of the development of mathematical analysis — a precursor to modern calculus.

Basic to modern trigonometry is the sine function. It was introduced into the Islamic world from India, probably through the astronomical text, *Suryasiddhanta*, brought to Baghdad during the eighth century CE. There were two types of trigonometry available then: one based on the geometry of chords best exemplified in Ptolemy's *Almagest*, and the other based on the geometry of semichords which was an Indian invention. The Arabs chose the Indian version. It is possible that two other trigonometric functions — the cosine and versine functions — also originated from the Indians.[18]

One of the important problems of ancient astronomy was the accurate prediction of eclipses. In India, as in many other countries, the occasion of an eclipse had great religious significance, when rites and sacrifices were performed. It was a matter of considerable prestige for an astronomer to demonstrate his skills dramatically by predicting precisely when the eclipse would occur.

In order to find the precise time at which a lunar eclipse occurs, it is necessary first to determine the true instantaneous motion of the moon at a particular point in time. The concept of instantaneous motion and the method of measuring that quantity is found in the works of Aryabhata I, Brahmagupta and Manjula (*fl.* 930 CE). However, it was in Bhaskaracharya's (or Bhaskara II's) attempt to work out the position angle of the ecliptic, a quantity required for predicting the time of an eclipse, that we have early notions of differential calculus: namely, a concept of an "infinitesimal" unit of time, an awareness that when a variable attains the maximum value its differential vanishes, and also traces of the "mean value theorem" of differential calculus, the last of which was explicitly stated by Paramesvara (1360–1455) in his commentary on Bhaskara's *Lilavati*. Others from Kerala (South India) continued this work with Nilakantha (1443–1543) deriving an expression for the differential of an inverse sine function and Acyuta Pisharti (*c.* 1550–1621) giving the rule for finding the differential of the ratio of two cosine functions.[19]

However, the main contribution of the Kerala School of mathematician-astronomers was the study of infinite-series expansions of trigonometric and circular functions and finite approximations for some of these functions. A motivation for this work was the necessity for accuracy in astronomical calculations. The Kerala discoveries include the Gregory and Leibniz series for the inverse tangent, the Leibniz power series for π, the Newton power series for the sine and cosine, as well as certain remarkable rational approximations of trigonometric functions, including the well-known Taylor series approximations for the sine and cosine functions. And these results had been obtained about 300 years earlier than the mathematicians after whom the results are now named. Referring to the most notable mathematician of this group, Madhava (*c.* 1340–1425), Rajagopal and Rangachari (1978, p. 101) wrote: "(It was Madhava who) took the decisive step onwards from the finite procedures of ancient mathematics to treat their limit-passage to infinity, which is the kernel of modern classical analysis". The growing volume of research into medieval Indian mathematics, particularly from Kerala, has refuted a common perception that mathematics in India after Bhaskaracharya made "only spotty progress until modern times" (Eves, 1983, p. 164).

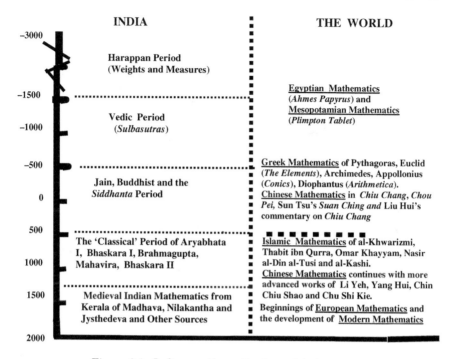

Figure 2.1. Indian mathematics in a global context.

Figure 2.1 below shows a highly summarised version what were the notable developments in mathematics in other parts of the world in relation to a backdrop of the history of mathematics in India. It is an extension of the earlier figure (Figure 1.3) presented in the last chapter. However this diagram is not a picture of the circulation of global mathematical knowledge. For that we now turn to the next section.

2.2 Indian Mathematics in World Mathematics: Global Interactions[†]

Figure 2.2 emphasises three features of global mathematical activity through the ages: (i) the continuity of mathematical traditions until

---

[†]This section is a summary of the discussion in Joseph (2011, pp. 12–22) which should be consulted for further details.

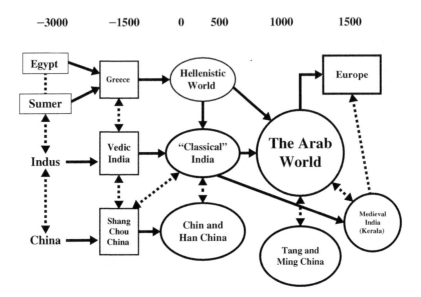

Figure 2.2. Indian mathematics in world mathematics.

the last few centuries in most of the selected cultural areas; (ii) the extent of cross-transmissions between different cultural areas which were geographically or otherwise separated from one another, and (iii) the relative ineffectiveness of cultural barriers (or 'filters') in inhibiting the transmission of mathematical knowledge. The lines of transmission are shown by the direction of the arrows (one-way or two-way) and whether broken or unbroken (unconfirmed or confirmed).

2.2.1   Egypt–Mesopotamia–Greek nexus

In both Egypt and Sumer (Mesopotamia) there existed well-developed written number systems as early as the third millennium BC. The peculiar character of the Egyptian hieroglyphic numerals led to the creation of special types of algorithms for basic arithmetic operations. Both these developments and subsequent work in the area of algebra and geometry, especially during the period between 1800 and 1600 BC are well known.

Figure 2.2 brings out another impressive aspect of Egyptian mathematics — the continuity of a tradition for over three thousand years, culminating in the great period of Alexandrian mathematics around the beginning of the Christian era. There is a tendency to view Alexandrian mathematics as a mere extension of Greek mathematics, in spite of the distinctive character of the mathematics of Archimedes, Heron, Diophantus and Pappus, to mention a few notable names of the Alexandrian period.[20]

The other early contributor to mathematics was the civilisation that grew around the twin rivers, the Tigris and the Euphrates, in Mesopotamia. There, mathematical activity flourished, given impetus by the establishment of a place-value sexagesimal (i.e. base 60) system of numerals, which must surely rank as one of the most significant developments in the history of mathematics. However, the golden period of mathematics in this area (or at least the period for which considerable written evidence exists) came during the First Babylonian period (*c.* 1800–1600 BCE), which saw not only the introduction of further refinements to the existing numeral system, but the development of an algebra more advanced than that in use in Egypt.

There is growing evidence of mathematical links between Egypt and Mesopotamia before the Hellenistic period, which we would expect given their proximity and the records we have of their economic and political contacts. There is also evidence of the great debt that Greece owed to Egypt and Mesopotamia for her earlier mathematics and astronomy. Friberg (2005, 2007) has discussed the 'unexpected links between Egyptian and Babylonian mathematics' and the 'amazing traces of a Babylonian origin in Greek mathematics'. Several of the Greek mathematicians listed earlier showed an easy familiarity with what Friberg describes as Babylonian 'metric algebra' (i.e. an approach that combines geometry, metrology and the solution of quadratic equations). The transmissions to Greece from the two areas, Egypt and Mesopotamia are shown in Figure 2.2. All three areas then became part of the Hellenistic world, and during the period between the third century BCE and the third century CE, and due to the interaction between the three mathematical traditions, there emerged one of the most creative periods in mathematics.

2.2.2   Indian–Mesopotamian–Greek nexus

The geographical location of India made her throughout history an important meeting-place of nations and cultures. This enabled her from the very beginning to play an important role in the transmission and diffusion of ideas. The traffic was often two-way, with Indian ideas and achievements traveling abroad as easily as those from outside entered her own consciousness. Archaeological evidence shows both cultural and commercial contacts between Mesopotamia and the Indus Valley. While there is no direct documentary evidence of mathematical exchange between the two cultural areas, certain astronomical calculations of the longest and shortest day included in the *Vedanga Jyotisa*, the oldest extant Indian astronomical/astrological text, as well as the list of 28 *nakshatras* (lunar mansion) found in the early Vedic texts have close parallels with those used in Mesopotamia.[21] And hence the tentative link, shown by a broken line in Figure 2.2. between Mesopotamia and India.[22]

The relative seclusion that India had enjoyed for centuries was broken by the invasion of the Persians under Darius around 513 BCE. In the ensuing six centuries, except for a century and half of security under the Mauryan dynasty, India was subjected to incursions by the Greeks, the Sakas, the Pahlavas and the Kusanas. Despite the turbulence, the period offered an opportunity for a close and productive contact between India and the West. Beginning with the appearance of the vast Persian Empire which touched Greece at one extremity and India at the other, tributes from Greece and from the frontier hills of India found their way to the same Imperial treasure houses at Ecbatana or Susa. Soldiers from Mesopotamia, the Greek cities of Asia Minor and India served in the same armies. The word "indoi" for Indians began to appear in Greek literature. By the time Ptolemaic Egypt and Rome's Eastern empire had established themselves just before the beginning of the Christian era, Indian civilisation was already well developed, having founded three great religions — Hinduism, Buddhism and Jainism — and expressed in writing some subtle currents of religious thought and speculation as well as theories in science and medicine.

There is little doubt that the Mesopotamian influence on Indian astronomy continued into the Hellenistic period. The astronomy and mathematics of the Ptolemaic and Seleucid dynasties became important forces in Indian science, readily detectable in the corpus of astronomical works known as *Siddhantas*, written around the beginning of the Christian era. Evidence of such contacts (especially in the field of medicine) has been found in places such as Jund-i-Shapur in Persia dating from between 300 and 600 CE. Jund-i-Shapur was an important meeting-place of scholars from a number of different areas, including Indians and, later, Greeks who sought refuge there with the demise of Alexandria as a centre of learning and the closure of Plato's Academy. All such contacts are shown in Figure 2.2 by lines linking India to Greece.

2.2.3   India–Islamic world nexus

By the second half of the first millennium CE, the most important contacts for the future development of mathematics were those between India and the Islamic world. This is shown by the arrow connecting India to the Arab world in Figure 2.2. The other major influence on the Islamic world was from the Hellenistic cultural areas, and the nature of these influences has been discussed in Joseph (2011). As far as Indian influence via the Islamic world on the future course of development of mathematics is concerned, it is possible to identify three main areas already highlighted earlier: (i) the spread of Indian numerals and their associated algorithms; (ii) the spread of Indian trigonometry, especially the use of the sine function; and (iii) the solutions of equations in general, and of indeterminate equations in particular.

2.2.4   India–China–Islamic world nexus

Figure 2.2 shows another important cross-cultural contact, between India and China. There is fragmentary evidence of contacts between the two countries before the spread of Buddhism into China. After this, from around the first century CE, India became the centre for pilgrimage of Chinese Buddhists, opening up the way for a scientific and cultural exchange which lasted for several centuries. In a catalogue of publications during the Sui Dynasty (*c.* 600 CE), there appeared Chinese translations of Indian works on astronomy, mathematics and medicine. Records of the Tang Dynasty indicate that from 600 onwards Indian astronomers were hired by the Astronomical Board of Chang'an to teach the principles of Indian astronomy. The solution of indeterminate equations, using the method of *kuttaka* in India and of *qiuyishu* in China, was an abiding passion in both countries. The nature and direction of transmission of mathematical ideas between the two areas is a complex but interesting problem. The two-headed broken arrow linking India with China is a recognition of the existence of such transmission. Also, there is some evidence of a direct transmission of mathematical (and astronomical) ideas between China and the Islamic world, around the beginning of the second millennium CE.[23] Numerical methods of solving equations of higher order such as quadratics and cubics, which attracted the interest of later Islamic mathematicians, notably al-Kashi (*fl.* 1400), may have been influenced by Chinese work in this area. There is every likelihood that some of the important trigonometric concepts introduced into Chinese mathematics around this period may have an Islamic origin.

2.2.5   India–Europe–China nexus

There are other broken lines of transmission in Figure 2.2 which need some explanation. One of the conjectures posed and elaborated in a later chapter (Chapter 10) is the possibility that mathematics from medieval India, particularly from the present-day southern state of Kerala, may have had an impact on European mathematics of the 16th and 17th centuries. While this is not substantiated at present by the existing direct evidence, the circumstantial evidence has become much stronger as a result of some recent archival research discussed in Joseph (2009a). The fact remains that around the beginning of the 15th century Madhava of Kerala derived infinite series for π and for certain trigonometric functions, thereby contributing to the beginnings of mathematical analysis about 250 years before European mathematicians such as Leibniz, Newton and Gregory were to arrive at the same results from their work on infinitesimal calculus. The possibility of medieval Indian mathematics influencing Europe is indicated by the arrow linking India with Europe.

During the medieval period in India, especially after the establishment of Mughal rule in North India, the Arab and Persian mathematical sources became better known there. From about the 15th century onwards there were two independent mathematical developments taking place: one Sanskrit-derived and constituting the mainstream tradition of Indian mathematics, then best exemplified in the work of Kerala mathematicians in the South and the other based in a number of Muslim schools (or *madrassas*) located mainly in the North. We recognise this transmission by constructing an arrow linking the Arab world to India.

The medieval period also saw a considerable transfer of technology and products from China to Europe which has been thoroughly investigated by Lach (1965) and Needham (1954). The 15th and 16th centuries witnessed the culmination of a westward flow of technology from China that had started as early as the first century CE. It included, from the list given by Needham (1954, pp. 240–241): the square-pallet chain pump, metallurgical blowing engines operated by water power, the wheelbarrow, the sailing-carriage, the wagon-mill, the crossbow, the technique of deep drilling, the so-called Cardan suspension for a compass, the segmental arch-bridge, canal lock-gates, numerous inventions in ship construction (including watertight compartments and aerodynamically efficient sails), gunpowder, the magnetic compass for navigation, paper and printing and porcelain. The conjecture here is that with the transfer of technology went certain mathematical ideas, including different algorithms for extracting square and cube roots, the 'Chinese remainder theorem', solutions of cubic and higher-order equations by what is known as Horner's method, and indeterminate analysis. Such a transmission from China need not have been a direct one, but may have taken place through

India and the Islamic world. And that completes the discussion of the nature and extent of the multi-directional transmissions implied by Figure 2.2 in which India was an important participant.

In conclusion, what this section highlights is the importance of emphasising circulatory histories over linear and bounded national or civilisational histories. Transmissions and exchange of ideas are in their way as important as the movement of goods and services between different cultures and nations.

Endnotes

[1] Gordon Childe, in his book *New Light on the Most Ancient East* (1935), wrote: "India confronts Egypt and Babylonia by the 3rd millennium with a thoroughly individual and independent civilisation of her own, technically the peer of the rest. And plainly it is deeply rooted in Indian soil. The Indus civilisation represents a very perfect adjustment of human life to a specific environment. And it has endured; it is already specifically Indian and forms the basis of modern Indian culture".

[2] Histories of Indian mathematics usually begin with a description of the geometry contained in the *Sulbasutras*. However, recent research into the history of Indian mathematics show that the essentials of this geometry could be older, contained in the altar constructions described in the Vedic texts, notably in the *Satapatha Brahmana* and the *Taittiriya Samhita*. The word '*samhita*' (or 'continuous recitation') indicates the strength and durability of the oral tradition. Even earlier, there existed *mantras* (words) which were required to be repeated a specified number of times, such as the *gayatri* from the *Rgveda* (*c*. 1000 BCE) whose *mantras* were required to be repeated 108 times. To keep track of counting, the joints of the two hands were assigned the numbers 1–9 (right hand) and 10–90 (left hand). Contained in the gestures are an early appreciation of the value of the place value principle.

[3] It is worth quoting Seidenberg in full, as recorded by van der Waerden (1983, p. 11)

> As Seidenberg has pointed out that in Greek texts as well as in the *Sulbasutras,* geometrical constructions were regarded important for ritual purposes, namely for constructing altars of given form and magnitude. In Greece this led to the famous problem of "doubling the cube", whereas in India it was not the volume but the area of the altar that was considered important. In both cases, one essential step in the altar constructions was the solution of the problem: to

construct a square equal in area to a given rectangle. To solve this problem, exactly the same construction was used in Greece and in India, a solution based on the Theorem of Pythagoras.

[4] In an early word-numeral system going back over two thousand five hundred years ago, the number 60,799 is denoted by the Sanskrit words, *sastim* (sixty), *sahsara* (thousand), *sapta* (seven) *satani* (hundred), *navatim* (nine ten times) and *nava* (nine).

[5] The term "Indian numerals" is used for the indigenous numerals from which the so-called "Hindu-Arabic" numerals evolved. Again, the term "Indo-Arabic numerals" is preferred to "Hindu-Arabic" numerals, for the former does not imply any 'exclusions', given the important contributions made by the Jains and the Buddhists to the development of these numerals, as discussed in Chapter 4.

[6] As late as the 14th century, when a more efficient method of forming word-numerals (*katapyadi*) was in place, a translation of a verse attributed to the Kerala mathematician, Madhava, read (with numbers given in parenthesis for sake of clarity):

> Gods (33), eyes (2), elephants (8), snakes (8), fires (3), trinity (3), qualities (3), *vedas* (4), *nakshatras* (27), elephants (8) and arms (2) — the wise say that this is the measure of the circumference when the diameter of a circle is 900 000 000 000.

Reading from right to left, the circumference is:

> 2 827 433 388 233 or an implied value for π of 3.141 592 653 59 correct to 11 decimal places.

[7] Most written scripts of India (including the *Devanagari* in which Sanskrit and Sanskrit derived languages are written) are not alphabetic in the strictest sense of the term, but semi-syllabic. Only the initial vowels have special characters, and the characters for consonants carry the vowel 'a' unless a special diacritical sign is used to remove it. Where two consonants precede a vowel, they are condensed into a single character. Vowel sounds other than 'a' are represented by a variety of diacritical marks attached to the character for the preceding consonants. [In this book the diacritical marks have been left out so as not to complicate matters further for uninitiated readers.] The letters of most Indian scripts are arranged in a phonetic and logical order in which groups of sounds formed against the back and front palates (guttural and palatal), gums (lingual), teeth (dental) and lips (labial), follow each other in sequences. This has formed the basis of the argument that phonetics and grammar may have developed as scientific studies in India even before writing was

introduced. A strong reason in Vedic India (i.e. India before 500 BCE) for the study of phonetics and grammar was to ensure perfect accuracy in pronouncing every syllable in a prayer or sacrificial chant. It is believed even today among certain orthodox sections that the written word is inferior to the spoken word. And possibly for that reason, the bulk of Indian scriptures, notably the Vedas, are still orally transmitted. Hence, the importance of mnemonics for easier recall of knowledge.

[8] The syllables of *cayagiyinusuchlr* represent the numbers 6, 30, 300 and so on to denote the number 57 753 336 reading from right to left.

[9] This system of notation can also be adapted for other languages. Indeed, its most common use was in Malayalam, the regional language of Kerala. While the subdivisions into vowels and consonants are similar to that of Sanskrit, the script is different. Sanskrit is written in the *Devanagiri* script while Malayalam has its own script.

[10] This was a system which facilitated easy memorisation since memorable words were made up using different chronograms. For example, if such a system is applied to English, the letters $b, c, d, f, g, h, j, k, l, m$ would represent the numbers zero to nine. So would $n, p, q, r, s, t, v, w, x, y$. The last letter, z, also denotes zero. The vowels, a, e, i, o, u are helpful in forming meaningful words, but have no numerical values associated with them. Thus 'I love Madras' is represented by the numbers '86 9234'.

[11] Certain definitions are in order. *Artificial* Sanskrit is a language intentionally created to deal with scientific works and deviate in some important respects from natural Sanskrit. By *scientific* Sanskrit is meant the use of the language for expressing scientific statements or laws. In the chronology of development, *grammatical* Sanskrit came first achieving a degree of artificiality that neither logical nor mathematical Sanskrit ever attained. Indeed, the efficacy of *logical* Sanskrit, that developed next was less to do with its artificiality and more to do with the expanding of the existing structures of Sanskrit syntax, especially the use of nominal composition and suffixes to express abstraction. In the case of mathematical Sanskrit, which was the last to develop, the artificiality arose not in the creation of a distinct language but more from the introduction of notations and technical terms that remained outside the natural language of Sanskrit. For further details, see Staal (1995).

[12] *Nyaya*: One of the six schools of Indian philosophy, particularly important for its analysis of logic and epistemology. *Navya–Nyaya* means literally the "more recent *Nyaya*".

[13] For an interesting discussion of this issue, see Surukkai (2008).

[14] The Mayan mixed base system arose from calendrical considerations. For a further discussion of the Mayan, Mesopotamian and Chinese place-value number systems, see Joseph (2011, pp. 66–73, pp. 136–142, 198–206).

[15] In the second century CE, Yavanesvara, played an important role in popularising astrology when he translated a Greek astrology text dating from 120 BCE. If he had merely made a literal translation it is doubtful whether it would have been of interest to more than a few. However, he popularised the text, by resetting the whole work in the context of Indian culture, using Hindu imagery and incorporating the caste system into his text.

[16] An example of an indeterminate equation in two unknowns (x and y) is $3x + 4y = 50$, which has a number of positive whole-number (or integer) solutions for (x, y). For example, $x = 14$, $y = 12$ satisfies the equation as do the solution sets $(10, 5)$, $(6, 8)$ and $(2, 11)$.

[17] What constitutes 'algebra' has been a vexed question and much effort and heat has been generated in discussion. There is a widely prevalent view that only mathematics from Descartes (or Fermat or Viete or Cauchy) can be legitimately considered as 'algebra'. This would mean 'Indian algebra' (or 'Babylonian' or 'Islamic') are at the very worst 'mindless recipes or at the best 'naïve'. Hoyrup's (2010, pp. 1–24) analysis regarding Babylonian algebra is apposite in the case of Indian algebra: "The prescriptions [of the Babylonian scribe] turn out to be neither renderings of algebraic computations as we know them nor mindless rules to be followed blindly; they describe a particular type of algebraic manipulation, which like modern equation algebra is analytical in character, and which displays the correctness of its procedures without being explicitly demonstrative"... "Whether all of this is sufficient to include Babylonian technique in an extended family of 'algebras' is a matter of taste and epistemological convenience." The author would add that it is also a matter of ideological predisposition.

[18] The etymology of the word "sine" is interesting. Because of the resemblance of a circular arc and its chord to an archer's bow and bowstring Indian astronomers referred to these objects as *capa* (bow) and *samasta-jya* (bowstring). The half-chord, in terms of whose length the Indians stated what we would refer to as a sine, was called *jya-ardha*, eventually abbreviated to *jya*. When the sine function was introduced into Arab astronomy from India, the Arabic term *jiba* was used to approximate the Sanskrit *jya*. Later European mathematicians, translating the Arabic works into Latin, mistook this borrowed term for another Arabic word *jaib* (spelled the same except for the omitted vowel markings), which signifies (among other things) the opening of a woman's garment at the neck, or bosom. Accordingly, *jiba* was translated into the Latin 'sinus' (which can variously signify a fold, a bosom, a bay or inlet, or a curve).

[19] We will be discussing some of these important developments in Chapter 11 which is devoted to Indian Trigonometry.

[20] Of the names mentioned, Diophantus' *Arithmetica* is an interesting synthesis of Greek, Egyptian and Mesopotamian mathematics. It was not only one of the earliest number-theoretic and algebraic texts, but contains a blend of rhetorical and symbolic exposition which we now would describe as syncopated mathematics. It owes less to Greek mode of thinking and closer to the Babylonian, Indian and Chinese traditions of mathematics. The text was essential in the development of Islamic algebra and European number theory and notation, especially in the development of the theory of indeterminate or Diophantine equations and inspired modern work in both abstract algebra and computer science.

[21] The ecliptic is divided into 27 *nakshatras*, which are variously called lunar houses or lunar mansions or asterisms. These reflect the moon's cycle against the fixed stars, of 27 days and $7\frac{1}{4}$ hours, the fractional part being compensated by the insertion of a 28th *nakshatra*.

[22] In the case of Indian astronomy and the mathematics associated with it, the early influences from Mesopotamia came through the mediation of the Greeks. Probably in the fifth century BCE, India acquired Babylonian astronomical period relations and arithmetic (e.g., representing continuously changing quantities with 'zig-zag' functions). Around the early centuries AD, the Babylonian arithmetical procedures were combined with Greek geometrical methods to determine solar and lunar positions, as reported in the Indian astronomical treatises, *Romaka-siddhanta* and *Paulisa-siddhanta*. For further details, see Pingree (1981).

[23] An exchange of astronomical knowledge took place between the Islamic world and the Yuan dynasty in China in the latter part of the 13th century when both territories were part of the Mongol empire. A few Chinese astronomers were employed at the observatory in Maragha (set up by Hulegu Khan in 1258) and probably helped in the construction of the Chinese-Uighur calendar (a type of a lunisolar calendar or a calendar whose date indicates both the phase of the moon and the time of the solar year). This calendar was widely used in Iran from the late 13th century onwards. There were at least ten Islamic astronomers working in the Islamic Astronomical Bureau in Beijing founded by the first Mongol emperor of China, Khubilai Khan in 1271. At this Bureau, continuous observations were made and a *zij* (or astronomical handbook with tables) was compiled in Persian. This work was then translated into Chinese during the early Ming dynasty (1383) and together with the influential Islamic text, Kushayar's *Introduction to Astrology*, served for a number of years as important sources for further research and study by Chinese scholars.

Ancient India: A Chronology to the Beginning of the Common Era

Pre-historic Era to 3000 BCE: Beginnings of Agriculture at Mehrgarh during the New Stone Age.

3000–1600 BCE: Emergence of the Harappan (or Indus Valley) civilisation around the banks of the Indus. The two most important sites uncovered so far by archaeologists, Harappa and Mohenjo-daro, show advanced urban development including multi-level houses and city-wide plumbing. Cultural and commercial contacts were maintained with Mesopotamia. The Harrapan civilisation probably collapsed because of some unknown natural disaster that altered the course of the Indus River.

1600–1000 BCE: The 'Aryans' (consisting of nomadic groups from present-day Afghanistan and Iran and further) moved into the Indus Valley region. The fusion of the newcomers and indigenous population led to the unfolding of the early Vedic culture.

1000–600 BCE: The later Vedic period continued the gradual merging of the Vedic population into the indigenous culture resulting in the emergence of the caste system. The codification and recording of the first Veda, *The Rg Veda* is followed by the recording of other *Vedas* and then the *Vedangas* (including the *Vedanga Jyotisa* and the *Sulbasutras* which are of mathematical and astronomical significance respectively). The Epics (*Ramayana* and *Mahabharata*) and the *Upanishads* were recorded around this time.

600–100 BCE: Certain Vedic practices were challenged by the rise of religious dissent in the form of Buddhism and Jainism which objected to the caste system, animal sacrifices, Brahmin dominance and the *Vedas* as sacred texts.

Some Notable Dates

563 BCE   Birth of Gautama Buddha, the founder of Buddhism.

540 BCE   Birth of Mahavira, the founder of Jainism.

537–509 BCE   Cyrus leads the Persian invasion into the west of the Indus River. Dairus conquers the Indus Valley region, making the area a province of the Persia. India open to the outside world and vice versa.

500–200 BCE   The written version of the *Mahabharata* (of which the *Bhagavat Gita* is a part) put into final form.

***c.* 500 BCE**   Composition of Panini's Sanskrit grammar, *Ashtadhyayi.*

327–326 BCE   Alexander's invasion and establishment of Greek rule in the Indus Valley.

323 BCE   Foundation of the Mauryan Empire by Chandragupta, with its capital at Pataliputra. In 304 BCE, Chandragupta exchanges 500 war elephants with Seleucus for the Indus Region and regions immediately to the West.

273–232 BCE   The Reign of Ashoka, grandson of Chandragupta.

184 BCE   End of the Mauryan Empire.

Chapter 3

The Harappa-Vedic Nexus: A Seamless Story or Two Separate Episodes?

3.1 The Beginnings: The Harappan Story

Of the many exhibits in the National Museum in Delhi there is one that captures immediate attention. This is the bronze figurine of the 'Dancing Girl of Mohenjo-Daro' (Figure 3.1). Known by that name because of her graceful and elegant pose, she takes us back to a period between 2300 BCE and 1750 BCE, when the Harappan Civilisation, containing the two great cities of Mohenjo-Daro and Harappa, reached its maturity. Referred to also as the Indus Valley Civilisation, the life line of this civilisation was the River Indus with its many tributaries, rising high in the Himalayas and meandering its way through the present-day lands of the Punjab and Sindh to reach the Arabian Sea. The area encompassed was vast, comprising almost one million square kilometers with more than 1,000 settlements identified so far. It covered practically the whole of modern Pakistan, North-West India and the Kathiawar-Saurashtra peninsula (see Figure 3.2).

The discovery of the Harappan Civilisation during the early decades of the twentieth century is a fascinating story. After its disappearance due to unknown causes, sometime between 2000 BCE and 1700 BCE, there was virtually no mention of or interest in the ruins that remained, except for rare references to the large number of very fine and standard size bricks scattered around the major

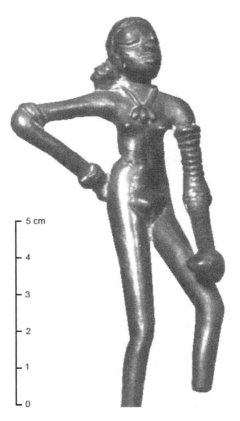

Figure 3.1. Dancing girl.

centres, notably in Harappa and Mohenjo-Daro. When Cunningham, Director of the Indian Department of Archaeology, visited the sites in the 1870s, he noticed a great depletion of bricks, owing to their use as ballast in laying a local railway track. In the early 1920s, the Harappa site was excavated by Cunningham's successors, Marshall and Sahni. They were joined by Banerji, who had earlier excavated the great mound of Mohenjo-Daro, a site on the River Indus itself, nearly 350 miles south-west of Harappa. When the three archaeologists began to compare notes they found that the two cities were sites from a common historical civilisation that flourished about 5,000 years ago. During the 1940s the British archaeologist, Mortimer Wheeler carried out excavations at both Harappa and Mohenjo-Daro; and since then

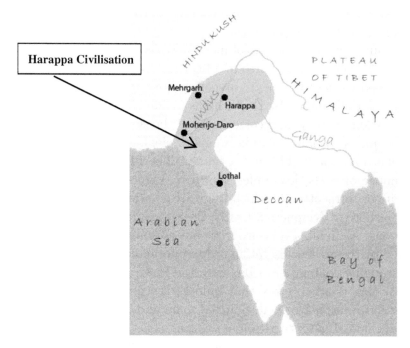

Figure 3.2. The Harappan civilisation.

work has proceeded apace on numerous sites in India and Pakistan, the results of which have added to our understanding of the entire civilisation.

The material evidence of this Harappan Civilisation, found in the ruins of many sites but principally at Mohenjo-Daro and Harappa, may be grouped into six categories. The first consists of the great cities themselves and their monuments. The higher citadels and the lower cities, the bathing pool of Mohenjo-Daro, whose walls were made water tight using bricks and gypsum mortar, the large well-ventilated granaries of Harappa, the brick walls and ground plans of houses, the elaborate drainages and brick-lined drain channels, and a dock at Lothal in Saurashtra, all testify to the engineering skills of the builders of that period and the excellent quality of the kiln-baked bricks used. There are also artefacts of everyday use, such as razors, bowls, cups, vases, and spindles, bronze and stone tools, terracotta models of toys, including mechanical ones, dress, jewellery and beads and an assortment of potteries.

A third category consists of just a few pieces of sculpture in stone or bronze, including the torso of a bearded man and the dancing girl mentioned earlier. The next category relates to the fauna and flora of that period, shown by the remains of the seeds of wheat, barley, peas, melons, sesame and cotton, and of the skeletal images found of such animals as bull, dog, pig, monkey, elephant, rhino, deer, tiger, scorpion and snake, but *not* a horse.[1] The fifth category includes a number of seals, [examples shown in Figure 3.3(b)] made from steatite or terracotta, on which there are a variety of pictograms and a script not yet deciphered. Lastly, there are a few remains of human skeletons. All in all, a treasure trove but nothing comparable to those offered by the ancient Egyptian or Mesopotamian Civilisation, which were roughly contemporaneous with the Harappan Civilisation.

Writing is often a key that unlocks the mind of a society. The Indus script has not been deciphered so far and is of little help for our purpose here. If a text written in two languages, one using that script and another which is already known, were to be found on a single seal or a document, the mystery of the Harappan script might yet be solved, just as the discovery of the Rosetta Stone in Egypt, in three languages of the same text, solved the puzzle of the hieroglyphs. At present, we can only judge the Harappan Civilisation from its material remains.

From the elaborate grid plans of cities, the standard brick sizes and the uniform set of weights and measures, we can infer that this was a highly organised culture with a strong numerate bent. The great urban granaries, the dock at Lothal and the seals, found as far away as Mesopotamia, suggest an extensive network of internal and external trade. The division between the citadel and the lower city and the variations in the burial practices, with some graves empty and others filled with jewellery, indicate a well-formed class structure and social hierarchy. From the meticulously constructed Great Bath of Mohenjo-Daro, measuring approximately 12 meters long, 7 meters wide and of a maximum depth of 2.4 meters, and the lingam-like stone structures and figures of divinity found on seals there, we have but a bare glimpse of the totems and rituals of that ancient culture. Although no clear purpose for the seals has been discerned, some scholars believe that they could be records of merchandise or amulets or even identity cards of the workers. The images of unicorn and other animals have been interpreted as symbols of dominant lineages and clan families. The cotton seeds of the area have been dated as the earliest anywhere in the world and these, along with spindles, permit us to say that Harappa contained significant textile production. The overall picture that we can draw from the material remains is that of a fairly sophisticated urban environment fed and enriched by extensive agriculture and commerce.

How such a civilisation came to be where it has been found — in the plains, deserts, valleys and coasts of the present-day regions of the Punjab, Sind, Baluchistan, Rajasthan and Gujarat — is a subject inviting both speculation and controversy. At first, it was thought that it must have been an extension of the Mesopotamian culture; but, over the decades since the 1920s, as growing number of new sites yielded artefacts and other evidence from deeper strata, there was the realisation that this culture was home-grown. The clinching argument came with the dig at Mehrgarh in Baluchistan, approximately 150 miles north-west of Mohenjo-Daro. There, an international team of archaeologists dug to the deepest levels under the surface, revealing a pre-Harappan site of mud-brick walls, stone tools and burial sites, an early Harappan site with specialised potteries and tools, a mature Harappan level with urban artefacts similar to those at Mohenjo-Daro and Harappa and; and finally, the post-Harappan site with evidence of material decline, nearer to the present day ground surface. The Mehrgarh results decisively demonstrated the indigenous roots of the Harappan Civilisation.

Controversy also surrounds the subject of the disappearance of this civilisation. For a long time, Mortimer Wheeler's judgement was considered the last word on the matter, which was that an invading Aryan forces had destroyed it. This theory is now discredited and, owing to detailed work done by the historians of climate and environment and the geologists, it is now accepted that a variety of natural and ecological factors contributed to its decline. Decline did not mean entire disappearance. Although the specific urban aspects of the civilisation withered away, some of its rural features continued and merged into the making of the future societies on the Indian sub-continent. A historically conscious traveller exploring today's Punjab, Sindh and Baluchistan may still observe shadows of the Harappan past.

3.2 'Frozen' Mathematics of the Harappan Civilisation

Early evidence of numeracy is found among the material ruins of the Indus Valley civilisation. Archaeological finds include an elaborate system of weights and measures. A total of over 500 plumb-bobs of uniform size and weight have been found throughout the vast area of the Harappa culture that conform to two series (binary and decimal) in the ratio of 1, 2, 4, 8, 16, 32, 64 and 10, 20, 40, 160, 200, 300, 640, 1600, 6400, 8000 and 12,800.[2] Equivalent weights have been in use in parts of India until recently, with conversion rates identical to the above ratios, thus forming the basis for an elaborate system of barter of one commodity for another.

Scales and instruments for measuring length have been excavated at major urban centres of this civilisation, notably in Mohenjo-Daro, Harappa and Lothal. The Mohenjo-Daro scale [Figure 3.3(a)] is a fragment of a shell 66.2 mm long, with nine carefully sawn, equally spaced parallel lines, on average 6.7056 mm apart, found during the excavation of the Great Bath in Mohenjo-Daro.[3] The accuracy of the graduation is remarkably high, with a mean error of only 0.075 mm.

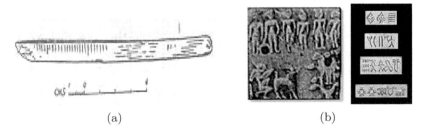

(a)   (b)

Figure 3.3. (a) The 'indus inch'. (b) Indus seals.

One of the lines is marked by a hollow circle, and the sixth line from the circle is indicated by a large circular dot. The distance between the two markers is 1.32 inches (33.53 mm), and has been named the 'Indus inch'.

There are certain interesting connections between this unit of length (if indeed that is what it was) and others found elsewhere. A Sumerian unit of length, the *sushi*, is exactly half an Indus inch. This would support other archaeological evidence of contacts between the two ancient civilisations. In North-West India, a traditional yard known as the *gaz* was in use until recently. The *gaz*, which is 33 inches (or 5840 mm) by our measurement, equals 25 Indus inches. Furthermore, the *gaz* is only a fraction (0.36 inches) longer than the megalithic yard, a measure that seems to have been prevalent in North-West Europe around the second millennium BCE. This has led to the conjecture that a base 5 scale of measurement, originating somewhere in Western Asia, spread widely as far as Britain to ancient Egypt and the Indus Valley (Mackie, 1977).

A notable feature of the Harappa culture was its extensive use of kiln-fired bricks and the advanced level of its brick-making technology. The bricks were exceptionally well baked and of high quality, and could still be used over and over again provided some care was taken in removing them in the first place. They contain no straw or other binding material. While 15 different sizes of Harappan bricks have been identified, the standard ratio of the three dimensions — the length, breadth and thickness — of each brick is close to 4:2:1. Even today this is considered the optimal ratio for efficient bonding required for a robust brick technology.

A correspondence between the Indus scales (from Harappa, Mohenjo-Daro and Lothal) and brick sizes has been noted by Mainkar (1984). Bricks of different sizes from the three urban centres were found to have dimensions that were integral multiples of the graduations of their respective scales.

Our argument on the possibility of development of mathematics (and of particularly geometry) in the Harappan Culture is not based on any direct evidence, nothing like the Mesopotamian clay tablets or the Egyptian papyri testifying to their origins in these centres of early civilisation. From the standpoint of an engineer, Kulkarni (1971) has argued, mainly on the earlier evidence of archaeologists such as Marshall and Mackay (i.e. without noting significant discoveries from the excavation of Kalibhagan and the more spectacular ones from the dock-yard at Lothal), that the elaborate constructions excavated cannot be understood by us today without attributing the knowledge of a number of geometrical propositions to the Harappan architects, engineers and masons: propositions relating to the shapes and mensuration of rectilinear figures and circles.[4]

3.3   Indus Writing: A Numerical Record?

Notwithstanding these connections between material objects and the 'frozen' mathematics embodied in them, which have served until recently as the only basis for making inferences about the numerate character of the Harappan culture, a critical piece of evidence, the written script [Figure 3.3(b)], has so far thrown no light on this subject. This script, usually referred to as the Indus script, was a system of writing employed in the Harappan Civilisation, more or less contemporaneously with two other civilisations, the Mesopotamian and Egyptian. As is well known and mentioned earlier, the writings of these other two civilisations have been deciphered, and hence the archaeological findings are richly supplemented by written records to obtain a better understanding of these civilisations.

The Indus script remains unread but hotly debated, despite years of ingenious attempts to do so. The script poses certain problems not present in the writings of other two ancient scripts. It is only available through objects of a very restricted medium, typically in the form of seals made of steatite, each seal on the average containing a text of only five graphemes (or signs). Also, the language or the language family of the Indus script texts is still unknown. A common assumption made in the past was that it is some form of proto-Dravidian language that had "disappeared" sometime before the middle of the second millennium BCE. There is apparently a long hiatus between this disappearance and the emergence of the so-called historical period of the Indian sub-continent (the Vedic period), conservatively dated as starting around 1500 BCE, causing thereby a big "hole" in the chronology of Indian mathematics right from its inception.

There have been a number of attempts to read the texts ever since a substantial collection of them became available around the 1920's. Many of the early attempts were phonetic interpretations based on unverifiable *a priori* linguistic speculations that involved mythological elements of the later Hindu traditions. A notable "objective" attempt was that of Hunter (1934) who carried out a positional and functional analysis of the signs of the Indus script and suggested methods for splitting the texts into sign combinations that constituted "words" independent of their linguistic attributes. Following Hunter's work, more recent investigations have involved detailed structural analysis of the texts with the aim of classifying the signs or sign combinations into linguistic units, such as root morphemes, attributes and other grammatical suffixes, and then read the texts phonetically, adapting a form of Dravidian as the underlying language. However, what has been achieved so far in the deciphering of the Indus script is summed up by the striking title of the 1988 Presidential Address to the Indian History Congress by the doyen of scholars studying the Indus script, Iravatham Mahadevan: "What Do We Know about the Indus Script? **Neti Neti** (Not This, Not That)".

Any fresh approach to the deciphering of the Indus script needs to take account of three distinctive features that have been identified in earlier studies. First, there exist rich structural regularities in the texts which make them distinct from other ancient writings. Second, the texts occur in almost all cases on seals so that the purposes of these seals become a matter of crucial importance. Finally, a closer examination should be made of the nature and significance of a number of animal and other motifs, named by archaeologists as 'field symbols' that occur on many of the seals together with the writing. And the examination of all three features should be made within the historical context of the emergence of the Harappa culture and its aftermath. We will consider just one study that has particular relevance for the history of Indian mathematics, namely that of Subbarayappa (1993).

It is believed that the Indus seals may have been records of administration and of internal as well as external trade. Indus seals have been found on sites in West Asia indicating periods of commercial contact between the Harappans and the wider neighbouring areas. There is also general agreement among archaeologists of the existence of an efficient and centralised administration, governing the vast area which constituted the Harappa culture, ensuring a degree of uniformity, whether it was in the construction of houses and public amenities, promotion of arts and commerce or other activities.

So what do the inscriptions on seals represent? To attempt an answer, certain features of the Harappa culture need to be highlighted. The wide geographical spread of this culture makes it highly unlikely that the language was the same throughout the length and breadth of that culture. Even the Mesopotamian civilisation spread over a smaller area than that of the Harappa culture had significant regional variations in language. Such variations are commonly observed throughout history. India has had a multilingual culture from ancient times. There have been marked variations not only in spoken but also in written languages. As to a common script, whatever may have been the situation in other ancient civilisations, the picture could well be different on the Indian sub-continent. Even the earliest extant written records from the first half of the first millennium BCE used two scripts — Kharosthi and Brahmi — as vehicles of the same language.

Are we therefore justified in assuming that there was one language throughout the Harappa culture — in the urban centres such as Mohenjo-Daro, Harappa, Chanhu-Daro, Lothal, Kalibangan and in the rural settlements? Yet this assumption of a similar literary script is necessary since the seals and other inscribed objects found in different parts of the Harappa culture have more or less identical forms with a noticeable uniformity of their own. Such continuity is usually more plausible in the case of a well-established system of numerical notation. Our universal system of numerals and numeration bears testimony to that fact.

Subbarayappa's monograph (1993) is an interesting example of a numerical interpretation of the Indus Script. It begins with an examination of why a purely literary interpretation of the script has failed, notably because of the implausibility of an assumption of a single language prevalent over the vast area encompassed by the Harappan Civilisation. The solution offered by Subbarayappa is that the Indus signs employed a 'ciphered system involving additive–multiplicative approach to arrive at and express the desired numbers'. He suggests that the Indus number system is a base 10 (or decimal) system, containing different symbols for the numbers 1 to 9, for 10, 100 and 1000, and for their multiples. Two types of numerical representation are then identified.[5] The first contains an ordered sequence of strokes representing numbers 1 to 12, with number 10 represented by a circle. The second is more complex in which pictorial symbols or geometric forms represent various numbers. For example, any sign with four lines, whether it is a square, or oblong or diamond or a cross, represents the number 4. Higher numbers are represented by additional strokes attached to the basic signs. Many of these numbers are identified from their apparent resemblance to numbers in contemporary or later number systems such as Babylonian, Chinese, Attic Greek, Kharosthi and Asokan Brahmi. The complexity of the system is increased further as many of the numeral signs are 'condensed in an artistic way' or 'embellished' to look like pictorial depictions.

If the Indus writing is seen as a string of numbers written from right to left, a practice that Subbarayappa characterises as characteristically Indian, the formation of larger numbers is achieved *either* by repetition *or* by addition of strokes *or* by ligaturing (i.e. 'bundling') of signs *or* by using special additive conventions. For example, a square with a cross inside {⊠} is $4 \times 4 = 16$; A lozenge enclosing five smaller squares and a circle is: $4 \times 4 \times 4 \times 4 \times 4 \times 10 = 10,240$. By such means very large numbers can be represented.

The Indus texts then become "exclusively quantitative records with no words or ideograms interposed in between." Subbarayappa has an interesting answer to the question as to what the quantitative records represent. He believes that the animal motifs depicted on the seals, shown in Figure 3.3(b), represent various agricultural commodities, the quantities of which are specified in the numbers indicated in the texts which accompany them. For example, the 'unicorn' represents symbolically the three important field crops of barley, wheat and cotton, each crop being specified by the variations in the standard-like object placed in front of the animal. Some of the other important identifications include short-horned bull, six-rowed barley, ox-antelope, two-rowed barley, elephant, wheat, rhinoceros, peas, buffalo, sesamum (herb), gharial (a fish-eating crocodile), rape-seed, tiger, date fruit, humped bull, cotton threads, hare and a ball of cotton. The purpose of the inscriptions was probably to maintain an account of the grains and cotton available to the people under a centralised dispensation. Duplicates of the inscriptions like those on the seals indicated that so many bundles or packages were sent from one place to another. What about the texts without any animal or other pictorial motifs? Subbarayappa points to the perforations present on most seals and explains that the text-seals were tied to other seals having pictorial motifs, the latter identifying the commodity.[6]

A practical objection to Subbarayappa's solution raised by Mahadevan (2002) is that "it is highly unlikely that the large and beautifully carved stone seals, apparently very expensive to make, would be used to record quantities of commodities varying with each transaction." It would have been much simpler to make use of cheaper and readily available material like cloth, palm leaves or clay for daily accounts not required to be preserved for posterity. A further objection is that "the great complexity of the system (proposed by Subbrayappa) would render it quite unsuitable for unambiguous recording of transactions for handing out daily rations or despatch of goods" (Mahadevan, 2002, p. 18).

Despite the failures of the past it would be a counsel of despair to suggest that any attempt to decipher the Indus Script is a futile pursuit unless some form of a Rosetta Stone is discovered. And this has not always been necessary in the past the since decipherment of the Linear-B and the Mayan scripts were achieved without the aid of such objects. However, there is a need to develop objective criteria for testing the validity of competing claims and as a guide for would-be decipherers. Mahadevan (2002) suggests some simple tests for 'a preliminary screening of the claims'.[7] They are discussed on pages 18–19 of his article, although it should be borne in mind that the proposed tests provide necessary but not sufficient conditions for claiming a successful decipherment.

3.4 Mathematics of the Vedic Age

3.4.1 Introduction

There was a long period from the very beginnings of the Vedic Age when writing was rarely used even though it was known to have existed. Indeed, there may have been contacts between the latter part of Harappan civilisation and the beginning of the Vedic period. However, given that the Indus script has not yet been deciphered and that we do not even know what family of languages it belonged to, it would be idle to speculate any further. This did not mean that knowledge and texts were not transmitted from generation to generation from the very beginning of the Vedic Age. There was a highly developed form of oral transmission and an efficient form of preservation by generations of professionals (*pandits*) whose techniques of recitation, memorisation and preservation can be assessed even today by observing present-day *pandits* at work.[8]

Over centuries, with the growth of commentaries on the sacred texts, there gradually emerged the need to establish the state of knowledge of subjects that had been orally transmitted in the Vedic poems. This gave birth to six disciplines (or *Vedangas*), four of them dealing with languages [grammar (*vyakarana*), etymology (*nirukta*), phonetics (*siksa*) and metrics (*chanda*)] and two others [astronomy (*jyotisa*) and ritual (*kalpa*)]. The primary functions of the three branches of *Vedangas* were the need to conserve oral texts in its pristine form, the need to conduct the rites at the auspicious date and time and to need to carry out the rituals in the correct manner.

The manner of oral recitation and later writing of Vedic knowledge was through a mnemonic form *par excellence*: the *sutra*. The term *sutra,* used both to describe a whole text or a single rule,

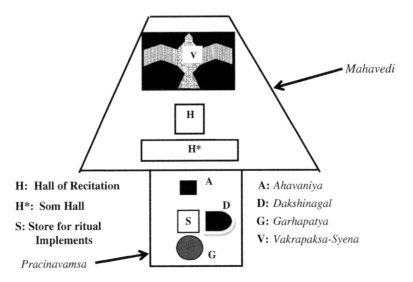

Figure 3.4. The *Agnicayana* ritual.

had be concise, focus on the essential point, link rule to rule and generalise wherever possible. The best exemplar of *sutras* was to be found in Panini's *Astadhyayi*. It contained rules to build up words and sentences from the roots and suffixes. In the next chapter we will discuss how Panini's grammar may help us to understand the character of Indian mathematics.

From early Vedic times, on propitious days, an elaborate 12-day rite was performed on a specially-constructed *Agnicayana* (i.e. 'piling up of altars').[9] The site for the public spectacle (Figure 3.4) was a large trapezoidal area, the *Mahavedi* (or 'Great Altar') and a smaller rectangular area to the west of it known as the *Pracinavamsa*. Situated in the *Mahavedi*, was a brick altar shaped in the form of a "falcon about to take wing" (*Vakrapaksa-Syena*). In the *Pracinavamsa* were three types of *agnis* (or fire altars) located in specified positions, known as the *Garhapatya*, the *Ahavaniya, and* the *Daksinagni*. The design, construction and the location of the *Mahavedi*, the *Pracinavamsa* and the structures had to be done exactly according to certain specifications which called for a certain amount of geometrical skills.[10]

The earliest written record of this geometrical knowledge is found in the *Taittiriya Samhita* and the *Saptapatha Brahmana* both of which conservative chronology places around 1000 BCE.[11] As Michels (1978) points out, these two texts[12] already contain important preliminaries for the development of Vedic geometry; found later in the *Sulbasutras*[13] which were recorded variously around 800–200 BCE. These practical manuals showed knowledge of:

- Rules for constructing specific geometrical figures of given dimensions;
- Dividing a given area into a specific number of smaller areas;
- Enlarging a given altar area step by step by 'piling up' successive layers according to the principle of $n^2, (n+1)^2, (n+2)^2, \ldots$, and
- Transforming geometrical figures into other shapes of equal areas.

The *Sulbasutras* also contained instructions for the construction of the sacrificial altars (*vedi*) and the location of sacred fires (*agni*) that had to conform to clearly laid-down instructions about their orientation, shapes and areas if they were to be effective instruments of sacrifice. There were two basic classes of sacrificial rituals — *Nitya* (perpetual or daily) and *Kamya* (optional). The sacrifice of the first class was obligatory and it was supposed to bring well-being for the family. The sacrifice of the second class was seasonal and was performed at special times — at new or full moon, at the time of the winter and summer solstices and so on. Each sacrifice had to be made on an altar of prescribed shape, size and orientation.[14] It was believed that even a small irregularity and variation in the form and size of the altar could annul the whole ritual. So it was necessary to abide by exact rules for construction of each altar.

The daily rituals were performed on altars of different shapes but conforming to an area of one square *vyama* (i.e. a standardised measure represented by the height of a man from his sole to the hair on his forehead). On certain occasions the 'optional' rituals were performed on a bigger area of $7\frac{1}{2}$ square *purushas* (i.e. a *purusha* is the height of a man with his arms lifted).[15]

Each smaller altar was constructed with five layers of bricks which together came up to the height of a knee and each layer contained a definite number of bricks of specified shapes. Each layer of the greater altar consisted of 200 bricks. The ritual required that the altar be constructed as layers of bricks of different shapes such that the interstices between bricks in contiguous layers did not coincide except at the centre, from where a deity was invoked through the chanting of verses and hymns.[16] Once invoked, the deity was believed to fly to the site of the sacrifice in his *ratha* (or chariot), partake of the sacrificial offerings placed on the altar and then depart. At the completion of the ritual the altar was abandoned or even destroyed. Of the altars used for the ritual, the falcon-shaped ones (or those that conformed to the shadow of a bird on the ground) were of particular importance.[17] There were a number of possible representations of falcons, some of which bore little or no resemblance to the bird, such as the one shown in Figure 3.5.

Figure 3.5 shows one of these simpler representations. This altar, known as a *Caturasra-Shyena* (a 'four-sided falcon'), was shaped as a 'falcon' with a square body and head, two rectangular wings, and a rectangular tail. The first, third and fifth layers of this altar were built with square bricks of four sizes which numbered: 24 bricks of *caturti* ('fourth') type, 120 bricks of *pancame* ('fifth') type, 36 bricks of *sasthi* ('sixth') type and 20 bricks of *dasmi* ('tenth') type where the numbers represented fractions of the area of a standard brick.

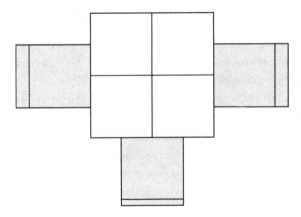

Figure 3.5. A square falcon altar.

The second and fourth layers contained square bricks of three sizes: 12 bricks of *caturti* ('fourth') type, 125 bricks of *pancame* ('fifth') type and 63 bricks of *sasthi* ('sixth') type. Note that the total number of bricks used in constructing each layer was 200. Other altar shapes required different combinations of brick sizes and shapes for each layer.[18]

The first step in the construction of the altar was to determine the size of the *angula*, the basic unit of measure used to determine the dimensions of the altar. The *Manava Sulbasutra* gave two ways to determine the size of the *angula*. In the first, the worshipper stood with his feet flat on the ground and his arms raised. It was assumed that the height reached by this individual with his arms stretched was 120 *angulas* (or approximately 90 inches so that 1 *angula* ≈ 0.75 of an inch). If he stood on his toes, it was then assumed that the height was 125 *angulas*. The unit of length of 120 *angulas* was known as a *purusha* (or a 'man'). The second method involved measuring the distance between the lines on the middle finger of the person offering the sacrifice which was assumed to be equal to one *angula*. If this person was unusually small, then one *angula* was taken to be equivalent to 64 sesame seeds laid together with their broadest faces touching.

The second stage in the construction of the altar was to determine its location and orientation. The orientation had both spatial and temporal dimensions. The orienting procedure determined not only the direction of the ritual object, but also the time of its construction. The simplest procedure for determining time and orientation utilised gnomons and ropes. The shadow of a gnomon was used to determine the east-west line; and pegs, ropes and rods are used to lay out different constructions.

There is a section in the **Satapatha** *Brahmana* that deals with the construction of an altar to carry out the twelve-day Agnicayana ritual, mentioned earlier in the chapter and shown on Figure 3.4. The constructions consisted of:

(i) The *Mahavedi* (Figure 3.6): Shaped like an isosceles trapezium, the two parallel sides were located in such a way that

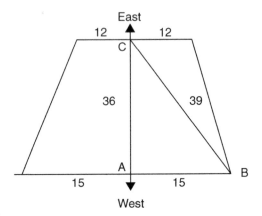

Figure 3.6. The *Mahavedi*.

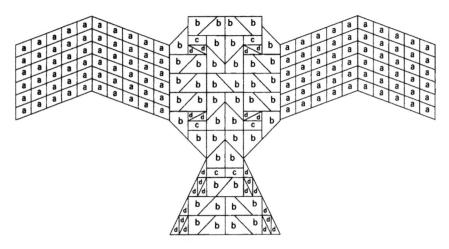

Figure 3.7. *Vakrapraksa-Syena*.
(The alphabetical labels refer to different shaped bricks).

the larger side measured 30 *prakramas* on the west and the smaller side 24 *prakramas* on the east with the altitude of the trapezium being 36 *prakramas*.[19] Contained in the *Mahavedi* was the complex falcon-shaped brick altar (*Vakrapraksa-Syena*) shown in Figure 3.7 below and may have represented time.[20]

The geometrical aspects relating to the construction of this altar will be discussed in the Appendix of this chapter. The other constructions in the *Mahavedi* had functional and ritual significance but little of mathematical interest.

(ii) To the west of the *Mahavedi*, shown in Figure 3.4, was the smaller rectangular area called *Pracinavamsa* in which, in specified positions, were three fire altars, previously referred to as *Garhapatya* (usually of circular shape symbolising the earth), the *Dakshinagni* (of semi-circular shape representing space) and the *Ahvantya* (a square representing the sky). In the construction manuals (*Sulbasutras*), there may have been the suggestion that the areas of the three fire altars were equivalent and equalled one square *purusha*.[21]

The last of these fire altars (i.e. the *Ahvantya* or sky altar) was a representation of the universe to include both space and earth. It was laid out in five layers, with the first representing the earth, the second the joining of earth and space, the third the space, the fourth the joining of space and sky and the fifth the sky. The need to maintain equivalence of area and of shapes in the case of different altars raised a number of geometric problems, the solutions of which led to early Indian geometry.

The instructions for the design of the *Mahavedi* (Figure 3.6) provide an insight into practical nature of the texts of the period. The *Apastamba Sulbasutra* (Verse 2) contains the following elaborated version of the original cryptic instruction:

> To a cord of length 36 *prakramas*, add 18 *prakramas*. Make two marks on the cord, one at 12 *prakramas* and the other at 15 *prakramas* from the western end. Tie the ends (of the cord) to pegs on the ends of the east-west (i.e. *prsthya* or the edge which runs along the back of the altar) line of length 36 *prakramas*. Take the cord by the mark at 15 *prakramas* and stretch it to the south and mark the point with a peg. Do likewise to the north. These are respectively the south-west and the north-west corners of the *Mahavedi*. Now untie the ends of the cord from the east-west line and retie the end that was previously fastened to the peg on the east end to the west end and vice versa. Repeat the previous procedure

but using the mark at 12 *prakramas* to obtain the south-east and north-east corners of the *Mahavedi*.

The augmented cord is $36 + 18 = 54$ **prakramas**. From the other end, the 12th mark is half of the smaller parallel side while the 15th mark forms the base (AB) of a right-angled triangle (ABC in Figure 3.6), with its hypotenuse (BC) being the remainder of the cord (i.e. $36 + 3 = 39$ *prakramas*) and the other side being the east-west line (AC) which measures 36 *prakramas*.

Apastamba, one of the composers of the *Sulbasutras*, gave other methods of designing the Mahavedi which is suitable for any length of a given side of right-angled triangles satisfying the measurements required by the **Mahavedi**. The method was based on taking the basic Pythagorean triples $(3, 4, 5)$ multiplied by 4 or 5; $(12, 5, 13)$ multiplied by 3; or triples $(15, 8, 17)$ and $(12, 35, 37)$. All these were chosen to ensure that at least one of the sides had the length of one side of the **Mahavedi**. Apastamba also introduced a slight modification.[22]

The most elaborate of the public altars, found in the **Mahavedi** and constructed also for the **Agnicayana** ceremony, was shaped like a giant falcon just about to take flight (Figure 3.7). It was believed that offering a sacrifice on such an altar would enable the soul of a supplicant to be conveyed directly by a falcon to heaven.

Generally, falcon-shaped altars were constructed with five layers of 200 bricks[23] each which reached to the height of the knee. For special occasions 10, 15, up to a maximum of 95 layers of bricks were used in the construction of the falcon-shaped altar. The top layer of the basic altar (Figure 3.7) had an area of 7.5 square *purushas*.[24] As mentioned earlier, a *purusha* was defined as the height of a man with his arms stretched above him, say $7\frac{1}{2}$ feet or 2.3 metres, which would give the altar an 'areal' measure of 56.25 square feet or 5.29 square metres. For the second layer from the top, the prescription was that one square *purusha* should be added, so that the total area of would be 8.5 square *purushas*.[25] Similarly, each successive layer area should be augmented by 1 square *purusha*, until with the 94th successive increase of 1 square *purusha*, the area of the base of this huge construction would be 101.5 square *purushas!*[26]

It is clear that if in the construction of larger altars they had to conform to certain basic shapes and prescribed areas or perimeters, two geometrical problems would soon arise: (i) the problem of finding a square equal in area to two or more given squares and (ii) the problem of converting other shapes (for example, a circle or a trapezium or a rectangle) into a square of equal area or vice versa. The constructions were achieved through a judicious combination of concrete geometry (the principle of dissection and reassembly),[27] ingenious algorithms and application of the so-called Pythagorean Theorem.

In the *Katyayana Sulbasutra* appears the following proposition:

> The cord (stretched along the length) of the diagonal of a rectangle makes an (area) which the vertical and horizontal sides make together. (3.11)

Using this version of the Pythagorean Theorem, the *Sulbasutras* show how to construct both a square equal to the sum of two given squares and a square equal to the difference of two given squares. Further constructions include the transformation of a rectangle (square) to a square (rectangle) of equal area and of square (circle) to a circle (square) of approximately equal area. The constructions "doubling the square" and "squaring the circle" lead naturally to the devising of algorithms for the square root of 2 and other numbers, for implicit estimates of π and for constructing similar figures in required proportions of a given figure. A few of these constructions and transformations are discussed in the Appendix of this chapter.

Another problem that led to some interesting mathematics arose from the need to ensure the precise distance and relative positions of the three fire altars, *Garhapatya, Dakshinagni and Ahavaniya* contained in the *Pracinavamsa* as shown in Figure 3.5. The general requirement was: *Dakshinagni* should lie south of the line joining the other two fire altars and be at a distance from *Garhapatya* of one third the distance between the other two fire altars. The concluding section of the Appendix contains a discussion of different ways of solving the same problem. None of these solutions are accurate; and the rules used are practical 'rules of the thumb' that an early surveyor may use, without mathematical considerations being predominant.[28] However, this did not mean that considerations of accuracy did not occur in early Indian geometry. For example, to calculate the square root of 2, the following instructions are given both by Apastamba (1.6) and Katyayana (3.13) who came after Baudhyana:

> "Increase the measure by its third and this third by its own fourth less the thirty-fourth part of that fourth. This is the value of a special quantity in excess (which needs to be deducted)."

If we take 1 unit as the dimension of the side of a square, the above formula gives the approximate length of its diagonal as:

$$\sqrt{2} \approx 1 + \frac{1}{3} + \frac{1}{3 \times 4} - \frac{1}{3 \times 4 \times 34} = 1.4142157.$$

The true value is 1.414216....

The *Sulbasutras* contain no clue as to the manner in which this accurate approximation was arrived at.[29] A number of theories or explanations have been proposed. Of these, a plausible one is that of Datta (1932), based on the "dissection and reassembly" principle and discussed in the Appendix of this chapter.

Apart from equivalence through area, the Vedic texts are believed to contain equivalences established between phenomena through numbers. The starting point is the centrality of the number 360 in the Vedic calendar. Parallels are drawn between human anatomy and planetary motions. Thus, the *Charaka Samhita*, an early medical text, counts the total number of *asthis* (or bones, teeth, nails, hard cartilages) in the human body to be 360, obtained from considering the 308 bones of the new born babe (before they fuse into a smaller number of 206 in the adult), 32 teeth and 20 nails where each of these *asthis* are associated with each day of the year. The parallel between the nominal year (360 days) and man (*purusha*) is carried further in *Satapatha Brahmana* where the basic falcon-shaped altar of 7.5 square *purushas* or 108,000 square *angulas*

(1 *purusha* = 120 *angulas*) is linked to 10 nominal years or 108,000 *muhurtas* (1 *muhurta* = 48 minutes).[30] It is interesting in this context that a total of 10,800 ordinary (*lokamprina*) bricks were used in the construction of the three fire altars found in the *Pracinavamsa* which is the same as the number of *muhurtas* in a nominal year. A number of other parallels based on the equivalence of numbers are found the Vedic literature of the period.[31] However, a word of caution should be sounded here. In the absence of direct evidence, conjectures about the mathematical pursuits of a bygone age have to be examined in the light of their plausibility, the existence of convincing alternative explanations and the quality of evidence available. It may well be the case that the parallels drawn here may well collapse under the heavy weight of conjectures piled upon it.

3.5 Restoring Continuity to Indian Mathematics

There is a widely-held view that Indian mathematics originated in the service of religion. The proponents of this view have sought their main support in the complexity of motives behind the recording of the *Sulbasutras*. Since time immemorial, they argue, the needs of religion have determined not only the character of Indian social and political institutions, but also the development of her scientific knowledge. Astronomy was developed to help determine the auspicious day and hour for performing sacrifices. The 49 verses of *Jyotisutras* (the *Vedanga* containing astronomical information) gave procedures for calculating the time and position of the Sun and Moon in various *nakshatras* (or signs of the zodiac). Also, a strong reason in Vedic India for the study of phonetics and grammar was to ensure perfect accuracy in pronouncing every syllable in a prayer or sacrificial chant. And the construction of altars and the location of sacred fires, as we have seen, had to conform to clearly laid-down instructions about their shapes and areas if they were to be effective instruments of sacrifice.

However, there is a danger that the magico-religious beliefs surrounding the Vedic rituals may be overly emphasised when considering the origins of Indian mathematics. We have seen the crucial role played by the *Agnicayana* ceremony in generating the geometrical concepts and techniques found in the *Sulbasutras*.

The rituals associated with the construction of fire altars may be looked at from two standpoints. The first is from the standpoint of beliefs connecting the shapes of altars with the specific desires to be fulfilled by their use in the rituals. The second is that of technology pure and simple: how exactly to construct altars with specific shapes, specified size and by using a specific numbers of bricks, and how to vary shapes of altars without affecting their size or area.

It is clear that the geometry originating in the *Sulbasutras* had more to do with the second standpoint. Thus, for example, whether a falcon-shaped altar provided for the entry into the sacrificed heaven or annihilation of enemies, is irrelevant to the problem of constructing it to conform to certain size and shape. As a matter of fact these problems would remain the same if somebody wanted that structure in their garden for ornamental purposes. In other words, the geometry developed in the *Sulbasutras* was required exclusively for solving the technical problems involved in constructing brick-structures. And it is this geometry that should be of interest to the historian of mathematics.[32]

Indeed, it may even be argued that since the *Sulbasutras* contain sections constituting a geometry textbook, and in particular those detailing the geometry necessary for designing and constructing the altars. Hence we may well have discovered the oldest applied geometry textbook in existence.[33]

Once the *Sulbasutras* are seen as mainly manuals for technicians, the question then arises where and when were the practical knowledge relating to bricks and brick technology acquired. Now references to bricks are conspicuous by their absence from the most sacred and earliest of Vedic literature, the *Rigveda Samhita*. When they do make an appearance in a revised text of a later *Veda*, the *Yajurveda Samhita*, bricks are viewed as marvellous and mysterious entities gathered for ritual purposes. It is, therefore, likely that the priests organising the rituals were acquainted with the burnt bricks from the same sites and would in course of time invest them with magico-religious properties.

In one of the last revised texts to *Yajurveda* appears the *Satapatha Brahmana* in which there is a discussion of the conducting of *Agnicayana*. The magico-religious elements of this ritual are accompanied by a short discourse on the construction of brick altars of various shapes and sizes. While the discussion lacks the geometrical sophistication of the *Sulbasutras*, it shows knowledge of brick technology which probably indicates that the Harappa culture was slowly percolating into the Vedic rituals. Staal's (1978) conclusion is particularly pertinent here:

> If we wish to understand more, we have to go beyond the texts and place the Agnicayana in a wider historical perspective. The techniques for firing bricks, which we met for the first time in the Yajurvedic texts dealing with the Agnicayana, could not have been imported by the Vedic nomads who had earlier entered the sub-continent from the north-west. Nomads have no need for bricks. The bricks, in fact occupy an exceptional position in the Vedic ritual, where all implements are made of perishable materials, and are taken away, destroyed, burnt or immersed in water after a ritual performance has been completed...

For the firing of bricks we have to look to a non-nomadic population. What springs to mind is that in the Harappan civilisation the use of fired bricks was normal and widespread. It is reasonable to suppose that knowledge of the techniques for firing bricks was preserved among the inhabitants of the sub-continent even after the Harappa civilisation had disappeared. So, where, then are we to look back for the origins of geometry in India? The common view is that the *Sulbasutras* are the source. However, a plausible hypothesis is that if the geometry embodied in the *Sulbasutra* texts is viewed as the outcome of a long and sophisticated tradition of brick technology, this geometry must have come into being when there was in fact an advanced form of brick technology with a long tradition behind it. This, in other words, would mean that whatever may be the time of the actual codification of the *Sulbasutras*, their contents came from a different period. The presumption, in short, is that the geometry which we find eventually codified in the *Sulbasutras* originated in the Harappan period. If this presumption is correct, the first and earliest of the discontinuity in the chronology of Indian mathematics hinted at in a previous chapter has been filled with the assistance of bricks.

3.6   Vedic Geometry and Greek Geometry: Possible Links

In 1997, Michael Witzel wrote: "Vedic sacred geometry is autochthonous (i.e. indigenous), and analogies between various cultures are not enough to prove actual historical exchange between them." This was a response to a long debate following a series of studies published by Seidenberg (1962, 1978, 1983) and Van der Waerden (1961, 1983) pointing to the similarities of the geometries of Greece and Vedic India both in terms of their content and their ritual applications. Van der Waerden (1983, p. 11) summed up the similarities appositely as follows:

Seidenberg has pointed out that in Greek texts as well as in the *Sulbasutras*, geometrical constructions were regarded as important for ritual purposes, namely for constructing altars of given form and magnitude. In Greece this led to the famous problem of "doubling the cube," whereas in India it was not the volume but the area of the altar that was considered important.

In both case, the one essential step in the construction was the solution of the problem: to construct a square equal in area to a given rectangle. To solve this problem, exactly the same construction was used in Greece and in India, a solution based on the Pythagorean Theorem. Also, the ideas about the religious importance of exact geometrical altar construction were very similar in both countries. Neither Seidenberg nor Van der Waerden based their theories on the common origin of Greek and Vedic geometry or on their discovery of similar elementary mathematical propositions in both traditions which could have led to the riposte of them being independent discoveries.[34] Instead, Seidenberg pointed to the similarity of certain specific altar constructions of which there were no other parallels found elsewhere in the ancient world. Van der Waerden went one step further and located the common source of the two geometries to an unspecified area in Eurasia.[35] It was left to Joseph Needham in his work on Chinese science to point out that as far as the corridor of communication of scientific ideas existed during the ancient and medieval periods, the Eurasian continent constituted an undivided unit. So the view that different cultures inhabited separate cognitive worlds in isolation from one another represented a misreading of history. The perspective that during the medieval period mathematics developed through complex interactions between the mathematics of Greece, India, China and West Asia,[36] reaching western Europe relatively late, the Americas still later and finally becoming global is now widely accepted.[37] However, the view that even "earlier scientific contacts are likely and worthy of study when they exhibit precise affinities even if historical relationships are not easy to document" (Staal, 2001), is only slowly gaining currency. The most telling argument against the 'Greek miracle' is therefore provided by the Indian geometry of the Vedic period as preserved in the *Sulbasutras* of the last millennium BCE.

3.7   From Ritual Geometry to Scientific Geometry

It is significant that the *Sulbasutras* were part of the *Srautasutras* which in turn was connected to the *Yajurveda*, the sacred texts describing the rules of Vedic rituals. Now it is argued that the relationship of the *Sulbasutras* to the *Srautasutras* is similar to that of Panini's work and the *Pratisakhyas*.[38] The *Pratisakhyas* consisted of four treatises pertaining to the creation of harmonic sound through permutation and combination of notes and the special characteristics of pronunciation as they were practised in various schools of Vedic studies. The compilers did not continue to restrict themselves to their own ritual *sutras*, but were interested in developing a more general discipline. The works of Panini were not just practical but exhibited the genesis of a linguistic science. The emergence of geometry from its ritual roots is likely to have happened earlier than that of linguistics from the analysis of recitation. We know that Baudhayana's *Sulbasutra* was compiled before 500 BCE and the Baudhayana *Srautasutra* was compiled in the seventh or eighth century BCE or even earlier.

Instead of modern algebra, the ancient Greeks and Vedic Indians possessed what has been called geometrical *algebra*, an algebra that makes use of geometrical methods. To illustrate, consider a problem posed by Seidenberg: construct a square of side c whose area is equivalent to a given rectangle of length a and width b. In the language of modern algebra, the relationship required can be expressed in one line: For a given rectangle of sides a and b, we seek a square of equal area to that of the rectangle, i.e. a square side c such that: $c^2 = ab$ or $c = (ab)^{1/2}$. This problem is solved by extracting the square root of ab.

In the Indo-Greek geometrical algebra, the construction of a square equal in area to a given rectangle is carried out in two steps. In the first, the rectangle is transformed into a difference of two squares. In the second, this difference is made equal to a square with the help of the Pythagorean Theorem. The Appendix of this chapter shows the solution suggested in the Baudhayana *Sulbasutra*, which also contains a discussion of how to construct a square equal to the difference between two squares.

It is Seidenberg's discovery of the similarity between the complicated constructions in Euclid's *Elements* and the *Sulbasutras*, based on the Pythagorean Theorem, which eventually led to Van der Waerden's argument of a common origin. However, as mentioned earlier, there was one important difference between the two sources: Vedic geometry emphasised equality of *area* whereas the Greeks emphasised equality of *volume*. Since volume is more complex than area, this led to the suggestion that the Vedic geometry might have been the earlier of the two, or a more direct survival of the original from which the two were derived.

Archaeological similarities or even contacts do not provide adequate grounds for speculating productively about scientific contacts. Often there has to be social contacts for there to be scientific contacts. It cannot be without significance that the general area of Bactria (sometimes independent but more often ruled by others, including Persians, Tocharians, Indo-Scythians and later Arabs) has been one of the most important junctions on the trade routes between West, East and South Eurasia. Contacts between the Indus and Oxus regions started early and continued through historical times. The importance of the entire area is unsurprising for it is the heart of Eurasia. It is true that there were links between ancient Mesopotamia and ancient Greece, India and China, especially in the area of astronomy; but the only really close similarities in mathematics pertain to the Vedic Greek geometry. That closeness is explained by the Indo-European background which also illustrates how ritual geometry was transmitted along with language and ritual. As Staal (1999, p. 123) points out: "Togolok [near Merv in present-day Turkmenistan] is about equally distant from Athens and Banaras: almost 2,000 miles as the crows flies. Our nomads and semi-nomadic pastoralists did not fly. They had to pass through deserts, circumvent rivers, lakes or seas and cross high passes, often carrying food and other necessities. Horses or camels assisted them on many occasions. They did not carry bricks, notebooks or paper. The transmission of their knowledge was oral for it preceded the invention of writing and numerals. That ritual geometricians went along on these expeditions and carried not only their language but also their science in their heads."

3.8   Conclusion

The geometry of the Vedic Indians (and of the ancient Greeks) was based on rituals which required that changes be made in the size and shape of altars and bricks. From that ritual background a more theoretical form of geometry developed, just as (in India) linguistics arose from the practice of recitation. Prior to attaining that more universal and scientific level, the geometry was inherent in rituals that wandering nomads or semi-nomads performed on their journeys. Rituals and ritual geometry went the same ways as the Indo-Aryan or Proto-Indo-Aryan languages: the majority of speakers went from Central to South Asia; but a tiny group reached the shores of the Eastern Mediterranean and left enough elements of ritual geometry behind for a kindred form of geometry to develop and flourish also there. Finally, there are two other issues which should be confronted in any discussion of Vedic geometry. They can be summed by two questions:

* Is Vedic geometry really geometry?
* Is Vedic geometry relevant in tracing the development of Indian mathematics?

 While the two questions are addressed, however tangentially in the Appendix that follows this chapter, it is important to recognise that there is a tradition of Western scholarship in the study of Chinese and Indian mathematics [such as Lloyd (1990)] that is constantly searching for parallels to Euclid's proofs that involve drawing logical deductions from axioms. Staal (1965) has argued convincingly that the Indian counterpart to Euclid's axiomatic deduction is not found in mathematics but in Panini's grammar. This is part of the larger argument that Indian philosophy was inspired just as much by grammar as the European was by mathematics. As for the common belief, that later Indian mathematics was only indirectly influenced by the *Sulbasutras*, this has been effectively refuted by Hayashi (1995) and Staal (2001).

Endnotes

[1] The absence of a horse among the animals featured on the Harappa seals has been noted and seen as indicative of the 'non-Aryan' nature of the Indus Civilisation. Against this background, Jha and Rajaram's (2000) discovery of a 'horse seal' from Mohenjo-Daro was a sensation until an exposé by Witzel and Farmer in their 'Horseplay in Harappa', showed it to be a computer-created picture that had allegedly converted the image of half a bull on Seal 453 into a full horse. What this incident illustrates is the ease with which extraneous considerations creep into unresolved issues regarding the origins of the Indus civilisation.

[2] 13.5 to 13.7 gm weight seemed to have been the standard weight used in the Indus valley.

[3] In Harappa, a fragment of a bronze linear measure graduated to one-half (9.34 mm) of a later Vedic unit, the *angula*, was also found, suggesting possible links between the two periods.

[4] Among the material remnants of geometrical knowledge are circle-drawing instruments found in excavations at Harappa and Mohenjo-Daro.

[5] Different notational systems co-existing side by side have characterised a number of historical cultures, notably the hieroglyphic and hieratic numerals in the ancient Egyptian culture and the 'head variant' and 'bars and dots' numerals in the case of Mayan culture. For further details of these different notational systems, see Joseph (2011).

[6] There are some interesting parallels with the early Sumerian systems of recording numbers. Whereas we use the same number signs regardless of their metrological meaning (the "three" for sheep is the same sign as the "three" for kilometres, or jars of oil), the Sumerians resorted to a wide variety of different symbols. Research has identified around 60 different number signs, which they group into a dozen or so systems of measurement. For example, the Sumerians used one system for counting discrete objects, such as people, animals or jars, and another for measuring areas. Each system had a collection of signs denoting various quantities. For further details of the Sumerian system of numeration, see Joseph (2011, 134–136).

[7] Briefly, one should (i) check the direction in which the inscriptions should be read which previous studies has definitively established as being right to left; (ii) whether the word boundaries established matches what is indicated by previous studies, and (iii) whether a frequency distribution patterns of the signs in the Indus inscriptions matches the sounds in the language proposed.

[8] The *'pandits'* had intensive training in the Sanskrit language and grammar, literary composition and interpretation, logic and recitation. Those who were the repositories of Vedic knowledge were trained from ancient times in three main types of recitations. The first involved taking words together according to rules governing their phonetic combination. The second involved recitation word by word with a pause after each word. The third involved grouping the words into pairs and reciting group by group or step by step. It was thanks to this group that a whole corpus of knowledge survived over thousands of years and can still be accessed today.

[9] The construction of the *Agnicayana* altar is discussed in the *Yajurveda*. It may be a prototype of a later model of a Hindu temple. Built with a 1,000 bricks, laid out in five layers (symbolically representing the five physical elements as well as the five senses), it is based on the Vedic division of the universe into three parts: earth, atmosphere (ether) and sky which are assigned numbers 21, 78 and 261 respectively, adding to 360 or the number of days in a year. An *Agnicayana* ceremony was carried out in Northern Kerala in 1975 and documented by Frits Staal (1983).

[10] A similar arrangement existed for many Vedic sacrificial ceremonies. The three fire altars were the 'householders' (*Garhapatya*) which had to burn continuously under the care of the head of the household; the 'oblation' (*Ahavaniya*) for solemn presentation to the Gods; and the 'southern' altar (*Daksinagni*).

[11] Chronologically, this period of Indian astronomy and mathematics is taken to commence from when the Vedic hymns began to be composed, which some date as going back as far as 4000 BCE! As mentioned in a previous chapter, issues regarding early Indian chronology have unfortunately become a tug-of-war between those who see themselves as the guardians and promoters of impartial scholarship and invariably adopt conservative dating and others who make excessive claims of antiquity for the early sources of Indian mathematics and astronomy. The tunnel visions of both groups make the task of incorporating recent discoveries, necessitating revisions to the conservative dating of the Vedic period, a more difficult one. What present evidence indicates is that the final forms of both *Satapatha Brahmana* and *Sulbasutras* should be placed earlier than the conservative dates attributed to these texts.

[12] The literature includes primarily four Vedic compilations (i.e. *Rigveda*, *Yajurveda*, *Samaveda* and *Atharvaveda*) in their various recensions. There are a set of commentaries called *Brahmanas* of which the *Satapatha Brahmana* is the most important for our purpose; a set of philosophical treatises called *Upanishads*; six *Vedangas*, written for the purpose of instilling the correct methods of recitation of the *Vedas* and performing Vedic rituals. Of the last set, the *Jyotisa* and the *Kalpa* are particularly important for our purposes since the first contains early knowledge of astronomy and the last contains the *Sulbasutras*.

[13] *Sulba* (*or sulva*) stands for rope, string or cord, and is derived from the root *sulv*, meaning 'to measure'. The word *sutra* means a thread and refers to a rule or a collection of formulas in the form of a manual. The *Sulbasutra* is therefore literally 'The Rules of the Cord'. To this day a cord is part of the basic equipment carried by an Indian mason in surveying or in laying out a structure in any construction activity. Three of the more mathematically important *Sulbasutras* were the ones recorded by Baudhyana (about 800 BCE), Apastamba (about 600 BCE), and Katyayana (about 200 BCE). The *Baudhyana Sulbasutra* is the biggest and divided into three chapters containing 525 *sutras*. The *Apastamba Sulbasutra* is divided into six sections containing 223 *sutras* in twenty one chapters. The *Katyayana Sulbasutra* is divided into two parts containing 130 *sutras* and presents the information in a more systematic way. A fourth, known as the *Manava Sulbasutra*, has a discussion regarding

a falcon-shaped altar which played an important part in ceremonies involving ritual practices. There exist other *Sulbasutras* written by Maitrayana, Varaha and Vadhula and later commentaries on some of the *Sulbasutras* are also available. Little is known about these *sulbakaras* (or composers of *Sulbasutras*) except that they were not just scribes but also priest-craftsmen performing a multitude of tasks including design, construction and maintenance of sacrificial altars.

[14] Square and circular altars were mostly sufficient for household rituals, while elaborate altars (such as the ones for the *Agnicayana* rites) contained shapes which were combinations of these basic figures and included rectangles, triangles and trapeziums.

[15] A *vyama* is equivalent to 96 *angulas* and a *purusha* 120 *angulas*, where an *angula* is the width of a finger approximately equal to $\frac{3}{4}$ inch or 1.8 cm.

[16] Altars for daily sacrifices were *Ahavaniya* (Square), Daksina (Semi-circular) and *Garhapatya* which could be a square or a circle. However, irrespective of the shape, the area of each had to be the same and equal to one square *vyama*.

[17] The falcon shape has a symbolic significance. The *Taittiriya Samhita* states: "He who desires heaven may construct the falcon-shaped altar; for the falcon is the best flyer among the birds; thus he [i.e. the sacrificer] having become a falcon himself flies up to the heavenly world." There were falcons with four-tipped, five-tipped and six-tipped wings. There were also altars resembling other birds (e.g., heron or *kanka*), geometrical shapes (such as *prayuga* or triangle), *ubhayatah prayuga* (or triangles on both sides of a rhombus), *drona* (or trough), *kurma* (or tortoise), *ratha-chakra* (or wheels of a chariot), etc.

[18] The most common fire altars for other optional sacrifices were of different shapes: a triangle (usually an isosceles triangle), rhombus, circle, isosceles trapezium and tortoise. Each of them had the same area of $7\frac{1}{2}$ square *purushas*). For a discussion of the construction of these altars in terms of the numbers, shapes and areas of the bricks required, see Staal (1983).

[19] The measures used in the *Saptapatha Brahmana* were the same as in the *Sulbasutras*. The important units of measurement were:

1 *pradesa* = 12 *angulas*
1 *pada* = 15 *angulas*
1 *prakrama* = 2 *padas*
1 *purusha* = 120 *angulas*
(A *prakrama* = 30 angulas is approximately 0.5 metres.)

[20] In Vedic mythology, time was represented by the metaphor of a bird. The year was divided into six seasons, with the head of the bird being the

vasant, the body being hemanta *and sisira*, the wings being **sarad** *and grishma* and the tail being *varsha*. It should be noted that although the word *syena* is generally used for a falcon, it is actually a comprehensive term for eagles, falcons and hawks, which constitute one of the three groups into which birds of prey were classified in ancient Indian texts. (All birds of prey are supposed descended from the primeval *Garuda*.) In fact *syena* is often used as a synonym for the *Suparna*, the celebrated golden eagle that is the strongest and fastest of the family.

[21] Seidenberg (1983, pp. 113–116) contains an interesting discussion of ambiguities in the Vedic texts relating to equivalences of area as well as the philosophical underpinnings of such a requirement.

[22] Apastamba's procedure can be interpreted as follows:

> Let the east-west line be x units in length. If the length of the cord is augmented to $x + x/12$, and the mark is at a distance of $5x/12$ and the other part of the cord is $13x/12$, then if the ends of the cords are tied to the ends of the east-west line, and the cord is stretched up to the mark, we get a right-angled triangle whose sides are x, $5x/12$ and $13x/12$. This relationship will hold for any integral value of x.

[23] Two different types of bricks were used in altar construction. There were ordinary bricks (*lokamprina*) and special (*yajushmatt*) bricks each of which had been consecrated and marked for purpose of identification. In the case of both types of bricks, they varied according to different size and shapes.

[24] Apart from minor variations, the body of the first-layer falcon-shaped altar was 4 square *purushas*. The wings and tail were one square *purusha* each plus the wing increased by 1/5 of a square *purusha* each and the tail by 1/10 of a square *purusha* so that the image would more closely approximate the shape of a falcon. Thus the total area of the top layer of *Vakrapaksa-Syena* altar is:

$$4 + (2 \times 1.2) + 1.1 = \textbf{7.5} \text{ square } purusha.$$

[25] In *Katyayana Sulbasutra* (5.4) appears the following instruction:

> For the purpose of adding a square *purusha* (to the original falcon-shaped altar), construct a square equivalent to the original altar together with the wings and tail, add to it a square of one *purusha*. Divide the sum (i.e. the resulting square) into 15 parts and combine two of these into a square. This will be the (new) unit of square *purusha* (for the construction of the enlarged figure).

[26] The instructions given in *Satapatha Brahmana* (X.2.3.11-14) for constructing a falcon-shaped altar consisting of 95 layers of bricks can be

interpreted as:

$$\text{Area of the body } (Atman) = 56 + \left(\frac{12}{7} \times \sqrt{56}\right),$$

$$\text{Area of two wings} = (2 \times 14) + \left(\frac{3}{7} \times \sqrt{14}\right) + \left(\frac{3}{7} \times \sqrt{14}\right)$$

$$+ \left(\frac{1}{5} \times \frac{1}{7} \times 3 \times \sqrt{14}\right),$$

$$\text{Area of tail} = 14 + \left(\frac{3}{7} \times \sqrt{14}\right) + \left(\frac{1}{10} \times \frac{1}{7} \times 3 \times \sqrt{14}\right).$$

The total area is about 116 square *purushas*, which is an over-estimate of the required 101.5 square *purushas*, arising in part from a rounding-off error from taking 14 rather $13 + 8/15$. *Baudhyana Sulbasutra* contains an explanation of how the last number is obtained. Expressed in modern notation:

Let the new unit after the mth augmentation be x.
Then,

$$x^2 = 1 + \left(\frac{2m}{15}\right),$$

where m runs from 1 to 94
For $m = 94$,

$$x^2 = 13 + \frac{8}{15}.$$

So that 14, the estimate used, is a rounding up of this number. The use of this more accurate figure gives the calculated total area as 110 square *purushas* which is closer to 101.5 square *purushas* than 116 square purushas.
[27] The essence of this method involves two common sense assumptions:

(i) Both the area of a plane figure and the volume of a solid remain the same under rigid translation to another place.
(ii) If a plane figure or solid is cut into several sections, the sum of the areas or volumes of the sections is equal to the area or volume of the original figure or solid.

The reasoning behind this approach was very different from that of Euclidean geometry, but the method was often just as effective, as shown in the Indian (and Chinese) "proofs" of the Pythagorean Theorem discussed in Joseph (1994). For example, the following sizes and shapes of bricks used to construct one of the layers of a falcon-shaped altar can be "dissected and reassembled" from a square.

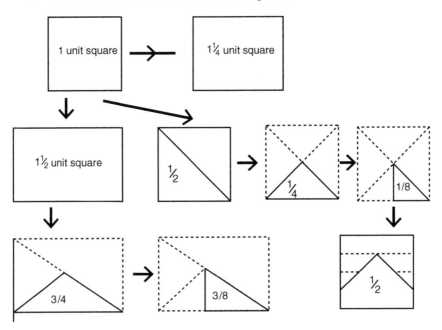

[28] However, it is interesting to note that the *Sulbasutras* contain computations of surds such as the square roots of 2, 3, 5 with attempts being made achieve computational accuracy, notably in the case of the square root of 2.

[29] Incidentally, some traditional Indian architectural texts suggest that the diagonal of a unit square is found by adding 5/12 of the side to itself which gives:

$$\sqrt{2} = 1 + \frac{5}{12} = 1 + \frac{1}{3} + \frac{1}{3 \times 4}.$$

[30] Various rituals required the day-time and night-time to be divided into two, three, four, five and fifteen equal parts. In the 15-fold division, each part was a *muhurta* or what would be equivalent to one-fifteenth of (12 × 60) or 48 minutes.

[31] For further examples, see Kak (1993) who gives examples of equivalences quoted here.

[32] There are indications in the texts that the authors of the *Sulbasutras* were aware of this distinction, for often one comes across expressions, "Thus

we are told" or "Such are our instructions," etc. The implication is that these instructions (say, on the sacrificial efficacy of different shaped altars or the astronomical codes to be adhered to) are not particularly relevant to the main purposes of the texts. These instructions are simply taken for granted, while the texts themselves pay exclusive attention to the technique of executing them. In fact, the texts are good exemplars of how exact science may grow directly out of applications.

[33] It is not our purpose to get into a debate on whether the *Sulbasutras* contained geometry or whether it predated Greek work, except to quote Thibaut, who provided the first translation and commentary in 1874. He wrote:

> In the first place they [the *Sulbasutras*] are valuable for the history of the human mind. In the second place, they are important for the mental history of India and for answering the question relative to the originality of Indian science. For whatever is closely connected with ancient Indian religion must be considered as having sprung up among the Indians themselves, unless positive evidence of the kind point to the contrary conclusion.

[34] It is useful to recognise an important difference between the Greek and Indian traditions of constructing altars. Ancient Greek civilisation left no living tradition of ritual and geometric knowledge while such knowledge still survives in India in the form of oral traditions in unbroken chains of transmissions over 1,000 of years.

[35] However, Van der Waerden (1983, p. 14) does mention vaguely 'the Danube region' as the common source of the geometries of both traditions without pursuing this matter any further.

[36] Mainly the area in West Asia that consists of present day Iraq, Iran and segments of neighbouring regions.

[37] The view that all mathematics (and philosophy) started around 600 BCE in ancient Greece is now less appealing in the face of mounting evidence accumulated over the last 100 years or so. However, there are still those who subscribe to the view of the uniqueness of Greece with the belief that the western science originated in Greece; and that in other traditions the knowledge that developed was not science as we understand it today. (Staal, 1999, p. 105).

[38] In the sixth century BCE, Panini completed his pioneering work in the field of Sanskrit grammar and linguistics. Besides, expounding a comprehensive and scientific theory of phonetics, phonology and morphology, Panini provided formal production rules and definitions describing Sanskrit grammar in his treatise called *Asthadhyayi*.

Appendix. Further Mathematics from the *Sulbasutras*

The *Sulbasutras* deal with construction of altars of different orientation, size, shape and area. They are not mathematical treatises. But implied in them are results relating to the geometry of different shapes and other geometrical truths which we are familiar with.

A.3.1   Theorem of the square of the diagonal (The 'Pythagorean' Theorem)

Knowledge of the Pythagorean result has a long history in India and elsewhere. As mentioned earlier, the *Satapatha Brahmana* (*c.* 1000 BCE), contains instructions for the construction of a *Mahavedi*, 36 units in length along the east-west (*praci*) line and 30 units along north-south line (see Figure 3.6). From that it is inferred that the *praci* and half the north-south line constitute two sides of a right-angle triangle (i.e. Δ ABC in Figure 3.6 has sides 36, 15 and 39 units). Besides, integer triples, such as $(5, 12, 13)$, $(12, 16, 20)$, $(8, 15, 17)$, $(15, 20, 25)$, $(12, 35, 37)$, ... rational triples such as $(2\frac{1}{2}, 6, 6\frac{1}{2},), (2\frac{1}{4}, 3, 3\frac{3}{4}), (7\frac{1}{2}, 10, 12\frac{1}{2}), (78\frac{1}{3}, 188, 203\frac{2}{3})$, ... were present in other *Mahavedis* whose areas were either integral multiples or fractions of areas of altars of the same or different shapes.[1]

Different versions of the Pythagorean result are found in the *Sulbasutras*. They include:

> The rope [stretched along the length] of the diagonal of a square produces an area double the size of the original square (Baudhayana and others).

Or more generally,

> The rope [stretched along the length] of the diagonal of a rectangle makes an area which (its) vertical and horizontal sides make together (Apastamba).[2]

Using these results the composers of the *Sulbasutras* were able to construct a square equal to the sum of two or more given squares or a square equal to the difference of two given squares. Further, they were able to 'turn' a rectangle into a square or vice versa, deduct one square from a larger square and even attempt to 'square a circle' and 'circle a square'.[3]

Two interesting aspects to the *Sulbasutra* versions of the Pythagorean Theorem, recognised by the authors of the *Sulbasutras* themselves, as the most fundamental theorem in geometry may be noted:

(a) The theorem is stated *separately* for rectangle and square where there is no mathematical distinction between the two; and
(b) The theorem focuses on properties of rectangles (or squares) rather than properties of right-angle triangles.

This is probably a reflection of an essential difference between a practical manual such as the *Sulbasutras* and a geometry textbook, such as Euclid's *Elements*, where the theorem makes its first written formal appearance in Greek geometry. However, as stated earlier, Seidenberg remarks on the similarity between the genesis of the two traditions which is particularly striking when geometric shapes are being transformed.

A.3.1.1 *Sulbasutra constructions*

As befits a surveyor's manual, there are a large number of geometric constructions in the *Sulbasutras*. Below is a list, by no means exhaustive, of these constructions with their sources given in brackets as abbreviations: (Baudhayana = BSS; Apastamba = ASS; Katayayana = KSS). A fuller list is found in Balachandra Rao (2004, 16–17).

- Divide a circle into any number of equal areas by constructing diameters (BSS ii, 73–74; ASS vii, 13–14).
- Divide a triangle into a number of equal and similar areas (BSS iii, 256).
- Draw a straight line at right angles to a given line (KSS i, 3).
- Draw a straight line at right-angles to a given line from a given point on it (KSS i, 4).
- Construct a square on a given side (All).

- Construct a rectangle of given length and breadth (BSS i, 36–40).
- Construct an isosceles trapezium of given altitude, face and base (BSS i, 41; ASS, v, 2–5).
- Construct a square equal to the sum of two different squares (BSS i, 51–52; ASS ii, 4–6; KSS ii, 22).
- Construct a square equivalent in area to two given triangles (All).
- Construct a square equivalent in area to two given pentagons (BSS iii, 68; KSS iv, 8).
- Construct a square equivalent in area to a given rectangle (BSS i, 58; ASS ii, 7; KSS iii, 2).
- Construct an isosceles trapezium having an area equivalent to a given square or rectangle (BSS i, 55).
- Construct a triangle equivalent in area to a given square (BSS i, 56).
- Construct a rhombus equivalent in area to a given square or rectangle (BSS i, 57; ASS xii, 9).

To get a flavour, consider the two instructions for constructing a square, using only a string or rope, found in the Baudhyana *Sulbasutra*.

(i) **Construct a square from two perpendicular diameters in a circle.** To draw a line, fix a pin at the middle of the string. Slip the tied end on to this pin and draw a circle with the mark (at the middle of the cord). Fix pins at the ends of the diameter. With the end-tie on the eastern pin draw a circle with the whole cord. Similarly with the western pin. The second diameter should be drawn through the points where these (circles) intersect. [The diameters are assumed to be vertical (N-S) and horizontal (E-W)]

(ii) **Construct a square on a given side.** [The given side is assumed horizontal (E-W)] To construct a square, tie up at both ends of a string as long as the desired side and make a mark at its middle. Draw a line (along the string) and fix a pin at its midpoint. Fixing the ties on this pin, draw a circle by the middle mark (of the string) and at the ends of the diameter fix pins. Fix one tie on the eastern pin and draw a circle with the other tie.

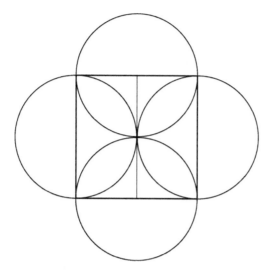

Figure A.3.1. Fish method.

Similarly, round the western pin. Through the points where they meet the second diameter should be drawn and pins should be fixed at this end. With the ties on the eastern pin a circle is to be drawn with the middle mark. Similarly, (do the same) round the southern, western and northern (pins). Their outer points of intersection form the square.

Construction (ii) (shown in Figure A.3.1) produces a pleasing geometrical pattern of four intersecting circles whose centres lie on the sides of the square. This is well-known to generations of school children as the 'fish method'.

A.3.1.2 *Combining areas*

(i) **Construct a square equal to n times the area of a given square.** The Katyayana *Sulbasutra* contains a simple method of doing so. Expressed in modern terminology, it requires the construction of an isosceles triangle ABC whose base AB is of the length $(n - 1)a$ where a is the side of a given square. Let AC $=$ BC $= \frac{1}{2}(n + 1)a$. The instructions in the *Sulbasutra* is

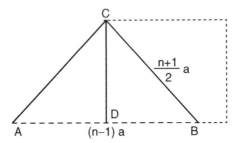

Figure A.3.2. A square from n squares.

summarised as follows (Figure A.3.2):

Drop the perpendicular CD to bisect AB since triangle ABC is isosceles. So BD $= \frac{1}{2}(n-1)a$. In right-angled triangle BDC,

$$CD^2 = BC^2 - BD^2 = \left[\frac{1}{2}(n+1)a\right]^2 - \left[\frac{1}{2}(n-1)a\right]^2 = a^2 n.$$

Thus a square with side CD $= a\sqrt{n}$ will have an area $= a^2 n$ or n times the area of the given square.

(ii) **Construct a square equal in area to two given unequal squares.** The instructions are: cut a section off equal to the width of the smaller square (side b) from the bigger square (side a). The rope which is stretched diagonally across the section unites both (squares). Figure A.3.3 is self-explanatory. The construction follow from the result $a^2 + b^2 = c^2$, where c is the side of the resulting square.[4]

(iii) **Construct a square equal in area to the difference of two squares.** The original instructions are somewhat convoluted and has been modified here for the sake of clarity:

To deduct one square from another square, cut off a piece from the larger square a [rectangular] segment equal to the side of the square to be deducted. Then draw a longer side of this segment diagonally across the other longer side and where it falls [on the other side] cut off that portion. The cut-off line is the side of a square the area of which is equal to the difference of the two

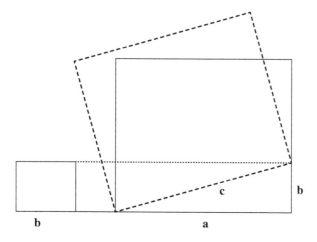

Figure A.3.3. Two squares in one.

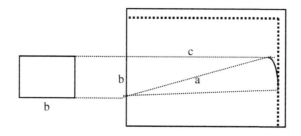

Figure A.3.4. Difference of two squares.

The two given squares are of side a and b. The piece of rectangular segment to be cut is the narrow strips at the top and the right of Figure A.3.4 whose total area is equal to the area of square side b. Applying the Pythagorean result $a^2 - b^2 = c^2$, the resulting square of side c equals the difference between the given squares of sides a and b respectively.

A.3.2 Transformation of figures

As mentioned earlier in the chapter, different shapes were prescribed for the fire-altars depending on the benefit sought: a falcon (*syenacit*) for attaining heaven, an isosceles triangle (*praugacit*) for destroying enemies, etc. However, the transformed shapes had to have the same area as the original figure. The *Sulbasutras* describe different methods of changing the shapes of figures while retaining the same areas.

A.3.2.1 *'Squaring' a circle*

All *Sulbasutras* contain cryptic 'rule of thumb' instructions on how to 'square' a circle [i.e. to find a square equal area to a given circle]. For example, Apastamba suggests: "Divide the [diameter] into fifteen parts and remove two of them. This gives the side of the [required] square."

Let the side of the square be a and the diameter of the circle be d. The instruction suggests that: $a = \frac{13}{15}d$.

So the area of the square is: $a^2 = \left(\frac{13}{15}d\right)^2$.

Now the area of the given circle $\approx \left(\frac{\pi d^2}{4}\right)$.

If area of the square is equated to area of the given circle, then $\left(\frac{\pi d^2}{4}\right) = \left(\frac{13}{15}d\right)^2$,

or $\pi \approx 3.0044$ which is a crude approximation for π.

A.3.2.2 *'Circling' a square*

The following diagrams (Figures A.3.5(a) and A.3.5(b)) are self-explanatory as to the method used in converting a square into a circle.

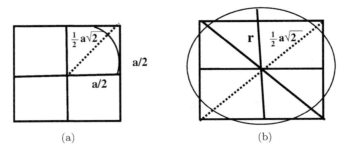

(a)   (b)

Figure A.3.5. Square into a circle.

If a is the side of the square and r the radius of the circle then it can be shown as (Joseph, 2011, 332–333):

$$r = \frac{a(2 + \sqrt{2})}{6} \Rightarrow a^2 = \frac{\pi}{36}[a(2 + \sqrt{2})]^2 \quad \text{which implies } \pi \approx 3.088.$$

For $a = 1$,

$$r = \frac{(2 + \sqrt{2})}{6} \approx 0.5690355.$$

or the area of the circle ($= \pi r^2$ and using Aryabhata's value of π as 3.1416) is 1.017252. This would seem a reasonably accurate conversion for that time of a unit square into a unit circle.

A.3.2.3 *Convert a rectangle into a square and vice versa*

The construction below is the basis for designing a number of different altars:

> Add a small piece (a small square) using the method taught on how to [cut off] the added piece (BSS, i, 54; ASS ii, 7; KSS iii, 2).

Figure A.3.6 and A.3.7 helps to clarify this vague instruction. The rectangle (ABCD) is partitioned into a square (ALKD) and two equal

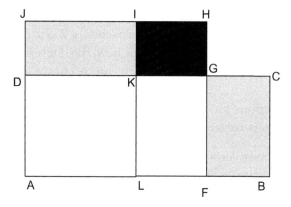

Figure A.3.6. Rectangle into square.

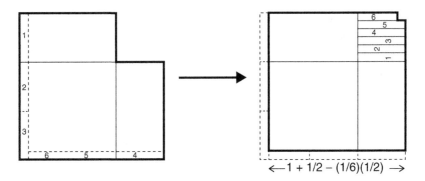

Figure A.3.7. Rectangle into a square.

rectangles (LFGK and FBCG). FBCG is inverted (moved) to DKIJ. The black square (KGHI) needs to be filled in. The last sentence of the instructions indicate in general terms how this can be done.

One way of doing so (see Figure A.3.7) is to take a strip from the left and bottom of the large square — the strips should be just thin enough to fill in the black square. The pieces filling in the little square will have half the length of the original square and six of these lengths will fit along the bottom and left of the large square. Note that there remains a tiny square slot at the top right-hand corner which needs to be filled in by the same procedure outlined above. Also, if the square ALKD is a unit square then it follows from Figure A.3.7 that the final larger square has the side *minus* the dimension of the tiny square slot. Successive application of this procedure will produce the required square whose side approaches $\sqrt{2}$. In a later part of this appendix there is a discussion of how the *Sulbakaras*, following this procedure, may have arrived at an accurate approximation for $\sqrt{2}$.

The instructions for the reverse process of transforming a square into a rectangle is clearer.

> Wishing to transform a square into a rectangle one should cut diagonally in the middle, divide one part again and place the two halves to the north and east of the other part. If the figure is a quadrilateral one should place together as it fits. This is the [desired transformation].

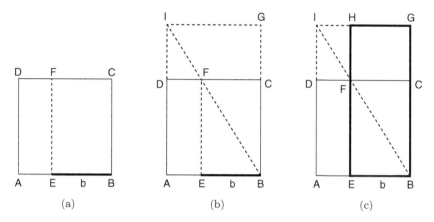

Figure A.3.8. Square into rectangle.

In Figure A.3.8(a), ABCD is the given square. Cut off length b from the sides AB and CD so that the rectangle EBCF is a rectangle with b as the smaller side. Join BF and produce it to cross AD extended at I. Complete the rectangle ABGI [Figure A.3.8(b)]. Produce EF to cross GI at H [Figure 3.8(c)] Then EBGH is the rectangle which is equivalent to the square ABCD. The proof is found from establishing the congruence of triangles ABI and GIB, triangles EBF and CFB and triangles DFI and HIF. Consequently the area of rectangle AEFD is equal to the area of rectangle FCGH. It would follow that area of the rectangle EBGH is equal in area to the given square ABCD. A similar proof can be provided in case the given side b of the rectangle is larger than the side of the square.

A.3.3 Areas of rectilinear figures and the circle

Methods of measuring areas of rectilinear and circular figures may have originated, as in many of the other ancient cultures, from the need to divide land and other immovable assets according to certain customary practices. In the Indian case, there was also a need as part of ritual to ensure that the altars such as *Garhpatya*, *Ahavaniya* and *Dakshinagni* which were circular, square and semi-circular respectively, were of equal area. A single example from the *Sulbasutras* is given as

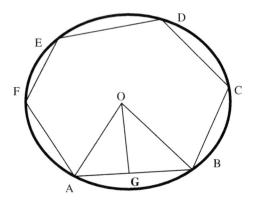

Figure A.3.9. Area of an inscribed hexagon.

A.3.3.1   *To find the area of a regular hexagon inscribed in a circle*

In Figure A.3.9, the area of the regular hexagon ABCDEF can be shown to be six times area of triangle AOB. This follows from the application of the Pythagorean Theorem.

$$OG^2 = OB^2 - \left(\frac{1}{2}AB\right)^2 = OB^2 - \left(\frac{1}{2}\right)^2 AB^2.$$

So area of hexagon ABCDEF $= 6 \times [(\frac{1}{2})AB \times OG] = 6 \times$ [area of triangle AOB].

A.3.4   Calculation of square root of 2

A Babylonian cuneiform tablet dating to about 1600 BCE in the Yale Collection shows a square with two diagonals [see Figure 4.2 in Joseph (2011, p. 145)]. The side and diagonal are labelled with the sexagesimal (i.e. base 60) values of 30 and 42; 25, 35 respectively. Below that diagonal is given the sexagesimal estimate of $\sqrt{2}$ as $1; 24, 51, 10$. In decimal notation, this approximates to: $1+60^{-1}(24)+ 60^{-2}(51) + 60^{-3}(10) \approx 1.41421297$ which is the square root of 2 correct to 5 decimal places.

A sexagesimal equivalent of the value of $\sqrt{2}$ from the *Sulbasutra* is $1; 24, 51, 10, 37, \ldots$, correct to 4 decimal places. The Babylonian procedure for the evaluation of $\sqrt{2}$ probably used an iterative numerical procedure which was the prototype to what came to be known later as the Heron's method (named after an Alexandrian mathematician who lived in the first century CE). This method resembles the procedure for extracting square roots used by the digital computer today.[5] On the other hand, the Indian derivation of square root was more likely to have been based on a geometrical (or 'constructive') method, following from the instructions for constructing a second square whose area was twice the area of the original square, or even from other constructions. In Baudhyana's *Sulbasutra* (i; 52) appears the statement that the diagonal of a square is the side of another square with two times the area of the first square. Now if we consider the side of the original square to be one unit, then the diagonal is the side (or root) of a square of area two, or simply the square root of two.[6] Verse 61 of the same *Sulbasutra* contains the following instruction for finding the length of the diagonal of a square: "Increase the length [of the side] by its third and this third by its own fourth less the thirty-fourth part of that fourth. The increased length (has) a small amount in excess". In modern notation, the rule may be expressed as:

$$\sqrt{2} \approx 1 + \frac{1}{3} + \frac{1}{4} \cdot \frac{1}{3} - \frac{1}{34} \cdot \frac{1}{4} \cdot \frac{1}{3}.$$

We are justified here in using the symbol $\approx$ instead of $=$ here since the above instructions show an awareness that the rule over-estimates the length of the diagonal by a small amount. This is borne out by a simple calculation. The values for the square root of 2 using a calculator and the above rule are 1.414213562 and 1.414215686 respectively where the latter is in excess of the former by 2.12426×10^{-6}.

Two related questions arise:

(a) How did the composers of the *Sulbasutras* derive such an accurate numerical value for the square root of 2?

(b) How did they know that the rule would overestimate the true value?

The *Sulbasutras* give no clues to the answers to either questions. So the suggested answers are merely conjectures. These conjectures fall into two main categories: the reasoning used was either *algebraic and computational* or *geometric and constructive.*[7] The algebraic/computational approaches involve trial and error methods such as the Heron method which is neither in keeping with the 'constructive' spirit of Vedic geometry nor consistent with the level of algebraic and computational skills prevalent in India of that period.

Consistent with the geometric techniques found in the *Sulbasutras* are conjectures that derive more recently from the work of Datta (1932),[8] but has a long history. Nilakantha's (*c.* 1500 CE) explanation is particularly noteworthy in this respect.[9]

Clearly, the method of deriving the approximation for square root of 2 has a geometrical basis. One of the more interesting and plausible ones of recent vintage is that of Henderson (2000) on which this exposition is based. Henderson claims that he can offer a "step-by-step" method (based on geometric techniques in the *Sulbasutras*) that will yield the approximation:

$$\sqrt{2} \approx 1 + \frac{1}{3} + \frac{1}{4} \cdot \frac{1}{3} - \frac{1}{34} \cdot \frac{1}{4} \cdot \frac{1}{3}.$$

In an earlier discussion, the method for constructing geometrically a square that has the same area as any given rectangle was outlined. It may be remembered that the first step was to take the shorter side of the rectangle for the side of a square, divide the remainder into two parts and, inverting, join those two parts to two sides of the square. This process changes the rectangle into a figure with the same area which is a large square with a small square cut out of its corner (see Figure A.3.6).

However, we can move directly to the first three terms of the approximation $\left(1 + \frac{1}{3} + \frac{1}{4} \cdot \frac{1}{3}\right)$ by considering the following dissection and re-assembly[10]:

One of the two squares of length, say 1 *pradesa*, is dissected into three sections of $\frac{1}{3}$ *pradesa*. The re-assembled large square shown on

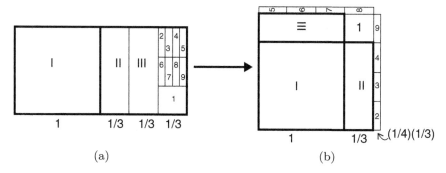

Figure A.3.10. The square root of 2.

the right of Figure A.3.10(b) has side equal to 1 *pradesa* plus $\frac{1}{3}$ of a *pradesa* plus $\frac{1}{4}$ of $\frac{1}{3}$ of a *pradesa*, or 1 *pradesa* and 5 *angulas* and the small square has side of 1 *angula*. [Note 1 *pradesa* = 12 *angulas*.] To make this into a single square we remove a thin strip from the left side and the bottom just thin enough that the strips will fill in the little square. Since these two thin strips will have length 1 *pradesa* and 5 *angulas* or 17 *angulas* we may cut each into 17 rectangular pieces each 1 *angula* long. If these are stacked up they will fill the little square if the thickness of the strips is $\frac{1}{34}$ of an *angula* (or $\frac{1}{34} \cdot \frac{1}{4} \cdot \frac{1}{3}$ *pradesa*). So we get the approximation for square root of 2 given in the Baudhyana *Sulbastura*:

$$1 + \frac{1}{3} + \frac{1}{4} \cdot \frac{1}{3} - \frac{1}{34} \cdot \frac{1}{4} \cdot \frac{1}{3} \approx 1.414215686 \quad \text{correct to 5 decimal places.}$$

Henderson continues this argument by suggesting that with a microscope it will be possible to observe that there is a tiny square still left out or our last approximation is still a little in excess of the *true value*. Follow the same procedure as last, remove a minute strip from the left and bottom edges from the large square and then cut them into $\frac{1}{34} \cdot \frac{1}{4} \cdot \frac{1}{3}$ *pradesa* lengths in order to fill in the left out square. If k is twice the number of $\frac{1}{34} \cdot \frac{1}{4} \cdot \frac{1}{3}$ lengths in $1 + \frac{1}{3} + \frac{1}{4} \cdot \frac{1}{3} - \frac{1}{34} \cdot \frac{1}{4} \cdot \frac{1}{3}$ *pradesa*, then the strips we remove must have width $\frac{1}{2k} \cdot \left(\frac{1}{34} \cdot \frac{1}{4} \cdot \frac{1}{3} \right)$ *pradesa*. k is easily calculated since we know that there were 17 segments of length $\frac{1}{4} \cdot \frac{1}{3}$ in the length $1 + \frac{1}{3} + \frac{1}{4} \cdot \frac{1}{3}$ and each of these segments was divided into 34 pieces and then one of these pieces was removed. Thus $k = 2[34(17) - 1] = 1154$. Therefore, even a finer approximation giving the answer correct to 11 decimal places may be obtained as:

$$\sqrt{2} \approx 1 + \frac{1}{3} + \frac{1}{4} \cdot \frac{1}{3} - \frac{1}{34} \cdot \frac{1}{4} \cdot \frac{1}{3} - \frac{1}{1154} \cdot \frac{1}{34} \cdot \frac{1}{4} \cdot \frac{1}{3}.$$

This procedure can be followed almost indefinitely with improving approximations for the square root of two.[11] The method also works for finding the square roots of other numbers and could be computationally more efficient than existing procedures. Henderson writes:

> The *Sulbasutra* contains many powerful techniques, which, in specific situations have a power and efficiency that is missing in more general techniques.[12] Numerical computations with the decimal system in either fixed point or floating point form has many well-known problems.[13] Perhaps we will be able to learn something from the (apparently) first applied geometry text in the world and devise computational procedures that combine geometry and numerical techniques.

A.3.5   Laying out the fires

An interesting problem that arose in Vedic geometry and referred to earlier is how to ensure the precise distances between and the relative positions of the three fire altars, *Garhapatya* (G), *Dakshinagni* (D) *and Ahavaniya* (A) contained in the *Pracinavams*, shown in Figure 3.4. The general requirement was: *Dakshinagni* should lie south of the line joining the other two fire altars and be at a distance from *Garhapatya* of one third the distance between the other two fire altars.[14]

The *Baudhyana Sulbasutra* contains three different versions of solving the same problem. To state the relevant passage in the first version given in Datta (1932, pp. 203–205), with slight modifications made for sake of clarity:

> "With the third part of the length [i.e. the distance between *Garhapatya* (G) and *Ahavaniya* (A)] describe three squares closely following one another (from west towards the east); *Garhapatya*

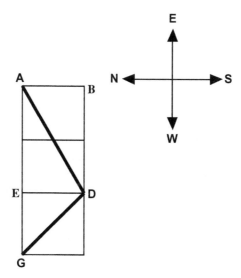

Figure A.3.11. Location of altars.

is located at the north-western corner of the western square; *Dakshinagni* (D) is at its south-eastern corner; and *Ahavaniya* at the north-eastern corner of the eastern square" (BSS i; 67)

Datta (1932, pp. 203–205) proceeds to use these instructions to construct Figure A.3.11 to obtain an estimate of the relative distances between the fire-altars. Given $AG = x$, $AB = x/3$ and $BD = 2x/3$. Using the Pythagorean result it can be shown that

$$AD = \sqrt{AB^2 + BD^2} = \sqrt{\frac{x^2}{9} + \frac{4x^2}{9}} = \frac{\sqrt{5}}{3}x$$

and

$$GD = \sqrt{ED^2 + EG^2} = \sqrt{\frac{x^2}{9} + \frac{x^2}{9}} = \frac{\sqrt{2}}{3}x.$$

Endnotes

[1] The *Soutramani* altar uses a right-angled triangle with sides $5\sqrt{3}$, $12\sqrt{3}$, $13\sqrt{3}$ and the *Aswamedha* altar a right-angled triangle with sides $15\sqrt{2}$, $36\sqrt{2}$, $39\sqrt{2}$. These irrational numbers must have arisen because of the

requirement that an altar be constructed accurately whose area is a multiple of another altar of the same shape.

2 The *Manava Sulbasutra* expresses this theorem in a different way from the other *Sulbasutras* and conforms closer to today's representation. It amounts to saying that the hypotenuse is obtained from taking the square root of the sum of the square of the length and the breadth of a triangle, being the smaller two sides.

3 This is an attempt to construct a square *equal* in area to a given circle or a circle *equal* in area to a given square. This is of course impossible.

4 This form of 'visual proof' is clearly consistent with the way that the *Sulbakaras* proceeded with their demonstration. However, this has led some historians of mathematics to wonder whether the whole result was no more than guess-work. This is to misunderstand the nature of demonstration contained in Indian mathematics in general and in the *Sulbsutras* in particular. The demonstration here involved no explicit proof but a proof implicit in the construction.

5 Let x be the number whose square root you want to find and the positive number a being your initial guess of the answer. Then $x = a^2 + e$ where the difference or error (e) can be positive or negative. We now try to find a better approximation for the square root of x which we denote by $a + c$. It is obvious that the smaller the error e, the smaller is c relative to e. Thus we impose the following condition on c:

$$x = (a+c)^2 = a^2 + e,$$

from which we get

$$2ac + c^2 = e.$$

Now if you made a sensible guess for a in the first place, c^2 will be much smaller than $2ac$ and can therefore be ignored. Therefore,

$$c \approx \frac{e}{2a}.$$

Hence, an approximation for the square root of x is:

$$a + c \approx a + \left(\frac{e}{2a}\right) = a_1.$$

Now taking $a_1 = a + \left(\frac{e}{2a}\right)$ as the new 'guess-estimate' the process can be repeated over and over again to get $a_2, a_3, a_4, \ldots$, as better and better approximations. Applying this method to find the square root of 2 up to the third approximation (a_2) gave the Babylonian sexagesimal value of $1; 24, 51, 10 \cong 1 + 60^{-1}(24) + 60^{-2}(51) + 60^{-3}(10) \approx 1.414221297$ correct to five places of decimals.

[6] The Sanskrit word for this length is *dvi-karani* or, literally, "that which produces 2".

[7] This is part of an old debate regarding the methodology of Indian mathematics which is believed to be predominantly algebraic and computational and hence neglected proofs and rationales. This ignores the fact that there was even a School (the Kerala School in Chapter 10), that emphasised proofs and delighted in demonstrating algebraic results geometrically. This theme is taken up in the concluding section of the last chapter.

[8] A short discussion of Datta's approach is found in Joseph (2011, pp. 334–336).

[9] In *Aryabhatiyabhasa*, Nilakantha writes:

> How does Baudhyana explain his method that does not involve the process of taking square roots (*vargamu likaranam vina*) to find the diagonal of a square (*samakarana*)? Baudhyana states that the side of the square itself added to the sum of a third part of the side of the square (*samacaturasra bahu svatryamsam*) and its own fourth part and lessened by its thirty fourth part gives the diagonal (*karna*). It is not his fault that he has given only a practical (or approximate) value.... (It) is better than the value (normally) used in the construction of sacrificial altars. To get this approximation he assumes a square of side twelve (*dvadsa bahukam samacaturasram*) and then proceeds as follows. If the square of twelve is doubled it will be eighty eight more than 200 (i.e. $2 \times 12^2 = 288$). This (result) is equal to one less than the square of seventeen (i.e. $17^2 - 1 = 288$). Therefore the square root is less than seventeen by a fraction of one with denominator twice seventeen. How is it known that it is less by a fraction of unity with denominator twice seventeen? The square root multiplied by 2 is to be removed from the remainder and the argument is the same. Thus the unit square (that is to be removed) is a multiple of a thirty-fourth part of one and so it is divided by thirty four. Seventeen minus thirty-fourth part of one will be the diagonal (of the square) having (in area) two hundred and eighty eight units and so it is stated that to twelve one third of it is to be added which in turn to be added to its one fourth and lessened by one thirty-fourth (of it). Twelve added to its one third (i.e. 4) gives sixteen which added to one fourth of one third gives seventeen. From this if one thirty-fourth of that one fourth is removed, what will be obtained is the 'approximate' diagonal (*asanna karma*).

[10] See also Joseph (2011, pp. 234–236) where the starting point is the construction of a square twice the area of a given square.

[11] At the fourth stage the method is accurate to about 23 decimal place with the only new calculation needed being $(1154)^2 = 1331716$.

[12] The most common method taught today to find square roots is the "divide-and-average" method. To find the square root of N, start with an initial approximation a_0 and then take as the next approximation the average of a_0 and N/a_0. In general, if a_n is the nth approximation of the square root of N, then $a_{n+1} = \frac{1}{2}[(a_n + (N/a_n)]$. The interested reader may check that if you start with $[1 + (1/3) + (1/12)] = [17/12] = 1.416666666667$ as your first approximation of the square root of 2, then the succeeding approximations are numerically the same as those given by *Sulbasutra's* geometric method. However, the efficiency of the *Sulbasutra's* approach is conditional on not changing the representation of the fractions from unit fractions into decimals or into standard fractions.

[13] Traditional units of measurement have a concrete basis as one would find in the *Sulbasutras*. For example, a traditional carpenter working with $2\frac{7}{16}$ inches is likely to work with it as $[2 + \frac{1}{2} - \frac{1}{8} \cdot \frac{1}{2}]$ *or* 2 inches plus a half inch minus an eighth of that half, which produces a clearer concrete image to hold onto and work with rather than 2.4375. Similarly, with the *Sulbasutra* approximation, it is easier to have a clearer image of the length of $\sqrt{2}$ than it is with the 'dividing and averaging' approximation of $\frac{8867310889}{6270135604}$.

[14] There was also a social prescription for the location of these fire altars. The *Apastamba Sulbasutra* (4.1) contains the following instruction (modified here for clarity): "In setting up the sacrificial fires, the prescribed distances between *Garhapatya* (*G*) and *Ahavaniya* (*A*) are... (for) the Brahmin: eight double-paces from; for the Kshatriya eleven double paces; and for the Vaishya twelve." Each pace is assumed to be 15 *angulas* or approximately one foot.

Chapter 4

From Zero to Infinity: Mathematics in Jain and Buddhist Literature

4.1 Introduction

The rise of Buddhism and Jainism was in part a reaction to some of the excesses of Vedic religious and social practices that had crystallised into an inflexible caste system.[1] The decline in offerings of fire sacrifices that had played a central role in Vedic rituals meant that occasions for constructing altars requiring practical skills and geometric knowledge became few and far between. There was also a gradual change in the perception of the role of mathematics: from fulfilling the needs of rituals to becoming both an abstract and practical discipline. The Jain contribution to this change is particularly notable, especially the role it played in bridging the gap between 'ancient' Indian mathematics of the Vedas and the mathematics of the 'Classical period', heralded by the work of Aryabhata I at the end of fifth century CE.[2]

The main Jain texts on mathematics date from a period between 300 BCE to 400 CE, although the Jain religion was born three centuries earlier.[3] Even for this period the material available is limited. There is some uncertainty as to how the later developments during the Vedic period had a noticeable impact on the Jain mathematics and vice versa. This is not something unique, for throughout the history of Indian mathematics it seemed as if each 'leap' took place independently of previous discoveries. This may have been due to the large size of the Indian sub-continent and the imperfect channels of communication within it.

A number of early Jain texts of mathematical significance are either no longer extant or have yet to be studied. What we know of them is based almost entirely on later commentaries.[4] Of particular relevance are the old canonical literature: *Surya Prajnapti, Jambudvipa Prajnapti, Sthananga Sutra, Uttaradhyayana Sutra, Bhagwati Sutra, Trilokasara* and *Anuyoga Dwara Sutra*. The first two texts date to the third or fourth century before the Common Era and the others are from at least two centuries later. There is also mention of individuals such as Bhadrabahu (*c*. 300 BCE) who wrote two astronomical works yet to be unearthed, and Umasvati, a commentator from 150 BCE, known more for his work on Jain metaphysics. The latter composed *Tattavarthadhiga Sutra Bhashya*, containing some mathematics, including a number of mensuration formulae. Some of them are listed at the end of this chapter. In all these texts, it is evident that the Jains regarded mathematics as an integral part of their religion to the extent that a section of their religious literature was named as *Ganitanuyoga* (calculation methods).

A development that would have a significant indirect impact on Indian mathematics hereafter was the pioneering work by Panini (sixth century BCE) in the field of Sanskrit grammar and linguistics.[5] Panini is believed to be the earliest linguist to present a comprehensive and scientific survey of the phonetics, phonology and morphology of Sanskrit. His *Astadhyayi* contains eight chapters, each subdivided into four sections. In it he distinguishes between the language of sacred texts and daily language of communication. He provides formal production rules and definitions of Sanskrit grammar. Basic elements such as vowels and consonants, parts of speech such as nouns and verbs are placed in categories. The construction of compound words and sentences follow ordered rules operating on underlying structures in a manner similar to present-day formal language theory.[6] On the basis of just under 4000 *sutras* (i.e. rules expressed as aphorisms) and just around 7000 words, he summarised virtually the entire structure of the Sanskrit language.

Panini's constructions have been compared to modern definitions of a mathematical function.[7] And it is even argued that the algebraic nature of Indian mathematics arose as a consequence of the structure of the Sanskrit language. In particular, the Indian way of representing numbers by words, and ultimately the development of modern number systems in India are all linked (however indirectly) to one another through the structure of the Sanskrit language.[8] The inference here is that Panini's work is an example of a scientific notational model that may have propelled later mathematicians to use abstract notations in characterising algebraic equations and presenting algebraic theorems and results in a scientific format.[9]

The use of symbolism with a mathematical underpinning also appeared around this period in the *Chandahsutra* of Pingala.[10] *Chandah* (i.e. prosody or the pattern of rhythm and sounds in poetry) consisted of identifying various combinations of Sanskrit syllables where, in a poetic composition the number of syllables is known and each syllable could be either be 'short' (*laghu*) or 'long' (*guru*). Later in this chapter there is a discussion of the mathematics resulting from different combinations of the two sounds in a three-syllabic metre.

During this highly creative period, there appeared another individual in the second half of the fourth century BCE. Vishnugupta Chanakya Kautilya composed the *Arthashastra* ('Science of wealth and welfare'). Containing 150 chapters in 15 volumes, it linked three complementary themes: (i) *Arthaniti* (promotion of economic growth), (ii) *Dandaniti* (administration of justice) and (iii) *Videshniti* (management of foreign policy to promote national interest). It would be misleading to think of this text just as an administration manual or as a treatise on statecraft which the often-made comparison with Machiavelli implies. Sihag (2014) has argued that Kautilya should be seen as the founder of economics as a theoretical discipline in that in that his treatise emphasises the 'scientific nature of economic analysis and its philosophical foundations, explaining the inner workings of the economic system along with the interactive effects of economics with society and politics'.[11] In the treatise are also the seeds of an early version of social arithmetic and applied mathematics in an increasingly numerate society.

4.2   Large Numbers: An Indian Obsession

In the *Rgveda*, the oldest of the four Vedas dated around 1000–1300 BCE, there appear nascent mathematical ideas involving astronomy and time reckoning embedded in hymns describing the creation of the world. Unlike any other culture, the Indian creation myths stipulate no beginning, with new worlds emerging endlessly from existing ones. From different parts of the body of the primeval man (*Purusha*), emerged new gods who themselves restart the clock of the universe's time cycle. It is the scale of the large numbers and preciseness of measurements that was notable.[12]

In the *Yajurveda Samhita* of the Vedas (conservatively dated around the beginning of the first millennium BCE), a list of number names is given as: *eka* (1), *dasa* (10), *sata* (10^2), *sahasra* (10^3), *ayuta* (10^4), *niyuta* (10^5), *prayuta* (10^6), *arbuda* (10^7), *nyarbuda* (10^8), *samudra* (10^9), *madhya* (10^{10}), *anta* (10^{11}), *parardha* (10^{12}). Each of these named denominations is ten times the preceding one, confirming a systematic mode of arrangement in naming numbers. In *Taittriya Samhita*, this list was extended up to *loka* (10^{19}); and in the Buddhist work *Lalitavistara* (100 BCE), there are series of number names based on the centesimal (i.e. 100) scale. For example, the sage Arjuna asks the young Gautama Buddha how the counting proceeds beyond *koti* (10^7) on the centesimal scale, and the Buddha replies: Hundred *kotis* are called *ayuta* (10^9), hundred *ayutas* is *niyuta* (10^{11}), hundred *niyutas* is *kankara* (10^{13}), and so on to *sarvajna* (10^{49}), *vibhutangama* (10^{51}) and *tallaksana* (10^{53}). After 10^{53} starts another series that went up to the enormous number 10^{420} named *asamkheya* — meaning innumerable or uncountable — a word that the *Lalitavistara* poetically defined as the number of raindrops falling on all the worlds for ten thousand years. The same text eventually reaches a number equivalent to 10^{421}!

The Jains also showed a similar propensity for extremely large numbers, but seem to favour powers of 2 rather than power of 10. In the *Anuyoga Dvara Sutra*, a Jain text from the beginning of the Common Era, the number of human beings in the world is given as 2^{96}. It is also in this work that there is the first mention of the word 'place' (*sthana*) used to define a class or group of numbers. A number of metaphysical significance, found in the Jain texts, is $(8400000)^{28}$ and given the name of *Sirsaprahelika*.[13] Thus, fascination with the enumeration of large numbers had eventually led to a number consisting of 52 digits followed by 142 zeroes![14]

An important objective of ancient mathematics the world over was to develop accurate ways of counting time. The ancient Indian calendar was from times immemorial a lunisolar[15] calendar with an extra day being inserted in the calendar in five-year cycles.

In Vedic literature, time is reckoned in *yugas*.[16] The four *yugas* are *Satya Yuga*, *Treta Yuga*, *Dwapara Yuga* and *Kali Yuga*, with the time-span of these four *yugas* being 1,728,000, 1,296,000, 8,64,000 and 4,32,000 years respectively in the ratio 4:3:2:1. The sum total of these four *yugas* constituted one *yuga* cycle (or *Mahayuga*), containing 4,320,000 years. Also a 1000 *yuga*-cycles (or 4,320,000,000 years) comprised one day in the life of *Brahma*, so that a period of a single day and night totalled 8.64 billion years! The numbers given above are extremely large; it is plausible that this preoccupation led to an early idea of infinity. The peace song in the *Isa* Upanishad of the Vedas mentions *purna* which has been taken to mean 'infinite fullness'.[17]

Across the literature, multiples of ten up to 10^{145} had individual names as though they were commonplace. In contrast, the Greeks had 'myriad' (10^4) and centuries later the words, billion, trillion and quadrillion came into existence. It is interesting that the Chinese as early as the beginning of the Common Era have names for numbers up to 10^{80}.

In India, such colossal numbers and attached names could have been of no practical use; they flowed from a contemplation of this infinite universe and a desire to reach out to that infinity.[18] They came naturally in a culture where people believed that they had to live through multiple cycles of existence spanning vast periods of time. A need was felt for an adequate way to express and calculate with astronomically large numbers, leading to the development of our decimal place value system with zero as a number, magnitude and place-holder.

4.3   The Evolution of the Indian Decimal Place Value System

There are four distinct elements that constitute our number system. They are: (i) recording on a base of ten, (ii) the use of zero, (iii) the notion of place value and (iv) the symbols for the nine digits and zero.

Consider counting with the base of ten first. We have already seen that in Vedic *Yajurveda* and other ancient texts, there frequently occur names for series of decuple (or ten-fold) terms starting from *eka, dasa, sata* reaching up to much larger powers of ten. And the Jain canonical literature and the Buddhist narratives inspired the construction of even much larger series. The terminology for increasing power series is unique to India; such large series did not exist anywhere outside India.

4.3.1   The history of zero: the Indian and global dimensions

The word 'zero' is a transliteration of the Arabic '*sifr*'. *Sifr* may in turn have been derived from the Sanskrit word '*sunya*' meaning void or empty. Introduced into Europe during Italian Renaissance in the 12th century by Leonardo Fibonacci (and by Jordanus de Nemore, a lesser known 13th century European mathematician) as '*zefiro*' from which emerged the word 'cipher'.[19]

The ancient Egyptians never needed (nor used) a symbol for zero in writing their numerals (Figure 4.1). Instead they had a word to denote value or magnitude. A bookkeeper's record from the thirteenth dynasty (about 1700 BCE) shows a monthly balance sheet for items received and disbursed by the royal court during its travels. On subtracting total disbursements from total income, no remainder was left in several columns. This zero remainder was represented by the hieroglyph, *nfr*, which also meant beautiful, or complete in ancient Egyptian. The same *nfr* also labelled a zero reference point for a system of integers used on construction guidelines at Egyptian tombs and pyramids. These massive stone structures required deep foundations and careful levelling of the

Egyptian Numerals (No 'Zero' Needed)

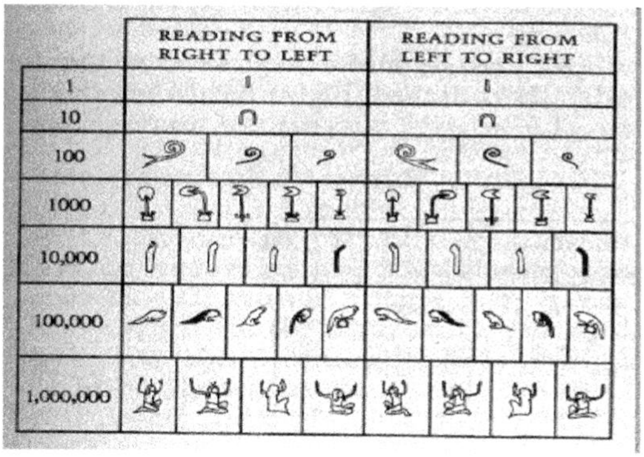

Figure 4.1. Egyptian numerals.

courses of stone. A vertical number line marked the horizontal levelling lines that guided construction at different levels. One of these horizontal lines, often at pavement level, was used as a reference and was labelled *nfr* (⚭) or zero.[20] Horizontal levelling lines were spaced 1 cubit apart. Those above the zero level were labelled 1 cubit above *nfr*, 2 cubits above *nfr* and so on. Those below the zero level were labelled 1 cubit, 2 cubits, 3 cubits and so forth, below *nfr*. Her zero was used as a reference for directed or signed numbers or as a direction separator.

It is quite extraordinary that Mesopotamian culture, which was more or less contemporaneous with Egyptian culture and had a fully developed positional value number system on base 60, did not use zero as a number. A symbol for zero as a place-holder (consisting of two slanted wedges, indicating 'nothing here') appeared late in the Mesopotamian culture (see Figure 4.2). If zero was represented as an empty space within a well-defined positional numeration system such a zero was present in Chinese mathematics centuries before the Indian zero (Figure 4.3). The absence of a symbol for zero in China did not prevent it from being properly integrated into an efficient computational tool capable of handling solution of higher order equations.

Babylonian Numerals

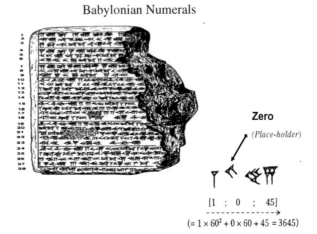

Zero

(Place-holder)

[1 ; 0 ; 45]
------------------→
(= 1 × 60² + 0 × 60 + 45 = 3645)

Figure 4.2. Babylonian numerals.

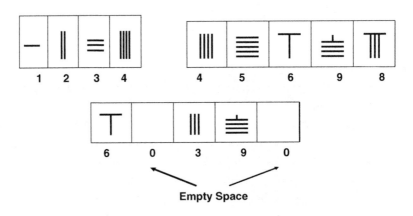

Figure 4.3. Chinese rod numerals and zero.

The early Greeks, who were the intellectual inheritors of Egyptian science, emphasised geometry to the exclusion of everything else. They did not seem interested in perfecting their alphabetic number notation system. They simply had no use for zero. This is because the Greeks thought geometrically representing numbers as lengths of line segments with no other abstract significance. They were not greatly interested in arithmetic, claiming that arithmetic should only be taught in democracies for it dealt with "relations of equality," an activity for slaves. On the other hand, geometry was the natural study for oligarchies for "it demonstrated the proportions within inequality" (Foucault, 1972, p. 219). Although a few Greek astronomers began using a symbol (the Greek letter *omicron*) to denote place holders, zero was not thought of as a number.[21]

In Indian mathematics, the oldest known text to refer to zero is a Jain cosmological text entitled the *Lokavibhaga* ("The Parts of the Universe"), originally composed in Prakrit[22] by a monk in the fifth century CE. It mentions the zero within a decimal place value system.[23] During the next few centuries preceding and following this text a place-value notation developed with the emergence of three successive types of numerals: Kharosthi (fourth century BCE — second century CE) and Brahmi (third century BCE) from which may have eventually evolved the Bakhshali Manuscript (date uncertain). Later, a written inscription dated 870 CE was discovered in Gwalior (in present-day Madhya Pradesh State). In a temple located on the path up to Gwalior fort there appears the number 270 (see Figure 4.4) where we find a circular zero in the terminal position.

Figure 4.4. The Indian zero.

This is widely quoted as the first evidence that India recognised zero as a number. Zero is not a "natural" candidate for acceptance as a number unlike other numerals. It requires a great leap of thought from the concrete to the abstract. With zero came the notion of negative numbers and along with all these comes a series of related questions about arithmetical operations on natural numbers, both positive and negative and zero.

In India, the zero as a philosophical concept predated zero as a number by hundreds of years.[24] The Sanskrit word for zero, *sunya*, means "void or empty". The word is probably derived from *suna* which is the past participle of *svi*, "to swell or enlarge". In the earliest Vedas, *Rgveda*, occurs another meaning: "lack or deficiency". It is possible that when a delicate balance was finally established, the two different meanings were fused into "*sunya*", a sense of "absence" with the potential for growth. It is in this last sense that the word *sunya* has now established itself to mean empty, void, vacant, vacuous, etc. Its derivative, *Sunyata* is the abstract noun of *sunya* describing emptiness, void, etc.

Sunyata is also the term used to describe the Buddhist doctrine of Emptiness, a spiritual practice of emptying the mind of all sensations, of which the Buddhist philosopher Nagarjuna (*c.* 150–250 CE) was the most notable exponent. This was a course of action prescribed for a wide range of creative endeavours. For example, the practice of *Sunyata* is recommended in projects involving writing of poetry, music composition, painting or any activity that springs from the mind of the artist. Traditional manuals of architecture (the *Silpa*) on building included instruction on the organisation of empty space, for "it is not the walls which make a building but the empty spaces created by the walls." The whole process of creation is captured in the following verse from an eigth century Tantric Buddhist text, *Hevraja Tantra* [1.32]:

First the realisation of the void (*sunya*),
Second the seed in which all is concentrated,
Third the physical manifestation,
Fourth one should implant the syllable.

A mathematical correspondence was soon established. 'Just as emptiness of space is a necessary condition for the appearance of any object, the number zero being no number at all is the condition for the existence of all numbers' (Datta, 1927)

A discussion of the mathematics of the *sunya* involves three related issues: (i) the concept of the *sunya* within a place-value system, (ii) the symbols used for *sunya* and (iii) the mathematical operations with the *sunya*. Material from appropriate early texts used as illustrations below.

It was recognised fairly early in India that *sunya* denoted notational place (place holder) as well as the "void" or absence of numerical value in a particular notational place.[25] Consequently all numerical quantities, however great they may be, could be represented with just ten symbols. A 12th century text (*Manasollasa*) states:

> Basically, there are only nine digits, starting from 'one' and going to 'nine'. By adding zeros, these are raised successively to tens, hundreds and beyond.

And in Patanjali's *Yogasutra* in the seventh century the following analogy appears:

> Just as the same sign is called a hundred in the "hundreds" place, ten in the "tens" place and one in the "units" place, so can one and the same woman be referred to as mother, daughter or sister.

The earliest representation of a symbol for zero occurs in the *Chandahsutra* of Pingala (*fl.* third century BCE) which discusses a method for calculating the number of arrangements of long and short syllable in a metre containing a certain number of syllables (i.e. the number of combinations of two items from a total of n items, repetitions being allowed).[26] The symbol for *sunya* began as a dot (*bindu*), found in inscriptions both in India and in South-East Asia around the seventh and eighth century CE and then became a circle (*chidra* or *randra* meaning a hole). However, the association between the concept of zero and its symbol was already well-established by the early centuries of the Christian era, as the following quotation shows:

> The stars shone forth, like zero dots (*sunya-bindu*) — scattered in the sky as if on a blue rug — the Creator reckoned the total with a bit of the moon for chalk.
>
> (Vasavadatta, ca. 400 CE)

Sanskrit texts on mathematics/astronomy from the time of Brahmagupta usually contains a section called '*sunyaganita*' or calculations involving zero. While discussions in the arithmetical texts (*patiganita*) were limited only to addition, subtraction and multiplication with zero, the treatment in algebra texts (*bijaganita*) covered such questions as the effect of zero on the positive and negative signs, division by zero and more particularly the relation between zero and infinity (*ananta*).

Take an example from Brahmagupta's seventh century text *Brahmasphutasiddhanta*. In it he treats the zero as a separate entity from the positive (*dhana*) and negative (*rhna*) quantities, implying that *sunya* is neither positive nor negative but forms the boundary line between the two kinds, being the sum of two equal but opposite quantities. He states that a number, whether positive or negative, remains unchanged when zero is added to or subtracted from it. In multiplication with zero, the product is zero. Likewise the square and square root of zero is zero. But problems arise when he tries to explain division. When a number is divided by zero, he gives no result but merely repeats that the quotient of any number and zero is a fraction with zero as a denominator. However, Brahmagupta's is the first attempt by any mathematician to explain the arithmetic operations on natural numbers and zero.

In the ninth century, Brahmagupta's attempts at defining operations using zero was taken up by Mahavira. Although he correctly states that a number multiplied by zero is zero, but wrongly adds that a number remains unchanged when divided by zero. The next attempt came from Bhaskara II in the 12th century who wrote in *Bijaganita*:

> A quantity divided by zero becomes a fraction, the denominator of which is zero. This fraction is termed an infinite quantity. In this

quantity consisting of that which has zero for its divisor, there is no alteration, though many may be inserted or extracted. [Similarly] no change takes place in the infinite and immutable God when worlds are created or destroyed, though numerous orders of beings are absorbed or put forth.

This, at its face value may seem approaching correctness, by suggesting that any number when divided by zero is infinity, but then Bhaskara II (b. 1114 CE) suggests that zero multiplied by infinity is any number, and hence all numbers are equal, which is certainly not correct.[27] Emancipated from the role of a mere place-holder, zero joined the family of numbers capable of arithmetical operations.

4.3.2 The history of the other zero: a digression

Evidence relating to Pre-Columbian Maya civilisation comes from three main sources: four screen-fold books called codices, a large number of stone monuments and thousands of ceramic vessels. The best account of the Maya culture around the time of the Spanish Conquest comes from a Franciscan priest, Diego de Landa who recorded the history and traditions of the Maya people around 1566. Piecing together these different strands of evidence, it is possible to construct an account of the social context in which the Mayan numerals and especially the Mayan zero emerged around the beginning of the Christian era.

The Mayan system of numerical notation was one of the most economical systems ever devised.[28] In the form that was used mainly by the priests for calendar computation as early as 400 BCE, it required only three symbols. A dot was used for one, a bar for five; and a symbol for zero which resembles a snail's shell [see Figure 4.5(a)].

With these three symbols, they were able to represent any number on a base 20. However, there was an unexpected irregularity in the operation of the place value system. Corresponding to our units, tens, hundreds, thousands, etc., the Mayans had units, 20's, (18×20)'s, (18×20^2)'s, (18×20^3)'s, etc. [see Figure 4.5(b)]. This anomaly reduces the efficiency in arithmetical calculations. For

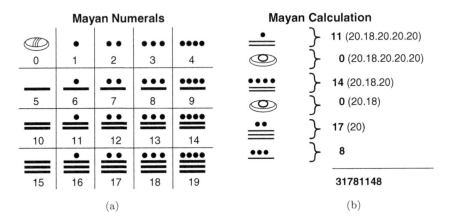

Figure 4.5. Mayan calculations.

example, a useful facility in our number system is the ability to multiply a given number by 10 by adding a zero to the end of it. An addition of a Mayan zero to the end of a number would not in general multiply the number by twenty because of the mixed base system employed. This inconsistency also inhibits the development of further arithmetical operations, particularly those involving fractions.

To understand this curious irregularity in Mayan numeration, it is important to appreciate the social context in which the number system was used. As far as we know, this form of writing numbers was used only by a tiny elite — a group of priest scribes who were responsible for carrying out astronomical calculations and constructing calendars.

At the top of the pyramid was a hereditary leader who was both a high-priest (*Ahau-Can*) and a Mayan noble. Under him were the master scribes who were priests as well as teachers and writers ("engaged in teaching their sciences as well as in writing books about them"). Mathematics was recognised as such an important discipline that scribes who were adept at that discipline were depicted in the iconography of Mayan artists (see Figure 4.6(a)) Their mathematical identity was signified in the manner in which they were depicted: either with the Maya bar and dot numerals coming out of their

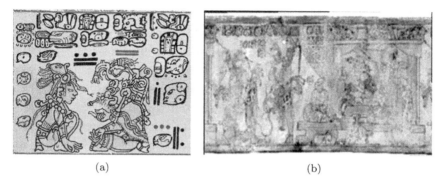

(a)   (b)

Figure 4.6. Mayan scribes at work.

mouths or with a number scroll carried under their armpit. The location of the scroll under the armpit with numbers written on it would seem to symbolise status.

In an interesting but faded illustration (Figure 4.6(b)) on another Maya vase from the beginning of the Christian era, there is a seated supernatural figure with the ears and hooves of a deer, attended by a number of human figures, including a kneeling scribe mathematician from whose armpit emanates a scroll containing the sequence of numbers 13, 1, 3, 3, 4, 5, 6, 7, 8 and 9. At the top right hand corner of this illustration there is the small figure of a scribe who looks female, with a number scroll under her armpit indicating that she is a mathematician and possibly the one who painted the scene and wrote the text on the vase. She is described as *Ah T'sib* ("the scribe"). Preceding this text is a glyph that has not been deciphered but which could be her name. Once the name is deciphered, and if the scribe is female, we may have the name of one of the earliest known women mathematician-scribes in the world. The existence of female mathematician/scribes among the Maya is further supported by a depiction found on another ceramic vase. The text on this vessel contains the statement of the parentage of the scribe in question: "Lady Scribe Sky, Lady Jaguar Lord, the Scribe". Not only does she carry the scribal title at the end of her name phrase but she incorporates it into one of her proper names, an indication of the importance she herself places on that reality.

Returning to the curious irregularity in the Mayan place value system, the general view is that it is tied to the exigencies of operating three different calendars. The first calendar, known as the *tzolokin* or 'sacred calendar', contained 260 days in twenty cycles of 13 days each. Superimposed on each of the cycles was an unchanging series of 20 days, each of which was considered a god to whom prayers and supplications were to be made. The second, known as a civil or secular calendar, was the one for practical use. It was a solar calendar consisting of 360 days grouped into 18 monthly periods of twenty days and an extra month consisting of five days. The last month was shown by a hieroglyph that represented disorder, chaos and corruption and any one born in that month was supposed to have been cursed for life. Finally, there was the third calendar of 'long counts' similar to the Indian 'Yuga' periodisation. The upper section of one of the oldest standing stelas at Ires Zapotes in Mexico shows the date of its construction in the calendar of 'long counts' as:

8 kins	$=$	8×1	$=$	8 days	(20 kins = 1 uinal)
16 uinals	$=$	20×16	$=$	320 days	(18 uinal = 1 tun)
0 tuns	$=$	$20 \times 18 \times 0$	$=$	0 days	(20 tuns = 1 katun)
6 katuns	$=$	$(18)20^2 \times 6$	$=$	7206 days	(20 katuns = 1 baktun)
16 baktuns	$=$	$(18)20^3 \times 16$	$=$	2304000 days	(20 baktuns = 1 piktun)
7 piktuns	$=$	$(18)20^4 \times 7$	$=$	20160000 days	(20 piktuns = 1 calabtun)

TOTAL   22,471,534 days which corresponds to 31 BCE

There were higher units of measurement, notably *kinchiltuns* (or kins) and *alautins* where 1 *alautin* equalled 23,040,000,000 days. Measurement of time constituted a central feature of the Mayan culture and the interest in measurement was carried into Mayan astronomy. We can only marvel at the high degree of accuracy that the Mayans achieved in their astronomical work. For instance, without any sophisticated equipment and with a deficient mixed base system, they deduced the mean duration of a solar year as 365.242 days (modern value: 365.242198 days) and the mean duration of a lunar month as equivalent to 29.5302 (modern value: 29.53059 days).

4.3.3 The two zeroes: common threads and differences

In the Prelude, we began with a discussion of the evolution of the Indian zero and this was then extended it to include the Mayan Zeroes. Why did the full use of zero within a well-established positional value system only emerge in two cultures? Were there any similarities between the two cultures that might provide an answer, however tentative?

From the existing evidence, much of it fragmentary, especially in the Mayan case, we are aware that both cultures were highly numerate with considerable interest in astronomy. The Indian culture from an early time showed interest and even fascination for large numbers and there is no contrary evidence to indicate that this was not so in the Mayan culture.[29] Both cultures were obsessed with the passage of time but in different ways. The Indian interest was tied up with the wide-spread belief in a never-ending cycle of births and rebirths that could only broken by the attainment of individual salvation. This was apparently aided during the Vedic times by sacrifices carried out on specially constructed altars conforming to specific shapes and sizes and on particular days chosen for their astronomical significance.

In the Mayan case, the preoccupation came about from fear that the world would come to an end unless the gods (and especially the Sun God) were propitiated by human sacrifice, undertaken systematically at certain propitious times of the year dictated by specific astronomical events. In both cases there was need for accurate measurement of time and hence the detailed calendars and the elaborate periodisation into eras. The need for such precise calculations may have stimulated the development of efficient number systems including a fully developed zero. And it was probably only an accident of history and geography that the Indian zero prevailed while the Mayan zero eventually disappeared into oblivion.

The consequences were far-reaching. As mentioned earlier, the spread of the Indian zero had to go through a number of cultural and linguistic filtering processes, the imperfect nature of which is evident even today. Culturally, our discomfort with the concepts of zero (and infinite) is often reflected in humour. Underlying such uneasiness is both a conceptual fuzziness regarding zero and a lack of confidence in the manipulation of mathematical expressions in which zero or infinity appear. A story told of youthful Srinivasa Ramanujan illustrates this point well. In an elementary mathematics class, the teacher was explaining the concept of division (or 'sharing') through examples. If three bananas were shared between three children, each child would get one banana. And similarly, the share would be one banana if four bananas were divided among four children, five bananas among five children and so on. And when the teacher tried to generalise this idea of sharing out x bananas among x boys, Ramanujan piped up with the question: If x equals zero, would each child then get one banana? There is no record of the teacher's reply. Hopefully, the teacher had the presence of mind to ask Ramanujan what is a zero banana!

Ask a mathematician whether zero is an even or an odd number? The answer would be: If you define evenness or oddness on the integers (either positive or negative), then zero should be taken to be even; but if you define evenness and oddness on the natural numbers, then zero would be neither. This is because we apply concepts such as 'even' only to 'natural numbers', in connection with primes and factoring, whereas 'natural numbers' are positive integers and so excludes zero. However, those who work on the foundations of mathematics consider zero a natural number, and for them the integers are whole numbers. From that point of view, the question whether zero is even just does not arise, except by extension. One may say that zero is neither even nor odd. Because you can pick an even number and divide it in groups, take for example 2, which can be divided in two groups of 1's, and 4 can be divided in two groups of 2's. But can you divide zero? This was Ramanujan's question. And the difficulty is basically caused by the fact that the concepts of even and oddness predated zero and the negative integers.

On the question of division by zero there seems to be confusion found even among even seasoned practitioners. In an example in Operational Research involving a 'Simplex Tableau', a division of 2 by zero occurred in performing the column ratio test. The result of

$2 \div 0$ is given as infinity (∞). However, the logic of this conclusion is not adhered to in the simplex calculations that followed. There was no awareness of the doubtful nature of this answer, for if one proceeded to ask the obvious question: Which number, when multiplied by zero, gives 2? Would it be 'Infinity'? But infinity is not a number; it is a concept. Would it be 'Undefined'?[30] For the correct answer, seek from the 'mouth of babes' or even from a pocket calculator: one cannot ever meaningfully divide by zero.

Finally, certain semantic issues relating to zero often remain unexamined. The concept of zero has often been associated with terms such as 'nothing' or 'emptiness' or 'void'. But is the presence of nothing (reflecting non-existence) different from the absence of something (reflecting non-availability) or even the absence of anything? "Not there" reflects that the number or item(s) exists but they are not just not available. "Nothing" reflects non-existence. To confuse the issue further, there are whole shades of meaning associated with the term 'zero' depending on whether it is used as a noun, a verb, an adverb, and even an adjective as in "zero possibility." Here we have a good illustration of how conceptual ambiguity in ordinary speech tends to make the comprehension of the mathematical meaning more difficult. Teachers of mathematics at all levels would be aware of the conceptual difficulties faced by their students at different levels.

The emergence of zero as numerical concept in Buddhist and Jain mathematics was possible because it could free-ride on certain cosmological and philosophical concepts. However, this does not explain its subsequent spread to other cultures. A possible explanation for this success is its violation of intuitive expectations of the number module. A numerical concept of nothing violates expectations provided by our evolved number sense. Cross-cultural experiments (Boyer and Ramble, 2001) show that people are prone to remember facts or narratives with an element that violates ontological expectations. Such counter-intuitive ideas are easier to remember than intuitive ideas. For example, a creature who can be in more than one place at the same time is more memorable than a creature that has to eat to sustain itself (Barrett and Nyhof, 2001).

Thus, zero as a numerical concept was long in the making because it is counter-intuitive. However, once accepted, it becomes appealing because it violates ontological expectations.

4.3.4   The spread of Indian zero eastward

The Egyptians needed no symbolic zero and neither did the Romans. The Babylonians had a place value sexagesimal (i.e. base 60) system, though not the zero. The Chinese too had a decimal place value system before the third century CE, but they did not have the zero (Figure 4.3). Nevertheless, could the Babylonian notion of place value and the Chinese decimal system have played any part in the development of the decimal place value in India? It is not impossible, although firm evidence is lacking.[31] This much, however, seems to be certain: the full development of the decimal place value in its modern form with the use of zero took place in India before the beginning of the Common Era, not in the fifth century as most of the current literature states. And it is from India that the system was transmitted to other countries.

It is one of the ironies of history of Indian science that the earliest inscriptions containing a symbol for zero were found not in India but

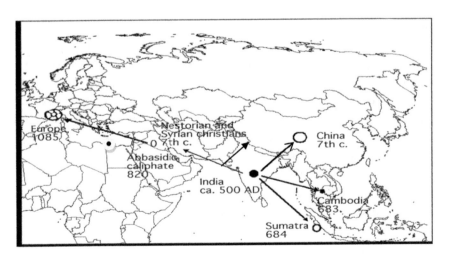

Figure 4.7. The spread of Indian numerals.

outside, notably the numerical expressions and symbols occurring in South-East Asia between the fifth and eighth centuries CE.[32]

For example, a Sanskrit inscription found at Bayan in Cambodia, the Saka year 526 (= 604 CE) records the year both using the place value system and as a chronogram *"rasa-dasra-saraih sak-endravarse"*. Two inscriptions have been found in South-East Asia that also contain the zero symbols. A Khmer inscription at Sambor in Cambodia from 683 CE records the *Saka* era 605, representing the zero with a thick dot (•); three years later, a Malay inscription at Palembang in Sumatra from 686 CE contains the numbers 60 and 608 *Saka*, where zero is shown by a small circle as we write today (o). Therefore, it has been argued, notably by Joseph Needham, that since these are the earliest occurrence of numerals within a consistent place value system with zero, the modern system of numeration originated in South-East Asia from where it spread to India (and China). Weighed against this argument is sparseness of supporting evidence (particularly in the social context in which zero developed), and the possibility of an earlier occurrence in India itself. On a charter found at Sankheda in Gujarat, the year is written in decimal notation as '346', the year corresponding to 594/6 CE.[33]

Further, if the symbol for zero was invented in South-East Asia in the seventh century under inspiration from China — why then did it not have a symbol of her own until the symbol was transmitted from India? A further question remains: Why were two symbols invented at the same time in South-East Asia, a small circle and a thick dot? Also, there are several inscriptions from this region composed in Sanskrit; which indicate the presence of Indian traders and settlers who brought with them the Sanskrit language, Indian calendar, the Indian *Saka* era, Sanskrit word numerals, and possibly the symbols for the digits 1–9. If so, they would also have brought the zero with them.

The fact is that these two (i.e. a dot and circle) symbols developed in India, not at the same time, but one after the other. Of these, the dot (*bindu*) is the earlier one. Later it was enlarged into a circle (*chidra* or *randhra*) for purely practical reasons; so that it could be clearly recognised as a symbol and not mistaken for an accidental dot.[34]

	0	1	2	3	4	5	6	7	8	9

○ ၅ ២ ៣ ៤ ៥ ៦ ៧ ៨ ៩

○ ៗ ២ ៣ ៤ ៥ ៦ ៧ ៨ ៩

○ ៦ ២ ៣ ៤ ៥ ៦ ៧ ៨ ៩

○ ៦ ២ ៣ ៤ ៥ ៦ ៧ ៨ ៩

Khmer Numerals

ℓ • ℓ = 605 Sambor (Cambodia) Inscription in CE 683

Figure 4.8. Changing Khmer numerals from South-East Asia.

Since these two symbols appear in inscriptions of South-East Asia in the seventh century, it is reasonable to conclude that both the symbols were in use in India at this time and that the circular symbol must have developed by this century at the latest. The change, however, was not simultaneous everywhere. The dot seems to have prevailed longer in Gandhara and Kashmir regions. And in the mathematical text known as the *Bakhshali Manuscript*[35] (discussed in the next chapter) and in an anonymous commentary on Sridhara's *Patiganita*,[36] both written in an early form of *Sarada* script some time after the ninth century, the zero is represented by a dot. Moreover, even after the dot was replaced by the circle, the symbol continued to be called *bindu* or *sunya-bindu*.[37] While the symbol for zero was gradually transforming itself from a dot to a circle, the symbols for the nine digits too underwent changes. These changes can be seen both in epigraphic records in India as well as in South-East Asia (Figure 4.8).

4.3.5   The spread to China

With the spread of Buddhism in China, there took place a significant exchange of scholars between India and China. Buddhist pilgrims visited India and carried back large quantities of manuscripts with them; Indian scholars went to China where they were active in translating Buddhist texts into Chinese. These Chinese and Indian scholars were primarily interested in religious texts, especially of Buddhism, and the reports generally talk only about religious and philosophical texts. But they probably also carried with them some notions or texts about Indian systems of numeration, mathematics and astronomy. For example, the catalogue of the Sui dynasty, completed in 610 CE, mentions Chinese translations of Indian works on astronomy, mathematics and medicine. These works are now lost, but their very existence shows that towards the end of the sixth century, the Chinese had gained some knowledge of Indian astronomy, mathematics and pharmaceutics.

Records of the Tang dynasty indicate that from 600 CE onwards Indian astronomers were employed at the Astronomical Board of Chang-Nan to teach the principles of Indian astronomy and calendar. One of the Indians named Gautama Siddhartha was reported to have constructed a calendar, based on the Indian *Siddhantas*. This calendar contains sections on Indian numerals and arithmetical operations and the construction of sine tables at intervals of $3°54'$ for a radius of 3,438 units, which are, as we will see in a later chapter, precisely the base values given in the Indian astronomical texts. There survives a block print text containing Indian numerals, including a dot to indicate zero.[38]

Strangely enough, neither the dot for zero is found in the calendar prepared by Gautama Siddhartha, nor the dot and circle for zero found in inscriptions in what Needham refers to as the "southern zone of the culture of the Chinese." It would seem that in the seventh century the Indian zero did not have any impact in China proper. The Chinese did not start using the zero symbol until the mid-thirteenth century when it appears for the first time in the work of Qin Jiushao. It is rather intriguing why the Chinese took such a long time (nearly six centuries) to use the zero symbol which was brought up to their doorstep, so to speak. Here is a case of clear transfer having no impact on the receiving culture until some centuries later. A similar phenomenon is noticeable in the case with of other elements of Indian mathematics and astronomy introduced into China during the Tang period.[39]

To conclude, there are extensive records about the transmission of Buddhist texts and ideas from India to China. There are also records of Sanskrit astronomical and mathematical texts being translated into Chinese. In the reverse direction, we have no written evidence of transmission from China to India, although such transmission is likely to have occurred with the movement of artefacts and technology. A number of parallel developments could have also taken place. But one looks in vain for firm evidence of interaction between these two cultures in the realm of mathematics. This then remains a grey area in the history of intellectual exchanges between these two cultures (and hence the representation of the interactions between the two cultures by broken arrows in Figure 2.3 in Chapter 2).

4.3.6   The spread of zero westwards

Fortunately, such uncertainties do not occur in the case of transmission to the Middle East or the Arab culture area. During the reign of the second Abbasid Caliph al-Mansur (753–774), the Indian province of Sindh passed under the control of the Caliphate and embassies from Sindh started visiting Baghdad. Sometimes these were accompanied by scholars. In his *India*, Al-Biruni states : "These star-cycles as known through the canon of Alfazari and Yakub Ibn Tarik, were derived from a 'Hindu' who came to Baghdad as a member of the political mission which Sindh sent to the Khalif Almansur, A.H. 154 (i.e. 771CE)."[40] Others also report about this mission. Ibn al-Adami states in the preface to his astronomical tables entitled *Nazm al-iqd* that "an Indian astronomer, well versed in the sciences, visited the court of al-Mansur, bringing with him tables of the equations of planets according to the mean motions, with observations relative to both solar and lunar eclipses and the ascension of signs. Abu-Masher of Balkh, an astrologer at the court of al-Mansur, mentions that he derived the knowledge of the Hindu great cycle of the '*kalpa*' from an Indian astronomer. The name of this Indian astronomer is given variously as 'Kankaraf', 'Kankah', or 'Cancah', 'Kenker'..."[41] It has also been suggested that the name was Kanaka and that the later Arabic writers slowly developed an elaborate mythology concerning Kanaka's role in the history of astronomy, attributing to this mythical figure scholarship and skills of diverse kinds.

Perhaps the real word was not *"Kanaka"* but *"Ganaka"* — not a proper name but a generic term for an astronomer or even for a group of astronomers. Hence, the word referred not to one particular astronomer who visited Baghdad in 771, but collectively to all Indian astronomers or learned people who visited Baghdad about this time. This would explain the diverse qualities attributed to this *Ganaka* who may have, in reality, represented different persons.[42] Be that as it may, the Indian decimal place value system reached the Middle East not through this visit of Kanaka/Ganaka but at least a century earlier. As early as 662 the Syrian Christian Bishop Severus Sebokht eulogised the Indian decimal numbers, in a passage that is often quoted.[43]

The embassies from Sindh to the court of Al-Mansur helped in the transmission of many more scientific ideas, besides the decimal system and arithmetical operations with this system. At the court of Al-Mansur, Brahmagupta's *Brahmasphutasiddhanta* and *Khandakhadyaka* were rendered into Arabic respectively by Al-Fazari and Yaqub ibn Tarikh. These are not literal translations but adaptations and came to be known under the names *Sindhind* and *Al-Arkand.* Through these works, Islamic scholarship became acquainted for the first time with mathematical astronomy, a few decades before the discovery of Greek astronomy. In the next century, around 820 CE, Muhammad ibn Musa al-Khwarizmi summarised the knowledge gained thus far in three works, one on arithmetic, another on algebra and the third on mathematical astronomy. In the first work on arithmetic entitled *Kitab al-jam'wal tafriq bi hisab al-Hind,* al-Khwarizmi explained the arithmetical operations of addition, subtraction, multiplication, division and the extraction of square roots according to the Indian system. Within a century, this treatise was superseded in the Eastern Islamic world by other introductions to Indian arithmetic. However, the work was still available in Moorish Spain in the 12th century where it was translated into Latin, under the title *Liber algorismi de numero Indorum*, (i.e. 'The Book of al-Khwarizmi on Indian numbers'). Soon European scholars recognised the value of al-Khwarizmi's treatise and there appeared more treatises in Latin elaborating al-Khwarizmi's treatise. What needs to be emphasised in this context is that the symbol for zero known to al-Khwarizmi and transmitted through him to Europe was a small circle.[44] Indeed zero was known in Arabic, not only as *alsifr*, but also by the expression *daira saghira* (small circle). And it is this small circle (*circulus*) which appears in early Latin manuscripts.[45] While zero retained its circular form in its transmission from the Eastern Islamic world to the Western Islamic world, the digits from 1 to 9 gradually underwent several changes, with the consequence that the western forms differed substantially from the eastern ones (see Figure 4.9). These western forms came to be known as *Ghubar* numerals. The eastern forms were transmitted to Italy and the Eastern Mediterranean basin where they were used by Latin authors in the 12th century. By the early 13th century these eastern forms were replaced in Europe by the western forms through the Latin translations made in

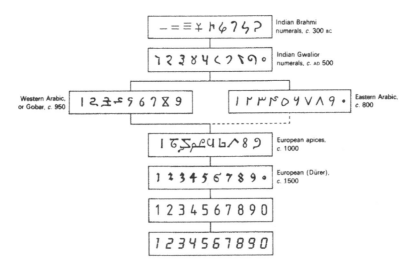

Figure 4.9. Evolution of Indo-Arabic numerals.

Moorish Spain. These then are the ancestors of what we call today 'Arabic' numerals (or what the Constitution of India, describes more accurately in article 343.1, as "the international form of Indian numerals").[46]

As mentioned earlier, the zero retained its form in Europe in the shift from eastern to the western numerals. But in the Arabic script, apparently at a later point, this circle was compressed into a dot, in order to distinguish the symbol for zero from that of five — a process that is the reverse of what happened in India.[47]

The westward transmission of Indian mathematical ideas was not limited to the number system alone.[48] The main areas which influenced the future course of development of mathematics are: (a) the spread of Indian numerals and their associated algorithms, first to the Arabs and later to Europe; (b) the spread of Indian trigonometry, especially the use of the sine function and (c) the solutions of equations in general, and of indeterminate equations in particular. We will discuss these influences in later chapters.

An intriguing question remains: Why was there an aversion to zero in Europe while in India (and later China) zero did not encounter any hostility but achieved acceptance and respectability from the time of its introduction as a number? The answer may lie in the differences of their respective philosophical underpinnings, with Europe being Aristotelian in its rejection of the possibility of absolute void and theologically preoccupied with the fullness of being while relegating nothingness to the realm of Satan. Indian culture nourished the Buddhist doctrine of *sunyata* and developed the idea of the mathematical void.[49]

4.4 *Sunyata* and the Emergence of the Mathematical Zero

To examine the role played by the concept of *sunya* in the evolution of the mathematical zero, it would be helpful to point again to the etymology of the term. The root of *sunya* is '*svi*' which has multiple meanings that includes 'to swell', 'to enlarge' and by a (semantic) extension the word 'hollow'. It is in the last sense that the word '*sunya*' has come to mean 'empty', 'void', 'vacant', 'absence', nonexistence and empty space. In Nagarjuna's[50] seminal text *Mulamadhyamakakarika* ('Fundamental Verses on the Middle Way') is a passage which examines the nature of reality:

> For him to whom emptiness (*sunya*) is clear,
> Everything becomes clear.
> For him to whom emptiness is not clear
> Nothing becomes clear
>
> (*Mulamadhyamakakarika* xxiv. 14)

Contained in that verse and inferences is an interesting scheme of parallelism showing the juxtaposition between three levels of comprehending zero, as shown by Gironi (2012, p. 35):

LEVEL 1 (Metaphysical)	**Emptiness** (i.e. lack of anything)	**Hole** (i.e. lack of something)
LEVEL2 (Mathematical)	**Zero as a numeral**	**Zero as a placeholder**
LEVEL3 (Linguistic)	**Undecidable** (i.e. something that cannot conform to either side of a dichotomy or opposition)	**Blank space**

Gabrioni proceeds to argue that the introduction of the concept of the 'hole' helps to move between the three levels where one compares (1) 'the metaphysical juxtaposition of 'emptiness' with the hole, with (2) a mathematical juxtaposition of zero as a numeral, with zero as a place-holder and with (3) the "role in a structural analysis of linguistic systems of a Derridan 'undecidable' and a mere blank space between linguistic signs". We began our discussion of the mathematical zero as fulfilling three functions as a measure of magnitude, as a place holder and finally as a number. We have therefore come full circle.

4.5   The Infinity: The Jain Preoccupation

There are innumerable soul units in a soul
There are an infinite number of space units in space
The number of units in clusters of matter may be numerable,
innumerable or infinite

(Umasvati's *Tattawarthadhiga Sutra Bhashya*)

A contemplation of very large numbers may have led the Jains to an early concept of infinity which, if not mathematically precise, was by no means simple.[51] All numbers were classified into three groups, enumerable, innumerable and infinite, each of which was in turn subdivided into three orders: (i) *Enumerable*: lowest, intermediate and highest; (ii)*Innumerable*: nearly innumerable, truly innumerable and innumerably innumerable and (iii) *Infinite*: nearly infinite, truly infinite and infinitely infinite.[52]

The first group, the enumerable numbers, consisted of all the numbers from **2** (1 was ignored) to the highest. What would be an idea of the 'highest' number is given in the following extract in the *Anuyoga Dwara Sutra* from the beginning of the Common Era:

Consider a trough whose diameter is that of the Earth (100000 *yojanna*) and whose circumference is 316227 *yojanna*. Fill it up with white mustard seeds counting one at a time. Similarly, fill up with mustard seeds other troughs of the sizes of the various lands and seas. Still the highest enumerable number has not been reached. (1 *yojanna* is about 9 kilometres.)

Of the third group Jain mathematics recognised five other kinds of infinity without any further elaboration: *infinity in one direction*; *infinity in two directions*; *infinite in area*; *infinite everywhere*; *and infinite perpetually*.[53] The Jains were among the first to discard the idea that all infinities were the same or equal, an idea that only gained currency in European mathematics with the work of Georg Cantor in the late 19th century. Further, the highest enumerable number of the Jains has some resonance (but no mathematical significance) with another concept developed by Cantor, *aleph-null* (i.e. the cardinal number of the infinite set of integers $1, 2, \ldots, N$), also known as the first transfinite number. It was Cantor who defined the concept of a sequence of transfinite numbers and devised an arithmetic of such numbers.[54]

The Jains distinguished between two basic types of transfinite number (i.e. the cardinal numbers of infinite sets). On both physical and ontological grounds, a distinction was made between *asmkhyata* and *ananta* (i.e. between rigidly bounded and loosely bounded infinities). With this distinction, the way was open for the Jains to develop a detailed classification of transfinite numbers and mathematical operations for handling transfinite numbers of different kinds.[55] However, unsurprisingly they did not do so given their limited technical and symbolic compass.

Fairly early in history of Indian mathematics, infinity acquired its own name: *khachheda*. The answer to any division by '*kha*' — the word numeral for 'space' or zero — had an intuitive mathematical appeal since such a division leads one towards infinity. The term *khacheda* for infinity was first used by Brahmagupta in his *Brahmasphutasiddantha* in the section on *shunya ganita* (or calculation with zero) which was touched on earlier but discussed in greater detail in Chapter 7. Bhaskara II in his *Bijaganita* who advanced the study of the mathematics of zero and infinity significantly preferred the term *khahara* for infinity.

Infinity is not found only at the top end of things: we find it at the bottom as well where it is known as the infinitesimal. In *Siddhantasiromani*, Bhaskara II defines the *nimesha* (literally, the blink of an eye) as $\frac{1}{972000}$th of a day, or about 89 milliseconds, and went on to divide it further and further till he reached the *truti*, a unit of time equal to $\frac{1}{2916000000}$th of a day!

The concepts of infinitesimal and infinite numbers in modern mathematics are now well incorporated into the differential calculus. Although they are now dealt with as a mathematical object, there are dangers lurking, the circumvention of which requires application of appropriate procedures. This should not make us forget that the introduction of infinity and the continuum in mathematics has historically aroused considerable uneasiness beginning with the Greeks (and especially Aristotle) and lasting for almost 2000 years until the work of Cantor (1845–1918) who developed the modern theory of transfinite numbers mentioned earlier.

4.6 Further Mathematics from the Jains

During the Jain and Buddhist period between 500 BCE and 500 CE, mathematics also fulfilled an important social role and hence more attention was paid to the study of the subject. According to Abhayadeva Suri (1050 CE), a notable commentator on the *Sthananga Sutra* (*c.* third century BCE), the topics of study in a mathematics curriculum numbered ten: *parikarma* (four fundamental arithmetic operations), *vyavahara* (measurement and computations), *rajju* (cord geometry such as contained in the *Sulbasutras*), *rashi* ('heap' or mensuration of solid bodies), *kalasavarna* (fractions), *yavat-tavat* ('as much as so much' or later, 'unknown quantity' in algebra), *varga* (square or product), *ghana* (cube or solid), *varga-varga* (square–square or fourth power) and *vikalpa* (permutation and combination).[56]

4.6.1 Sequences and progressions

Jain interest in sequences and progressions developed out of their philosophical theory of cosmological structures. Schematic representations of the cosmos constructed according to this theory contained innumerable concentric rings of alternate continents and oceans, the diameter of each ring being twice that of the previous one, so that if the smallest ring had a diameter of 1 unit, the next largest would have a diameter of 2 units, the next 2^2 units, and so on to the nth ring of diameter 2^{n-1} units.

Arithmetic progressions were given the most detailed treatment while there was little work on geometric progressions. The foundations for these subjects were laid then to be taken up by mathematicians whose works will be discussed in subsequent chapters. Rules were worked out for finding the first term a, the common difference d, the number of elements n in the series and the sum S of the terms. This is evident in a Jain text entitled *Trilokaprajnapti* of Yativrsabha (500 CE). One of the problems is to find the sum of a series consisting of 49 terms made up of 7 groups, each group itself forming a separate arithmetical progression. For the sum of any group within the series, the rule used corresponds (in modern symbolic notation) to:

$$S = n \left[a + \left(\frac{n-1}{2} \right)^2 d + \frac{n-1}{2} - d \right].$$

We shall not attempt to explain the rule here. For further details, see Sarasvati (1961/62).

It was the elaborate treatment of mathematical series by a later Jain, Mahavira (*c.* CE 850) that paved the way for some notable work by medieval mathematicians who came after him. We shall examine these developments in a later chapters.

4.6.2   Indices and logarithms

Without a convenient notation for indices, the laws of indices cannot be formulated precisely. But there are some indications that the Jains were aware of the existence of these laws, and made use of related concepts. The *Anuyoga Dwara Sutra* lists sequences of successive squares or square roots of numbers. Expressed in modern notation as operations performed on a certain number a, these sequences may be represented as:

$$(a)^2, (a^2)^2, [(a^2)^2]^2, \dots,$$
$$\sqrt{a}, \sqrt{\sqrt{a}}, \sqrt{\sqrt{\sqrt{a}}}, \dots.$$

In the same text, we come across the following statement on operations with power series or sequences: 'The first square root multiplied by the second square root, [is] the cube of the second square root; the second square root multiplied by the third square root, [is] the cube of the third square root.' Expressed in terms of a, this says that

$$a^{\frac{1}{2}} \times a^{\frac{1}{4}} = (a^{\frac{1}{4}})^3 \quad \text{and} \quad a^{\frac{1}{4}} \times a^{\frac{1}{8}} = (a^{\frac{1}{8}})^3.$$

As an illustration, the total population of the world is given as 'a number obtained by multiplying the sixth square by the fifth square, or a number that can be divided by 296 times'. This gives a figure of

$2^{64} \times 2^{32} = 2^{96}$, which in decimal form is a number of 29 digits! Does this statement indicate that following law of indices was familiar to the Jains?

$$a^m \times a^n = a^{m+n} \quad \text{and} \quad (a^m)^n = a^{mn}.$$

From the period around the eighth century CE, some interesting evidence in the *Dhavala* commentary by Virasena suggest that the Jains may have had early ideas of logarithms to base 2, 3 and 4 without using them for any computational purpose. The terms *ardhacheda, trakacheda* and *caturthacheda* of a quantity may be defined as the number of times the quantity can be divided by 2, 3 and 4, respectively, without a remainder. For example, since $32 = 2^5$, the *ardhacheda* of 32 is 5. Or, in language of modern mathematics, the *ardhacheda* of x is $\log_2 x$, the *trakacheda* of x is $\log_3 x$, and so on.[57]

4.6.3 Permutations and combinations

A *permutation* is a particular way of ordering some or all of a given number of items. Therefore, the number of permutations which can be formed from a group of unlike items is given by the number of ways of arranging them. As an example, take the letters a, b, and c, and find the number of permutations two letters at a time. Six arrangements are possible:

$$ab, ac, ba, ca, bc, cb.$$

Instead of listing all possible arrangements, we can work out the number of permutations by arguing as follows: the first letter in an arrangement can be any of three, while the second must be either of the other two letters. Consequently, the number of permutations for two of a group of three letters is $3 \times 2 = 6$. The shorthand way of expressing this result is $_3P_2 = 6$.

A *combination* is the number of selections of r different items from n distinguishable items when order of selection is ignored, unlike a *permutation* where order is taken into account. Therefore, the number of combinations which can be formed from a group of unlike items is given by the number of ways of selecting them. To take the same illustration as above, but of combinations of two letters at a time from a, b and c is three: ab, ac, bc.

Again, instead of listing all possible combinations, we can work out how many there are as follows: in each combination the first letter can be any of the three, the second letter has two possibilities and the third letter has just one possibility, so that there are six possibilities in total. But if you are not concerned about order of appearance of the two letters (i.e. although ab and ba are two different permutations, they amount to the same combination.), you must divide the total possibilities (6) by 2 to get 3 as the number of combinations. A shorthand way of expressing this result is $_3C_2 = 3$.

Permutations and combinations were favoured topics of study among the Jains. Statement of results, presumably arrived at by methods like the one just discussed appear quite early in the Jain literature. In the *Bhagabati Sutra* (c. 300 BCE) are set forth simple problems such as finding the number of combinations that can he obtained from a given number of fundamental philosophical categories taken one at a time, two at a time, and three or more at a time. Others include calculation of the groups that can be formed out of the five senses, and selections that can be made from a given number of men, women and eunuchs. The *Bhagabati Sutra* gives the corresponding values correctly for selections of up to three at a time. Expressed in modern mathematical notation, the results are[58]:

$$_nC_1 = n, \quad _nC_2 = \frac{n(n-1)}{2!}, \quad _nC_3 = \frac{n(n-1)(n-2)}{3!},$$

$$_nP_1 = n, \quad _nP_2 = n(n-1), \quad _nP_3 = n(n-1)(n-2).$$

Values are given for $n = 2$, 3, 4, and there is then the following observation: "In this way, 5, 6, 7, 10, etc., or an enumerable, unenumerable or infinite number of things may be specified. Taking one at a time, two at a time, ..., ten at a time, as the number of combinations are formed, they must all be worked out." Apart from the generalisations implied, the application of the principle to different kinds of infinities or different dimensions is noteworthy.

Even before the advent of Jainism there was some interest in the notion of permutations and combinations. Sushruta's great work on medicine in the sixth century BCE, contains the statement that 63 combinations may be made out of 6 different tastes (*rasa*) — bitter, sour, salty, astringent, sweet, hot — by taking the *rasa* one at a time, two at a time, three at a time, and so on. This solution of 63 combinations can easily be checked as follows:

$$_6C_1 + {_6C_2} + {_6C_3} + {_6C_4} + {_6C_5} + {_6C_6}$$
$$= 6 + 15 + 20 + 15 + 6 + 1 = 63.$$

Another interesting example relates to the number of ways of combining different metres (*chandas*) in a poetic composition. In a book entitled *Chandasutra* ('Rule of Metrics') from the second century BCE, Pingala, who was mentioned earlier, considered a method of calculating the number of combinations of short (*laghu*) and long (*guru*) sounds (or syllable patterns) in a given poetical composition. During this period, the music of sound variations (*varnasangita*) was based mainly on these two sounds. Pingala considered a three-syllabic metre, for which the following different combinations of the sounds of *guru* and *laghu* could result: three *guru* sounds will occur once, two *guru* and one *laghu* three times, one *guru* and two *laghu* also three times, and three *laghu* sounds once. The rule given in the original sutra is cryptic to the point of incomprehensibility. So we have to be dependent on the commentaries. In modern notation, the rule may be expressed thus:

If we represent *guru* by a and *laghu* by b, then the different combinations may be represented by the coefficients of the binomial expansion:

$$(a + b)^3 = a^3 + 3a^2b + 3ab^2 + b^3.$$

For a four-syllabic metre, different combinations of the two sounds can be found by the same representation:

$$(a + b)^4 = a^4 + 4a^3b + 6a^2b^2 + 4ab^3 + b^4.$$

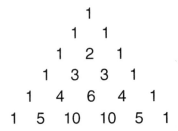

Figure 4.10. A pyramid of numbers.

This technique of finding the number of variations of sounds was useful as a means of testing the quality of different metres, and after Pingala it was used for this purpose.

Around the end of the tenth century CE, Halayudha produced a commentary on Pingala's *Chandasutra* in which he introduced a pictorial representation of different combinations of sounds, enabling them to be read off directly. Figure 4.10 shows Halayudha's *meruprastara* (or pyramidal arrangement) for a binomial expansion of $(a+b)^n$, where $n = 0, 1, 2, 3, 4, 5$. We come across the same triangular array of numbers, Pascal's triangle, in Chinese mathematics. (Joseph, 2011, 271–274). However, there is no evidence that the triangle was used for any other purpose, such as numerical solutions to higher order equations, as it was in China. Indeed, there is no evidence that the device was ever incorporated into Indian mathematics.[59]

4.6.4 Jain geometry

The term *rajju* was used in two different senses in the Jain texts. In cosmology it was a frequently occurring measure of length, approximately 3.4×10^{21} kilometres according to the Digambara[60] school. But in a more general sense it was the term used for geometry or mensuration, in which they followed on from the Vedic *Sulbasutras*. Their notable contribution was in the measurements of the circle. In Jain cosmography the Earth is a large circular island called the Jambu Island, with a diameter of 100,000 *yojanna*. While there are a number of estimates of the circumference of this island, including the rather crude 300,000 *yojanna*, an interesting estimate mentioned in both the *Anuyoga Dwara Sutra* and the *Triloko Sara*, from around the beginning of the first millennium of the Common Era, is

316 227 *yojanna*, 3 *krosa*, 128 *danda* and $13\frac{1}{2}$ *angula*,

where 1 *yojanna* is about 10 kilometres, 4 *krosa* = 1 *yojanna*, 2000 *danda* = 1 *krosa*, and *96 angula* (literally a 'finger's breadth') = 1 *danda*. This result is consistent with taking the circumference as $\sqrt{10}d$, where $d = 100\,000$ *yojanna*. The choice of the square root of 10 was quite convenient, since in Jain cosmography islands and oceans always had diameters measured in powers of 10. It is therefore no surprise that Jain mathematics contain some interesting results that relate to the circle. In Umasvati's work *Tattvarthadhigamasutrabhasya* (219 CE) where the following formulae (apart from the ones given above) make their first appearance.[61]

Circumference = $3 \times$ diameter (crude value),

Circumference = $\sqrt{10}d$ (accurate value),

Area of a circle = $\dfrac{1}{4}$ circumference × diameter,

Chord of a circle = $\sqrt{4\,sara\,(\text{diameter} - sara)}$,

Sara = $\dfrac{1}{2}[\text{diameter} - \sqrt{\text{diameter}^2 - chord^2}]$,

Arc of segment less than a semicircle = $\sqrt{6\,sara^2 + chord^2}$,

Diameter = $\dfrac{sara^2 + \frac{1}{4}chord^2}{sara}$.

Note: Sara is Rversine or $(R - R\cos\theta)$.

It would seem that there was some interest in exploring the relationship between the diameter of circle (d), the arc length (a), the corresponding chord (c), and the Rversine (v). The following relationships were given in Nemicandra's *Trilokasara* (10th century CE) which became by then a part of ancient Jain mathematics.

$$c^2 = 4c(d - c),$$
$$a^2 = 6v^2 + c^2.$$

Endnotes

[1] There were other reasons for the rising influence of Buddhism and Jainism. These included the growing power of the *Kshatriyas* (the second in the caste hierarchy) who had gained political ascendancy, and the increasing prosperity of the *Vaisyas* (the third in the caste hierarchy) having brought into commercial cultivation large areas of arable and pastoral land. Increased trade and urbanisation (including higher population density) and greater social mobility, all of which led to a more individualistic way of thinking. Vedic altars and rituals had been integrally linked to a rural 'communitarian' emphasis on repetitive seasonal cycles. It is also interesting that it was two Kshatriya princes who founded Buddhism and Jainism.

[2] A well-known eulogy of the mathematician, Mahavira (*c.* 850 CE) begins with an invocation of his illustrious name-sake, Lord Mahavira, who embodies, 'the lamp of the knowledge of numbers by whom the whole universe is enlightened', and then continues pointing to the versatility of using mathematics (*ganita*) in a whole range of practical fields. This eulogy is quoted in full at the beginning of Chapter 2.

[3] The sacred texts of the Buddhists and the Jains were mainly composed in the everyday vernaculars of Pali and Prakrit with some commentaries and other literature available in Sanskrit. These texts are difficult to date and in many cases their final versions were only available as late as the fifth century CE.

[4] The only complete extant text on mathematics authored by a Jain is *Ganitasarasangraha* of Mahavira (*c.* 850 CE) which became a mainstream text of the 'Classical period'. We will discuss the text and its author in Chapter 7.

[5] Panini was born in Shalatula, a town near the Indus river in present-day Pakistan and on the border of Afghanistan. The dates often attributed to Panini are little more than guesses. The uncertainty concerning the period he lived (sometime between the fourth and seventh centuries BCE) extend to the scope and relevance of his work. What is of little doubt is that he was one of the most innovative scholars who ever lived and has been justifiably described as the founder of linguistics.

[6] This resonates with the fundamental notion of using terminals, non-terminals and production rules in the modern study of computer science. For further details, see Rao and Kak (1998).

[7] Although the rules contained in Panini's *Astadhaya* should not to be considered as mathematics *per se*, it constitutes "a part of a quantitative, symbolic and abstract approach to understanding language that was unique in the modern world." (Plofker, 2009, p. 54). Staal (1965, p. 113)

has pointed out an essential difference between Euclid's geometry and Panini's grammar.

> In Euclid's geometry, propositions are derived from axioms with the help of logical rules which are accepted as true. In Panini's grammar, linguistic forms are derived from grammatical elements with the help of rules which were framed *ad hoc* (i.e. as *sutras*).

There is an essential difference in the way that the word 'derived' is used. In Euclidean geometry it means 'demonstrated' while in Panini's grammar it means 'generated'.

[8] Two strands may be identified in the birth and development of computer science. The first involved techniques developed to model the computational structure of nature and mind of which Alan Turing was the pioneer. The second was the creation of new computing algorithms and machines, including ways of representing and manipulating knowledge, inference and deduction. It has been argued, notably by Ingalls (1951), Staal (1988), Matilal (1968), Briggs (1985) and Kak (1987), that many developments in formal logic, linguistics, and computer science are but a rediscovery of the work of Panini and other early linguists. The formal structure of computer programming languages, introduced in the late 1950's by Backus and Naur — the inventors of the Backus Normal Form (BNF) — was used to describe the syntax of modern computer languages. They were searching for a language "suitable for expressing a large class of numerical processes in a form sufficiently concise for direct automatic translation into the language of programmable automatic computers." On this basis an analogy has been drawn between the BNF and Panini's grammar where there are similarities in formal techniques, including recursions, transformations and other rules. For this reason, Ingerman concluded that Panini's notation is equivalent in its power to that of Backus and Naur.

[9] The intricate system of conventions governing rule interaction and application, the linear arrangement of partially ordered sets in the form of linearised representation of hierarchical relationships, the use of markers to trigger the application of the *sutras* are some of the techniques that one would associate with the modern information theory. These are also concepts attributed to Panini by certain historians which others dispute. For example, Indraji (1876) claimed that the Brahmi numerals (discussed later in this chapter) developed out of using letters or syllables as numerals. He suggested that Panini was the first to come up with the idea of using letters of the alphabet to represent numbers. This is not totally convincing since the symbols for 1, 2 and 3 clearly did not come from Brahmi letters but more likely arose from representation

of these numbers with one, two and three lines respectively. However, there is a widespread acknowledgment of Panini's *Astadhyayi* as a characteristically Indian text whose symbolism and technical conventions have deeply influenced many other disciplines and especially Indian mathematics.

[10] Pingala, the author of the *Chandahsutra*, has been variously identified in popular literature as the younger brother of Panini, or as Pantajali, the author of *Yogasutra*, or of even the major treatise of Sanskrit grammar, the *Mahabhashya*, composed around second century BCE. The *Chandahsutra* contains eight chapters difficult to understand without a commentary. A notable and extensive commentary was supplied in the 10th century CE by Halayudha containing a description of the construction of the so-called Pascal triangle. The *Chandahsutra* also contains traces of an early summation of a geometric series of the form:

$$1 + 2 + 2^2 + \cdots + 2^n.$$

As we will see in a later chapter, the first systematic treatment of the geometric series is found in Mahavira (850 CE) *Ganitasarasangraha*. Earlier, a Jain mathematician Virasena (*c*. 810 CE) in his commentary *Dhavala* derived the volume of a frustum of a circular cone by a sort of infinite procedure:

$$\frac{1}{4} + \left(\frac{1}{4}\right)^2 + \left(\frac{1}{4}\right)^3 + \cdots = \frac{1}{3}.$$

The proof of this result had to wait for the Kerala astronomer/ mathematician, Nilakantha in *Aryabhatiyabhasya*. And this was a milestone in the work on infinite series and calculus by the Kerala School of Mathematics.

[11] Balbir Sihag, in his writings, attempts to highlight the contributions of this neglected savant. A special issue of the journal *Humanomics* (2009) is devoted to Kautilya's contributions to economics, with Sihag as the guest editor.

[12] Take the following excerpt from the famous creation hymn in *RigVeda* (Book 10, Hymn 90):

> On every side pervading earth he fills a space ten fingers wide
> A thousand heads has *Purusha*, a thousand eyes, a thousand feet;
> . . .
> All creatures are one-fourth of him, three-fourth eternal life in heaven
> When three-fourth Purusha went up; one fourth of him remained;
> . . .
> Forth from his navel came mid-air; the sky was fashioned from his head;
> Earth from his feet, and from his ear the regions. They formed the worlds.

What is clear from this passage is that as early as the second millennium BCE, there existed a precise metric sense and that included fractions and a sense of infinity of space.

[13] In calculating $(8400000)^{28}$, an initial count up to a number of eighty-four lakhs (8400000) gave the '*purvi*'. Squaring the '*purvi*' gave a '*purva*'. Successive multiplication by 84 lakhs gave the ultimate number '*Sirsaprahelika*'. That works out to a 196 digit number!

[14] It is difficult to provide a definitive reason for this fascination with large numbers, The explanation may be found in the complexities of cosmology or from observing nature that may have led to abstract thinking such as the comparison between the vastness of a high mountain and the smallness of a mustard seed *or* a contemplation of the number of grain sands on a vast beach.

[15] There are basically three different types of calendar: lunar, lunisolar and solar. Both lunar and lunisolar use 'lunation' (i.e. the average length of the time between the new moons or 29.53 days) as calendar months whereas solar calendars define calendar months as a fraction of the year (i.e. the average month length is $365.24 \div 12 \approx 30.44$ days). Our Gregorian solar calendar has either 30 or 31 days long (not counting February).

[16] According to the Vedic cosmology, life in the universe is created, destroyed once every 4.1 to 8.2 billion years, which is equivalent to one full day and night for Brahma. The lifetime of a Brahma himself was given as 311 trillion and 40 billion years. The *yugas* (cycles) are said to repeat like the seasons, waxing and waning within a greater time-cycle of the creation and destruction of the universe.

[17] The word '*purna*' has been used as a word-numeral for zero. It has also been interpreted as infinity. This follows from a mathematical representation of a verse from Upanishads where,

$$Purna - Purna$$
$$= Purna \text{ implies } 0 - 0 = 0 \text{ or infinity} - \text{infinity} = \text{infinity}.$$

[18] There is a legend from the Islamic world that illustrates well the Indian fascination with large numbers. Being shown the *chaturanga*, (an early version of chess) by its inventor, a raja promised any reward asked. The inventor requested that for his reward, one grain of the wheat be placed on the first square of the board, 2 on the second, 4 on the third, 8 on the fourth and so on up to the sixty-fourth square. The raja soon realised, after consulting the mathematicians at his court, what he earlier thought was an insultingly modest reward, could not be met even if he gave away all the grain in his kingdom. The total grain needed was $2^{64} - 1$! A wily court official suggested that the country's granaries should be opened and the inventor be asked to keep counting without stopping the number of

grains desired. The inventor got the message and was never heard of again!

[19] The word zero that we use today is believed ultimately to be derived from the original *sunya* in Sanskrit. Transliterated from Sanskrit into Arabic as *sifr* (meaning 'vacuum' or equivalently 'nothing'), the word travelled to Europe changing to different sounding Latin words *sifra, cyfra, zyphra, zephirum*. From 15th century , some of these words came to describe the set of Indo-Arabic numerals (*chiffre*) in different European languages. The word *zephirum* was taken by Fibonacci to Italy in the 13th century and transformed into *zefiro*, from which originated the English 'cipher' and the French 'chiffre'. The word was finally transformed into *zeuro* or *zero* and used to name the symbol in English, French and Spanish while the word in German means 'null'.

[20] It is interesting that the Egyptian hieroglyph '*nfr*' for zero also had a connotation meaning "beautiful, good, perfect", suggesting a positive association in contrast to its negative association in medieval Europe.

[21] It is important in this context to recognize that a place value system can exist without the presence of a symbol for zero. The Babylonian and the Chinese numeration systems were good examples. But the zero symbol as part of a system of numerals would never have come into being without a place value system. In neither the Egyptian nor Greek nor the Aztec cultures was there existed a place value system. A zero as a number in any of these systems would in any case have been superfluous (Menninger, 1977, 391–392).

[22] In ancient India, Sanskrit was spoken only by Vedic Brahmins. The common language was *Prakrit*, derived from Sanskrit. Jain religion and literature was promoted with different versions of *Prakrit*.

[23] And unlike many ancient texts, the *Lokavibhaga* provides detailed astronomical evidence to indicates that it was composed on 25 August 458 CE .

[24] The concepts *of lopa* in Panini, *abhava* in *Nyaya* and *sunya* in Buddhist philosophy are closely related to the idea of zero. For further discussion of the philosophical underpinnings of the Indian zero, see Sarukkai (2008).

[25] It may be noted that within a decimal place-value system there are three positions for zero:

- *Medial* which is the classical role of a blank space, e.g. 105. 1005.
- *Terminal* which has a more restricted role, e.g. 250, 2500.
- *Initial* which has a superfluous role and has become more important in this age of computers e.g. 025 = 0025 = 25.

[26] In this age of binary computation and their application to logical circuitry in computers, the origins of binary numbers may be traced

ultimately to Pingala (*fl.* 100–200 BCE) who wrote the influential text on prosody mentioned earlier entitled *Chandahsutra*. Prosody at that time in India was not just versification or study of speech-rhythms, but the complete technology of poetry with metrics, metre with syllables, syllables with sound, sound with vibration, etc.

[27] Bhaskaracharya's definition of infinity (*khahara*) as a fraction where the denominator is zero (i.e. $a/0 = 8$). Today, this formula would only be valid in the case of a type of transformation of the Euclidean plane where the inverted image of zero is a point at infinity. However, Bhaskara correctly deduces that it is infinity by an ingenious limit process. Indeed, by that time, the rules for operating with fractions were already known and Bhaskara proceeds to assert that since $\frac{1}{1/2} = 2$, $\frac{1}{1/3} = 3, \ldots$ and so on, it would follow $\frac{1}{1/n} = n$ as n tends to infinity. This would seem a profound discovery as important as the discovery of zero. In fact, it opens the way for a prototype notion of limits that would be taken up in greater detail in a later chapter.

[28] It should be remembered that in comparison with other numbers systems our number system is economical both in the space occupied by any written number, its ability to express an arbitrarily large number and above all its computational facility. For example, a simple 888 in our number system is equivalent to a 12-digit Roman number DCCCLXXXVIII.

[29] In his book *Cosmos* (1980), Carl Sagan writes "...The dates on Mayan inscriptions also range deep into the past and occasionally far into the future. One inscription refers to a time more than a million years ago and another perhaps refers to events of 400 million years ago, ... The events memorialized may be mythical, but the time scales are prodigious."

[30] There are a few mathematical expressions which are called undefined or indeterminate and all of these involve zero and infinity. Whatever the context, division by zero is meaningless so that such an operation is invalid. This does not preclude the fact that in higher mathematics we often encounter the so-called indeterminate forms that can only be evaluated through a limiting process. For further details, see Maor (1987, pp. 7–9).

[31] Lam Lay Yong (1986) has argued for an earlier Chinese origin of Hindu-Arabic numerals on the basis of similarities in the algorithms for multiplication, division and square root extraction found in the works of Islamic mathematicians al-Uqilidisi and al-Khwarizmi when compared to the mathematical work of a Chinese mathematician Sun Zi written five centuries before. This view has not received wide acceptance.

[32] For further details, see Coedès (1930/32, pp. 323–328).

[33] An extensive discussion is found in a number of different sources including Datta and Singh (1938, pp. 40–51) and Ifrah (2000, pp. 356–439).

[34] It is interesting to observe that an analogous shift occurred with the zero symbol (*anusvara* in Kannada-Telugu script). In early inscriptions, the *anusvara* was represented by a small dot, often hardly visible on the copper plates. Gradually it was replaced by a small circle identical to the present symbol for *sunya*.

[35] For more details see Hayashi (1995, p. 89).

[36] For more details, see *The Patiganita of Sridharacarya, with an ancient Sanskrit Commentary*, edited by Kripa Shankar Shukla, Lucknow, 1959, Introduction, p. xxix.

[37] A parallel situation obtains in the case of the expressions *nalika* and *ghatika*. The first expression meant originally an outflow water clock of cylindrical shape and time unit of 24 minutes measured by this clock. Around the fourth century CE, this type of water clock was replaced by the sinking bowl type of water clock which was called *ghatika*. The time unit of 24 minutes measured by this variety of water clock was called *ghatika*. But *nalika*, the older designation for this time unit also continued to be used. Often both are used as synonyms in the same text. In the South, however, *nalika* survives in Tamil and Malayalam (with slight phonetic modification) while *ghatika* (with slight phonetic change) survives in Kannada and Telugu.

[38] There is considerable literature on interactions of scientific ideas within the Asia. For further details, see Sen (1962, 1970, 1986) and Gupta (1982, 1989).

[39] Duan Yaoyong (1996, pp. 110–111) investigated this question in his master's thesis and came to the following conclusion:

> I undertook a systematic study and organization of this topic for the first time (some made fragmentary study about it before). [I] got that trigonometry knowledge from English translation of Indian Astronomy Corpus and tried to make it systematic (nobody did it before). I found an equivalent way that is equal to the method of trigonometry in China; made study of Chinese and Indian calendric calculation system, and tried to find the internal relations [between] them. I [tried to] answer Dr. J. Needham's question [viz. 'How much trigonometry did the Tang astronomers know and where did they learn it?'], with his reply that 'Chinese calculation had nothing to do with Indian trigonometry. Chinese algorithm specialist[s] solved the problem with 'the *gougu* theorem' which is equivalent way of trigonometry used by Indians'. I have tried to set out reasons

why Chinese never accepted trigonometry. [Thesis Abstract, *Ganita-Bhāratī: Bulletin of the Indian Society for History of Mathematics*, 20 (1998), pp. 110–111, with minor modification for the sake of clarity]

[40] A story found in *Alberuni's India*, translated and reprinted in Delhi in 1964 in Vol. 2, p. 15.

[41] For further details of this story, see Sen (1962, pp. 24–25).

[42] In the 14th century Arab navigators report of their meeting the "Kanakas" on the Malabar coast, who probably were *Ganakas*, (i.e. astronomers or more specifically navigators).

[43] See Joseph (2011, p. 462) and also Reich (2000, pp. 478–489).

[44] "Algorizmi said: [i.e. al-Khwarizmi mentioned earlier]

> When I saw that Indians composed out of IX letters any number due to the position established by them, I desired to discover, God willing, what becomes of those letters to make it easier for the student... Thus, they created IX. letters, figures of them are as follows ... The beginning of the order is on the right side of the writer, and this will be the first of them consisting of unities. If instead of unity they wrote .X and it stood in the second digit, and their figure was that of unity, they needed a figure of tens, similar to the figure of unity so that it became known from it that this was X. Thus they put before it one digit and wrote in it a small circle "o," so that it would indicate that the place of unity is vacant. [Cited in Rosenfeld (1990, p. 132)].

[45] See Burnett (2002, pp. 237–288) for a discussion of the first appearance of Indian numerals in Europe.

[46] This is no doubt an elegant solution for the nomenclature of the numerals of the form 1, 2, 3 etc. But how does one designate unambiguously the numerals associated with the Arabic–Persian script, such as 123? Again, when the Indian numerals were transmitted to Europe, they were first known as the "Indian" numerals. One wonders when Europe started calling them "Arabic" numerals.

[47] The introduction and acceptance of zero into Europe was by no means a simple process. Awe, fear, suspicion and rejection of the new number is well summed up by Menninger (1969, p. 422) in the following passage:

> What kind of crazy symbol is this which means nothing at all? Is it a digit or isn't it? 1, 2, 3, 4, 5, 6, 7, 8 and 9 all stand for numbers one can understand and grasp — but 0? If it is nothing, then it should be something. But sometimes it is nothing and then at other times it is something: $3+0 = 3$ and $3-0 = 3$, so here zero is nothing and when it is placed in a front of a number it does not change. $03 = 0$. So zero is

> nothing, *ulla figura*! But write the zero after a number, and it suddenly multiplies the number by 10: $30 = 3 \times 10$. So now zero is something — incomprehensible but powerful, if a few 'nothings' can raise a small number to an immeasurably vast magnitude.... (I)n short, the zero is nothing but 'a sign that creates confusion and difficulties'.

Thus the resistance to the Indian numerals by those who used the counting board (abacus) for calculations took two forms: some regarded them as the creation of the Devil, while others ridiculed them.

48 However, the spread of Indian numerals to Europe was a slow process. In the 16th century, before widespread adoption of decimal arithmetic in European commerce, a wealthy German merchant consulted a scholar regarding which European university offered the best education for his son. He was informed that if his son needed to learn addition and subtraction with the 'new-fangled method', any French or German university would be sufficient. But if the intention was to master multiplication and division and assuming that his son had the necessary ability, then he had to be sent to Italy (Ifrah, 2000, p. 577).

49 The exposition that follows is based on an article by Fabio Gironi (2012) entitled '*Sunyata* and the Zeroing of Being: A Reworking of Empty Concepts'.

50 Nagarjuna was a highly influential Buddhist philosopher of whom little personal details are known. One account suggests that he was an abbot of a monastery. Author of a number of Buddhist texts and inspiring a larger number of commentaries composed in India, China and Tibet, he was well known for his writings on the philosophy of emptiness (*Sunyata*). His philosophical texts were directed against logicians of non-Buddhist schools and the doctrines and assumptions of the non-Mahayana Buddhist schools. The core theme of his writing is the path to Buddhahood (enlightenment).

51 In Buddhist literature there is some awareness of indeterminate and infinite numbers. Numbers were deemed to be of one of three types: *Sankheya* (countable), *Asankheya* (uncountable) and *Anant* (infinite). But beyond that there was little development in Buddhist writings.

52 There exists another interesting concept of infinity. The Jains recognized different types of infinities: infinite in length (one dimension), infinite in area (two dimensions), infinite in volume (three dimensions), and infinite perpetually (infinite number of dimensions).

53 For a student of any age, the concept of countable infinity ($\aleph_0$), is a difficult one. Asked to explain the concept of infinity, the following were some of the responses of young children: "there is no larger number," "there is no ending to a number," "infinity is counting numbers that will

never end, "infinity is a word to say that a number does not change, "a number that keeps going and going and going,... forever, "that means there is lots of numbers and it just keeps going," and an "everlasting number." A more insightful statement: "No one can say infinity is a certain number because it goes on and on non-stop," implying the idea that infinity is a concept rather than a number.

54 Now when the highest enumerable number, call it N, is attained, infinity may be reached via the following sequence of operations:

$$N + 1, N + 2, \ldots, (N + 1)^2 - 1$$
$$(N + 1)^2, (N + 2)^2, \ldots, (N + 1)^4 - 1$$
$$(N + 1)^4, (N + 2)^4, \ldots, (N + 1)^8 - 1 \quad \text{and so on.}$$

It is not practical to examine this fascinating area of mathematics here. Simple introductions to the concept of transfinite numbers and operations with such numbers are given by Stewart (1981, pp. 127–143) and by Sondheim and Rogerson (1981, pp. 148–159).

55 Yet another imagery of infinity. In Umasvati's *Tattvarthasutra*, appears what seems an infinite chain of concentric circles:

> There are islands and oceans that bear names such as Jambu Island, Lavana Ocena and so on. The islands and oceans are concentric circles, the succeeding ring being double the preceding one in breadth. At the centre of these islands and oceans is the round island Jambu with a diameter of 100,000 yojanas and Mount Meru at its navel. There are seven continents on Jambu Island: *Bhara, Haimavati, Hari, Videha, Ramyake, Haianyavata* and *Airavata*. The six mountains that extend from east to west and divide the seven continents are.... Continuing the description of islands, the text states that the number of islands and oceans is innumerable.
>
> Some have argued that 'innumerable' in this context means 'countable infinite'. But since the author later mentions that the concentric circles of islands and oceans stop at an ocean called *Svayambhuramana*, the number of islands and oceans cannot be infinite.

56 There is a difference in opinion as to the precise meaning of these topics and we are frequently warned about the dangers of applying today's symbolic package and terminology in describing ancient mathematics, especially those emanating from the Non-Western world. There is a lively debate between those who believe that past can be interpreted only in the context of the past culture ('historicism') and those who attempt to understand such works in terms of the present ('presentism'). Whether one inclines to 'historicism' or ' presentism' in one's own interpretation

depends to an extent on how one sees present-day mathematics. Did it evolve from older mathematics where the older mathematics has been absorbed into our own, or whether one sees different mathematical traditions as being to a significant extent incommensurable? Ultimately those who condemn all attempts to interpret a text on the ground that the text itself does not employ our linguistic and conceptual structures should be reminded that no translation or interpretation is sacrosanct, and that the real challenge for anybody is to make their presuppositions conscious and to convey through their work both a sense of wonder and uncertainty when confronted by the original text.

[57] A cautionary note should be sounded here. The use of the word 'logarithm' in this context does not imply that it ever became a computational tool in Indian mathematics. Like all computational tools they have become redundant with the introduction of new and better methods such as calculating machines and computers.

[58] Again, a cautionary note should be sounded. It is not claimed that the Jains had the symbolic package or the conceptual compass for expressing the results as we have done. It is a form of transplantation which can only be justified on grounds of comprehension and clarity.

[59] However, there are other examples of combinations and construction of the *meru* (tabular figure) in Indian mathematics. In *Brhatsamhita* (Chapter 76, Verse 20) of Varahamihira (*fl.* 550 CE), a discussion of perfumery leads to the calculation of 1820 combinations from choosing 4 perfumes from a set of 16 perfumes (i.e. $^{16}C_4 = 1820$) and the construction of a *meru*. A commentator on Varahamihira's text, Bhattopala (*c.* 950), provides further elaboration.

[60] The Digambara are one of the two main sects of Jain monks whose members shun all property and wear no clothes. In their active practice of nonviolence, they use a peacock-feather duster to clear their path of insects to avoid trampling them.

[61] As far as we know Umasvati was no mathematician and it is more than likely he got these results from the mathematical works that was available to him in his time.

Chapter 5

Down to Earth and Reach for the Stars:
The *Bakhshali* and *Siddhanta* Episodes

5.1 The *Bakhshali* Manuscript

In 1881, near the village called Bakhshali on the northwest border of
India, a farmer digging in a ruined stone enclosure came across the
remains of a manuscript. A large part of the manuscript had been
destroyed and only about 70 leaves of birch-bark, of which a few were
only scraps, survived to the time of its discovery. Written in Sarada
(as opposed to Devanagari) characters, the find was described as
being as fragile as 'dry tinder', with a substantial part mutilated
beyond repair. What remained was put in order and parts of it
translated into English by Hoernle. It now resides in the Bodleian
Library at Oxford.

Kaye (1927) produced another translation of this manuscript
with a commentary. There are serious errors in both his translation
and interpretation — errors which have passed into histories of
mathematics that cite his work. There has been much controversy
over the manuscript's age, and here Kaye's pronouncements have
been particularly unfortunate. On the basis of doubtful literary
evidence, Kaye argued that the Bakhshali Manuscript belonged
to the twelfth century CE. An earlier view supported Hoernle's
assessment that the manuscript was a later copy of a document
probably composed at some time in the early centuries of the
Common era. Hoernle's dating was based on a consideration of a

number of aspects, including the mathematical content, the units of money used in some examples, the presence of the symbol + for the negative sign and the neglect of certain topics (in particular, the solution of indeterminate equations) which appeared in works known to have been written later. However, Hayashi (1995) argued that the date of the original is not earlier than seventh century CE although the Manuscript itself could be a later copy composed between the eighth and the twelfth centuries. Hayashi who produced an authoritative transcription, translation, and commentary of the entire work, gives convincing reasons for his dating of the manuscript, including palaeographical evidence, the material and the language of the manuscript, which was Gatha, a modified form of classical Sanskrit consistent with medieval northwest Indian vernaculars of that time. The Manuscript used a script, *Sarada*, popular during the Gupta era around 350 CE. And finally, the nature of the problems discussed indicates a period before Aryabhata I (pre-fifth century CE). Choosing between the differing scholarly evaluations is a highly specialised skill which the author of this book does not possess.

If Hoernle's assessment is correct, the Bakhshali manuscript would be the next substantial piece of evidence, after Jain mathematics, to bridge the long gap between the *Sulbasutras* of the Vedic period and the mathematics of the 'Classical period', that began around 500 CE. It is also the earliest evidence we have of Indian mathematics free of any religious or metaphysical associations. Indeed, one detects a certain resemblance between the Manuscript and the Chinese *Jiu Zhang Suan Shu* ('Nine Chapters on Mathematical Arts') from some centuries earlier both in the topics discussed and in the style of presentation of results. It should, however, be added that the premier Chinese text is far more wide-ranging and 'advanced' than the Bakhshali work.[1]

The Bakhshali Manuscript is a handbook of rules and illustrative examples together with their solutions. It is devoted mainly to what we would term as arithmetic and algebra, with just a few problems on geometry and mensuration. Only parts of the Manuscript have been restored, and so we cannot be certain about the balance between different topics. The examples from arithmetic cover fractions, square roots, profit and loss, interest and the rule of three, while the algebraic problems deal with what we would now describe as simple and simultaneous equations, quadratic equations, and arithmetic and geometric progressions. All these topics will be developed in detail in later chapters. There is no clue as to who was the author of the work.[2]

The subject-matter is arranged in groups of *sutras*, and presented as follows. In a typical case, a rule is stated and then a relevant example given, first in verse and then in notational form. The solution follows in prose, and finally we have the demonstration or 'proof'.[3] This method of presentation is quite unusual in Indian mathematics. The few texts arranged in this way are invariably commentaries on earlier works. Since, in terms of its content, the Bakhshali Manuscript is but a prelude to more substantial work in a later period, we confine our discussion to a few novel features in this text. We begin with an examination of the system of notation used, as it is a recognisable precursor of later systems.

5.2   Notations and Operations

The system of notation used in the Bakhshali Manuscript has some resemblance to those used by mathematicians such as Aryabhata I (*c.* 476–550 CE), Brahmagupta (b. 598) and even Bhaskaracharya (b. 1114). But there is one important difference. In the Bakhshali text we find that the sign for a negative quantity looks exactly like the present 'plus' (+) symbol used to denote addition or a positive quantity. This sign was placed after the number it qualifies. For example,

$$\left| \begin{array}{cc} 15 & 8+ \\ 4 & 3 \end{array} \right| \quad \text{means } 15/4 -8/3$$

Later, the + sign was replaced by a dot over the number to which it referred. Incidentally, this is one of the clues which tell us that the Bakhshali Manuscript must have originated before the twelfth century. Another interesting aspect of the notation shown above is the representation of fractions. It is similar to the

present-day representation in that the denominator is placed below
the numerator, but the line between the two numbers is missing.

This and other aspects of the notation and operations are brought
out if we take an example from the 25th *sutra*. There the following
representation appears:

$$
\begin{array}{|cccc|}
\hline
\bullet & 1 & 1 & 1 \\
1 & 1 & 1 & 1 \quad bha\ sesam\ 32 \\
 & 3+ & 3+ & 3+ \\
\hline
\end{array}
\quad phalam\ mula\ 108
$$

Here the black dot is used very much in the same way as we
use the letter x to denote the unknown quantity whose value we
are seeking. A fraction is, as we saw earlier, denoted by placing one
number under another, without a line between them. A compound
fraction is shown by placing three numbers under one another; thus
the second column of the representation above denotes 1 minus
1/3 (or 2/3). Without the + sign, it would denote 1 plus 1/3.
Multiplication is indicated by placing the numbers side by side. Thus
the representation above means $(2/3) \times (2/3) \times (2/3)$, or *bha* is an
abbreviation of *bhaga*, meaning 'part', and indicates that the number
preceding it is to be treated as a denominator. *Bha* is thus the symbol
for division. The representation above therefore means:

$$x = [(2/3)^3]^{-1} \times 32 = (27/8) \times 32 = 108$$

or, in words, 'the remainder (*sesa*) is divided by $1(1-\frac{1}{3})(1-\frac{1}{3})(1-\frac{1}{3})$;
the result (*phalam* or 'fruit') is 108'.

In the Bakhshali Manuscript the dot was also used to represent
zero. The use of the same symbol to represent both an unknown
quantity and a numeral is interesting. At the time the dot indicated
an empty place, as its Sanskrit name shows: *shunya* means 'empty',
or 'void'. It is this interesting 'dual' meaning that gives us another
clue to the age of the text.

On only two occasions the symbol for addition which is the
abbreviation *yu* (for *yuta*) is used. On almost all occasions, the two
numbers to be added are put side by side. The whole operation is

enclosed between lines, and the result is set down on the right of *pha* (abbreviation for *phalam*) Thus $3 + 6 = 9$ is represented as

$$
\begin{array}{|c c|}
\hline
3 & 6 \\
 & \quad yu \;\bigg|\; pha\,9 \\
1 & 1 \\
\hline
\end{array}
$$

5.2.1 The rule of three (*trairasika*): a brief survey

An early statement of this rule is in Aryabhata's *Aryabhatiya* and discussed in the next chapter.

> In the rule of three (*trairasika*), multiply the *phala-rasi* (fruit) by the *iccha-rasi* (desire or requisition) and divide by the *pramana* (measure or argument). The required result *iccha-phala* (or fruit corresponding to desire or requisition) will be thus obtained.

Given that f, i, p and m are *phala-rasi, iccha-rasi, pramana* and *iccha-phala* respectively, then $m = \frac{f \times i}{p}$.

This was the first time in Indian mathematics that the technical names for the 'Rule of Three' and for the four numerical quantities involved are given. However, the succinct manner in which the rule is stated would indicate that it was already well known and that Aryabhata was merely restating it as a prelude to its use in astronomical computations. The antecedents of this rule has been traced back about a thousand years to a verse in the *Vedanga Jyotisa* and discussed by Sarma (2002). This rule becomes a standard item in all texts after *Aryabhatiya*. Though normally used in solving commercial problems, the rule played an important role in other areas of Indian mathematics and astronomy. In arithmetic, as we will see below, it was also used as a means of verification in solving other problems. More importantly, it was employed in astronomical computations. For example, the rule is applied in the computation of the mean position of a planet from the number of its revolutions in a *kalpa* of 4,320,000,000 years.[4] Many of the problems of spherical trigonometry were solved by applying the rule to similar right-angled triangles such as those found in the *aksajaksetra* ('figure produced by the latitude').[5] Also, the rule forms the basis for computing trigonometric ratios. Bhaskara I's commentary on Aryabhata's work contains a detailed discussion of the 'rule of three' and points to how it can be extended to encompass Rules of Five, Seven, etc. Bhaskara also introduces the question of the logical sequence in which the three numerical quantities should be set down and the order in which the multiplication ($f \times i$) and division by p should be carried out. Later, Brahmagupta's formulation of the rule became a model for subsequent writers bringing out more explicitly the fact that the three quantities should be set down in such a way that the first and last be of like denomination and the middle one of a different denomination. This is reiterated by Sridhara ($c.$ 800), Mahavira ($c.$ 850) and Aryabhata II ($c.$ 950) without adding much to the principles underlying the rule.

However, Bhaskara II ($c.$ 1114–1200) in his *Lilavati* states that nearly the entire arithmetic is based on the 'Rule of Three' and that most of the topics dealt within *ganita* are but variations of this rule:

> Just as the universe is pervaded by Hari with His manifestations, even so all that has been taught [in arithmetic] is pervaded by the 'Rule of Three' with its variations." (p. 239)

Further, Nilakantha (1444–1544) declares in his commentary on the *Aryabhatiya* that the entire mathematical astronomy (*graha-ganita*) is pervaded by two fundamental laws: by the law of relation between the base, perpendicular and hypotenuse in a right-angled triangle — which goes today under the name of Pythagoras theorem — and by the 'Rule of Three'.

However, depending on the original date of composition of the Manuscript, the first systematic application of this rule may very well have been in here. The problem to which the rule is applied is of a type familiar to school children today: 'If 8 oranges cost 92 pence, what will 14 oranges cost?' The solution is $(92 \times 14)/8 = £\,1.61$.

The method indicated in the Bakhshali Manuscript, which is also found in later works, may be stated in the following terms: If p oranges (argument, or *pramana*) yields f pence (fruit, or *phala*), what will i oranges (requisition, or *iccha*) yield? It is suggested in the Manuscript that the three quantities be set down as follows:

where p and i are of the same denominations, and f is of a different denomination. For the required result the middle quantity is to be multiplied by the last quantity and divided by the first, to give the result as $\frac{f \times i}{p}$.

The following example from the Bakhshali Manuscript (Example 2 from *Sutra* 53) is taken to illustrate the 'Rule of Three', without in any way implying that the solution in the Bakhshali Manuscript follows the method we would use today.[6] For the fuller Bakhshali solution, the reader should refer to Hayashi (1995, pp. 385–386).

Example 5.1 Two page-boys are attendants of a king. For their services one gets 13/6 *dinaras* a day and the other 3/2. The first owes the second 10 *dinaras*. Calculate and tell me when they have equal *amounts*.

Suggested solution: Take the denominators 6 and 2, together with the number 10 that the first has to give. The lowest common multiple of 2, 6 and 10 is 30, so 30 is the *iccha* (requisition).[7] Now apply the rule of three (shown in the table below). Note that if the first page-boy gives the second 10 *dinars*, both will be left with 55 *dinars*.

	p (day)	f (*dinars*)	i (day)	**Required result** (fi/p)
1st page-boy	1	13/6	30	65
2nd page-boy	1	3/2	30	45

Consider a more apposite illustration from *sutra* 27 of the 'rule-and-example' type of exposition, beginning with the statement of the rule, followed by the problem expressed verbally and then in a numerical form, and ending with a verification using the 'Rule of

Three'. This example is a minor modification of what is contained in Hayashi (1995).

Example 5.2

Rule: Multiply [the weights of] the gold pieces by [their respective] impurities. One then divide their sum by [the weights of] the gold pieces added together. The [result] gives the loss [of gold] per unit [weight of the alloy].

Example [Four] gold pieces, the quantities of which are one, two, three and four *suvarnas* [respectively] are debased by one, two, three and four negative *masa* [per *suvarna* in that order]. [They are melted into] and formed into single alloy. What is the impurity [of that alloy]? (Note: 1 *suvarna* = 16 *masas*)

The Problem in Numbers

Weight in *suvarnas*	1	2	3	4
Impurity in *masas*	[−] 1/2	[−] 1/3	[−] 1/4	[−] 1/5

Computation

Multiply [the weights of each] gold piece by its own impurity. Add the fractions together by reducing each to the same denominator. Result 163/60. Divide this sum 'by the weights of the gold pieces added together (Result: 163/600). This is the loss [of gold] per *suvarna*'.

Verification

The calculation is verified by applying the 'Rule of Three' in obtaining the combined impurity of all four pieces which amounts to $1630/600 = 163/60$ *masas*. The calculations are shown in Table 5.1 below. In terms of the modern notation: If w_j = weight of the jth gold piece in *suvarnas* and c_j is the impurity (or non-gold component) of the jth gold piece in *masas* for $i = 1, 2, \ldots, n$, then the combined impurity of all n gold pieces, when melted together is

$$\frac{w_1 c_1 + w_1 c_1 + \cdots + w_n c_n}{w_1 + w_1 + \cdots + w_n}.$$

Table 5.1.

Type	Pramana (p)	Phala-rasi (f)	Iccha-rasi (i)	Iccha-phala (m) $m = (f \times i)/p$
1	10/1	163/60	1/1	163/600 *masas*
2	10/1	163/60	2/1	163/300 *masas*
3	10/1	163/60	3/1	163/200 *masas*
4	10/1	163/60	4/1	163/150 *masas*
Total				**1630/600 *masas***

5.2.2 Extracting square roots

The Bakhshali Manuscript extended the work on square roots in the *Sulbasutras*, which we discussed in a previous chapter, to give a more accurate formula for finding an approximate value of the square root of a non-square number. The relevant *sutra* (No. 18) may be expressed more clearly in the following terms:

> In the case of a non-square number, subtract the nearest square number; divide the remainder by twice the nearest square; half the square of this is divided by the sum of the approximate root and the fraction. This is subtracted, and will give the corrected root.

In symbolic form, this rule is:

$$\sqrt{A} = \sqrt{a^2 + r} \approx a + \frac{r}{2a} - \frac{\left(\frac{r}{2a}\right)^2}{2\left(a + \frac{r}{2a}\right)},$$

where a^2 is the perfect square nearest to A and $r = A - a^2$. For example:

$$\sqrt{41} \approx 6 + \frac{5}{12} - \frac{\left(\frac{5}{12}\right)^2}{2\left(\frac{6+5}{12}\right)} \approx 6.4031 \text{ (to 4 decimal places)}.$$

(The reader may wish to try using this rule to evaluate $\sqrt{3}$ and $\sqrt{5}$.)

The formula was applied in the Manuscript to calculate the approximate square root of 481 as 424642/19362 which is correct to four decimal places. The formula may be compared with the approximation procedure for finding the square root of a non-square integer, usually attributed to the Hellenistic mathematician Heron (*c.* 200 CE), although there is a close antecedent of the procedure in Babylonian mathematics. Heron's formula for finding the square root of *A* is

$$\sqrt{A} = \frac{1}{2}\left(a^* + \frac{A}{a^*}\right),$$

where a^* is a first approximation to the square root of A, and can be a non-integer. (This is not possible with the Bakhshali method).[8]

5.2.3   Indeterminate (diophantine) equations: their early appearance

The following is one of a number of similar problems found in the manuscript:

Example 5.3 Three persons possess 7 *asavs*, 9 *hayas* and 10 camels, respectively [*asavs* and *hayas* are two breeds of horses]. Each gives two animals, one to each of the others. They are then equally well off. Find the price of each kind of animal and the total value of the livestock possessed by each person.

Solution (in symbolic terms) Let x_1, x_2 and x_3 be the prices of an *asava*, a *haya* and a camel respectively. Then, from the information given in the question,

$$5x_1 + x_2 + x_3 = x_1 + 7x_2 + x_3 = x_1 + x_2 + 8x_3 = k$$

or

$$4x_1 = 6x_2 = 7x_3 = k$$

and we seek values of x_1, x_2, x_3 and k which are positive integers.

To get integer solutions, we take k to be any multiple of the lowest common multiple of 4, 6 and 7. In the Bakhshali Manuscript k is taken as $4 \times 6 \times 7 = 168$. Then the price of an *asava* is 42, the price of a *haya* is 28 and the price of a camel is 24. The total value of livestock in the possession of each person is 262.

From these humble beginnings, over the next 1,000 years there was to be a systematic development of indeterminate analysis, a subject which will be examined in sections in later chapters.[9]

5.2.4   Series

An early example of an arithmetic series, found in the Bakhshali Manuscript, relates to two people going on the same journey with the first travelling a *yojanas* (1 *yojana* is about 9 miles) on the first day and an additional b *yojanas* each successive day. The second person travels at a uniform rate of c *yojanas* per day, given a head start of t days. The question is: when would the first person overtake the second?

If x is the number of days after which the first person will overtake the second? In modern notation, the value of x can be obtained from solving the equation:

$$(t + x)c = a + (a + b) + (a + 2b) + (a + 3b) + \cdots \ x \text{ terms.}$$

A correct solution for x was provided although no explanation is given.

However, there is a related problem where two persons start with different initial velocities a_1 and a_2, travelling on successive days, increasing at different rates b_1 and b_2. But they cover the same distance after the same period of time. To find this period of time, the Bakhshali Manuscript gives the following rule:

> Twice the difference of initial terms divided by the difference of the common difference is increased by unity

Expressed in modern notation the solution is obtained from:

$$x = \frac{2(a_1 - a_2)}{b_2 - b_1} + 1.$$

Among the arithmetic series found in the Bakhshali Manuscript, there are some unusual ones. As an example, consider the next problem, reconstructed from a mutilated birch-bark strip, and found in Hayashi's translation and interpretation of Example 7 in Sutra N6 (Hayashi, 1995, pp. 286 and 395)

Example 5.4 Oh wise man! A certain king gave five horse-men a gift of 57 (monetary unit missing). Each person in order, I tell (you), obtained twice the amount of his predecessor and one more. What then was obtained by the first person and what by each of the others?

Solution: The problem may be expressed thus. Let x_i be the share of the ith horse-man. Then

$$x_{i+1} = 2x_i + 1 \quad (i = 0, 1, 2, 3, 4).$$

The solution is lost in the original manuscript. Expressed in modern notation, the problem may be expressed in terms of the following equations:

$$x_1 = x_0 + 0, \quad x_2 = 2x_0 + 1, \quad x_3 = 4x_0 + 3,$$
$$x_4 = 8x_0 + 7, \quad x_5 = 16x_0 + 15.$$

The solution is obtained as:

$$x_1 = 1, \quad x_2 = 3, \quad x_3 = 7, \quad x_4 = 15, \quad x_5 = 31$$

being the shares received by the five horse-men.

5.3   Early Indian Astronomy and the Emergence of the *Siddhantas*

The history of Indian astronomy goes far back into distant past. Controversies reign both on the nature and dating of this knowledge, fuelled by battle lines drawn between those promoting a form of chauvinism on one hand and excessive scepticism on the other. What we can say with some assurance is that both evidence of origins and information regarding early transmission to India of mathematical and astronomical knowledge from other cultures is at best spotty and offers opportunities for speculations and interpretations, especially of the Vedas and the supposed knowledge and dates encoded in them. What is a safe bet is that like many other ancient cultures, interest in astronomy was present at least by the middle of the millennium BCE and possibly earlier dealing especially with calendar periods.[10]

It was in the field of astronomy during the early centuries of the first millennium that India began to make her mark. While written records of such activity have not survived the ravages of time, there is enough to get a flavour of what was happening then. The fundamental objective of mathematical astronomy was to help to determine the solar, lunar and planetary positions from a particular place. Additionally, as astronomical knowledge increased, further objectives became important: to predict eclipses, to convert celestial coordinates, to find the gnomon shadow lengths[11] at a particular time for the purposes of constructing accurate calendars or to make astrological predictions or to determine geographical directions. It was soon recognised that a critical problem was determining the mean position of the celestial body in question, correcting this position for its orbital anomalies and then using the true position to predict the occurrence of sunrise, new and full moons, conjunctions, eclipses and so on. With the arrival of data on astronomical period relations from Babylonia probably in the 5th century BCE, there were some major changes in Indian astronomy, notably the incorporation of the arithmetic character of Babylonian astronomy, such as calculation of parameters and representing continuously changing characters with linear zigzag functions.[12] This was embodied in the emerging *siddhanta* (i.e. a 'doctrine' or a 'tradition').

The first of the major fully preserved astronomical texts which both integrated mathematical methods with astronomical explanations as well Hellenistic components with indigenous elements is found in the well-known work *Aryabhatiya* of Aryabhata (b. 476 CE).[13] Aryabhata accurately calculated celestial constants like earth's rotation per solar orbit, days per solar orbit and days per lunar orbit. Incidentally, his computation of the circumferences of a circle given its diameter (i.e. π) as 3.1416 and the length of a solar year as 365.358 days were extremely accurate by the standards of the next 1,000 years. Aryabhata was apparently quite skeptical of the widely held doctrines about eclipses and also about the belief that the Sun goes round the Earth. He did not think that eclipses were caused by Rahu but by the Earth's shadow over the Moon and the Moon obscuring the Sun. As early as the sixth century, he talked of the diurnal (daytime) motion of the earth and the appearance of the Sun going round.

A typical *siddhantha* contained not only an explanation of the methods involved but also a discussion of the technical instruments available then for measuring time and angles. Both in terms of intellectual and technological transfers across cultures, the *siddhanta* became an important tool for the advancement of mathematical astronomy in India. By the early centuries of the first millennium CE, a synthesis was emerging between indigenous traditions of astronomical and calendrical concepts and computations and the Hellenistic contributions in the form of plane trigonometry of chords and geocentric models involving spherical bodies with planetary eccentrics and epicycles.[14] The period of intellectual exchange seemed to have come to an end before the emergence of Ptolemaic astronomy in the third century CE, after which Indian astronomy developed in near isolation for the next few centuries.

The first of the major fully preserved astronomical texts which both integrated mathematical methods with astronomical explanations as well as Hellenistic components with indigenous elements is found in the well-known work *Aryabhatiya* of Aryabhata (b. 476 CE). A discussion of this work and its author will be postponed to the next chapter.

5.4   A Summing Up

The state of Indian mathematics around the middle of the first millennium CE, as exhibited by the Bakhshali Manuscript, may be summarised as follows.

1. While the mathematics of the Vedic age and of the Jains were principally inspired by religion and metaphysics, the mathematics of this period became more practical and secular, being applied to everyday problems. Examples of profit and loss, computation of the average impurities of gold, wages or gifts to be paid to subordinates, and speeds and distances to be covered formed the subject-matter of the Bakhshali Manuscript.

2. Whereas the writers of the *Sulbasutras* had already devised rules to find approximate values to transform and combine different shaped altars, these rules became more elaborate and were being used to obtain, for example the square root of any number to a greater degree of accuracy.

3. There is some evidence that work on mathematical series begun during the Vedic and Jain period was continued by later mathematicians, including the composer(s) of Bakhshali Manuscript.

4. This period marks the beginning of the great interest in indeterminate analysis. Such an interest did not arise solely from the demands made by astronomical calculations. Other problems, some of them of a recreational nature, were also catered for. The examples discussed in the Bakhshali Manuscript and the solutions offered are not difficult, but they mark the beginning of a study which was to reach an advanced level during the so-called Classical period of Indian mathematics.

5. There is evidence of the existence of a well-developed place-value number system which included zero (represented by a dot). The facility with which the system is used in the Manuscript suggests that the system predates the document by a few 100 years.

6. In contrast to the majority of Indian mathematical works composed before and after this manuscript, the method of exposition follows a systematic order, as illustrated by the examples discussed earlier. The solution conformed to the following order (i) statement of the rule (*sutra*), (ii) example(s) to apply the rule to, (iii) a solution using the rule and (iv) verification of the correctness of the solution Many of the other sources of Indian mathematics, until the emergence of Kerala mathematics in the late 14th century contain concise statements of the rules, usually without any attempt at deriving or demonstrating them. These were left to subsequent commentators or teachers to explain.

Endnotes

[1] For a more detailed discussion of the contents of the Chinese "Nine Chapters", see Joseph (2011: pp. 215–241).

[2] However, as pointed out by Montelle (2013, p. 1661), there is a colophon on the Manuscript that states that it was composed by a Brahmin who was the son of a Chajaka (i.e. king of calculator) who lived in the ancient city of Martikavati near present-day Peshawar. Since we do not know whether the individual was a scribe or the author and nothing about the date when it was produced, the author of the Manuscript remains for all purposes anonymous. There is, however, some evidence that the Manuscript was the creation of more than one scribe, for at one point between two lines a rule stated is dismissed as incorrect.

[3] It is interesting that the prose explanations do not always reflect the solution procedures offered, possibly suggesting that some of the verses stating the problem may have been recorded earlier than the explanations.

[4] For further details of this computation, see Plofker (2009, p. 68).

[5] I am grateful to Takao Hayashi for providing a precise definition of the term *aksajaksetra*. There is a short introduction to Indian spherical trigonometry in Chapter 11.

[6] We would solve this problem today by working out that the first page-boy gets $13/6 - 3/2 = 2/3$ *dinaras* more than the second each day. The former needs 20 *dinaras* more than the latter to return his 10 *dinaras* debt and for both to end with equal amounts. For that 30 days are required when each has $13 \times 30/6 - 10 = 55$ *dinaras*.

[7] The Bakhshali Manuscript shows evidence of knowing how to reduce the numbers of fractions to a common denominator to carry out further arithmetical operations. For example, to find the sum of 2, $1\frac{1}{2}$, $1\frac{1}{3}$, $1\frac{1}{4}$, and $1\frac{1}{5}$, you are asked to reduce each to a common denominator of 60 and then add the numerators to get the sum of the fractions as $437/60$.

[8] It is worthy of note that the Babylonian procedure, permits successive approximation not possible with the Bakhshali method. It is easily seen that the Bakhshali formula and Heron's formula produce the identical result for the square root of 3, for $a = 1$ and $a^* = 1.5$. In Heron's example in his book, *Metrica* 1.8, $A = 720$, $a^* = 27$ and $A/a^* = 720/27 = 26\frac{2}{3}$ eventually produces a square root approximation for 720 correct to 3 decimal places. For further details, see Joseph (2011, p. 365).

[9] In Bhaskaracharya' *Lilavati*, there appears a similar problem where there are four persons and the animals are replaced by precious stones, suggesting that such problems with 'indeterminate' solutions were known earlier than Bhaskaracharya's time.

[10] It is interesting that the Ishango Bone discovered in the Congo is reckoned to be one of the earliest records (going back to over 20,000 BCE)

of attempts to quantify time. One view is that it represents a six-month lunar calendar. For further details, see Joseph (2011, pp. 30–35).

[11] Probably the first astronomical measuring instrument was the gnomon, which in its simplest form is a straight stick stuck vertically into the ground. The gnomon was used to track both the day and the longer term movement of the Sun in the sky. A sundial is little more than a graduated gnomon.

[12] A linear 'zigzag function' is a line that proceeds by sharp turns in alternating directions. In early Babylonian astronomy we find the first linear zigzag functions describing the difference in the day duration. The same function was used for investigating the lunar and astral phenomena in the astronomical texts.

[13] In the next chapter (Chapter 6) we will be concentrating mainly on the mathematical content of Aryabhata's contribution. The astronomical work of Aryabhata I and the others who followed him will only be discussed in passing in terms of its contribution to the development of trigonometry which is surveyed in Chapter 12.

[14] The early Hellenistic mark on Indian astronomy is particularly noticeable in works such as *Romakasiddhanta* (meaning 'Doctrine of the Romans') and *Paulisasiddhanta* (meaning Doctrine of *Paulisa Muni*) where the names themselves are suggestive. It was only with Varahamihira's *Pancasiddhanta* in the 6th century CE that the combination of the indigenous, Babylonian and Greek procedures occurred.

Chapter 6

Heralding the Golden Age:
Aryabhata I and his Followers

6.1 Introduction

Of the major mathematicians/astronomers of the "classical" period of Indian mathematics and astronomy, Aryabhata is the best known. His name often appears at the head of a group of individuals who made notable contributions to the development of Indian mathematics and astronomy. His impact is more substantial than his known contributions, for he founded a School that profoundly influenced the development of both astronomy and mathematics in different parts of India, especially in Kerala, between the 14th and 17th centuries.

There are a number of unresolved questions regarding him, including his place of birth and indeed his very identity. There is some confusion arising from the fact that more than one mathematician bearing the name of Aryabhata lived over a short period of time. It is well-known there was one Aryabhata of Kusumpura who wrote the famous text *Aryabhatiya* in 499 CE; and another who composed an astronomical text *Maha Aryasiddhantha* around 950 CE. But then both the Indian Brahmagupta (b. 598) and the Persian Al-Biruni (b. 973) mention a third Aryabhata who lived before the Aryabhata of Kusumpura. In his history written in 1036 CE, al-Biruni talks about both an elder Aryabhata and the Aryabhata of Kusumpura. In 628 CE Brahmagupta has some harsh words in his *Brahmasphutasiddhanta* for certain planetary theories of another Aryabhata. But in an astronomical text he wrote 27 years later (*Khandakadyaka*), there are references in highly reverential terms to the work of an Aryabhata. Since the positions of planets and other heavenly bodies found in Brahmagupta's earlier text are very different from those that can be inferred from *Aryabhatiya*, one may conclude that the object of Brahmagupta's admiration could very well have been an older Aryabhata who lived before the Aryabhata of Kusumpura.

There is a tendency among historians of Indian mathematics to recognise the later two Aryabhatas and ignore the earliest one. The first, the Aryabhata of Kusumpura is usually referred to as Aryabhata I and his year of birth is now accepted to be 476 CE since he states that he was aged 23 when he wrote *Aryabhatiya* (499 CE). The second, a relatively minor figure known for only one commentary, *Maha Aryasiddhanta*, is now referred to as Aryabhata II whose mathematical work is briefly discussed in a later chapter.

The other unresolved question is regarding Aryabhata I's place of birth. This has become a matter of some contention because of the claims that he was born in Kerala. Kusumpura was not his place of birth but the place where he lived and worked; there are many references, including one in Nilakantha's commentary on *Aryabhatiya*, that he was born in a place called Asmaka. It is the location of these two places that has aroused controversy. One view is that while Kusumpura is a place in the North, probably near modern Patna, Asmaka could refer to a place in the South, probably Kerala. There is insufficient evidence to support either view.

The appearance of Aryabhata's *Aryabhatiya* (499 CE) and Varahamhira's *Pancasidhantha* (505 CE) may be seen as heralding the beginning of the 'classical period' of Indian mathematics and astronomy, a period notable for rapid developments in both interrelated disciplines. These texts established a particular mode of writing in concise verses, a practice that would continue on for over a thousand years. They were both compendiums, a particular genre of knowledge known at that time. And this was so of Varahamhira's *Pancasidhantha*, a name that implied a summary of five *siddhantas*.

Commentaries were recorded to help to understand the original texts which in some cases were cryptic to the point of incomprehension. Bhaskara I (usually referred to as such to distinguish him from a later Bhaskara) wrote the oldest commentary on *Aryabhatiya* entitled *Aryabatiyabhasya* (meaning a 'commentary on *Aryabhatiya*'). It was composed about a 100 years later and was the first of about 18 major commentaries, some written as late as the 19th century. It was completed a year after the highly influential *Brahmasphutasiddantha* of Brahmagupta whose work will be discussed in the next chapter.

There has been a tendency to ignore Bhaskara I and his contributions in both mathematics and astronomy despite the fact that his commentary is the first surviving source of a detailed discussion of mathematics in Sanskrit.[1] This is not to say that there was no significant mathematical activity in the post-Vedic period as evidenced from Bhaskara I's stray comments on texts and compositions which are no longer extant. However, in our discussion of the contents of *Aryabhatiya* we incorporate material from the commentary of Bhaskara 1 who provides useful illustrations and extensions to what is often terse and difficult verse rules to understand. This is not to ignore some of Bhaskara's own original contributions which will be discussed as well. These include not only his attempts at definition of the scope of mathematics but also unravelling the different meanings of terms that Aryabhata introduces, the most notable example being the reinterpretation of Aryabhata's 'pulveriser' algorithm discussed later in this chapter.[2]

Little is known about Bhaskara I apart from his works which suggest that he was a prominent member of the early Aryabhata School. Since Bhaskara often refers to the *Asmakatantra* instead of the *Aryabhatiya* that he may have been a member of a school of mathematicians/astronomers situated in Asmaka, believed by some, as stated earlier, to be the birthplace of Aryabhata. There are other references to places in India in Bhaskara I's writings. For example, he mentions Vallabhi (today's Vala), the capital of the Maitraka dynasty in the 7th century and a highly reputed centre of Buddhist learning, and Sivabhagapura, which were both in Saurashtra (found today in the state of Gujarat on the west coast of the Indian subcontinent). Also mentioned are Bharuch (or Broach) in southern Gujarat and Thanesar in the eastern Punjab ruled by Harsha for 41 years from 606 CE. This helps us to place Bhaskara I in the early part of the seventh century.

Bhaskara I was the author of three treatises, in chronological order, the *Mahabhaskariya*, the *Laghubhaskariya* (an abridgement of *Mahabhaskariya*), and the commentary *Aryabhatiyabhasya*. The *Mahabhaskariya* is a work of eight chapters on Indian mathematical astronomy and includes topics such as the longitudes of the planets, the conjunctions of the planets with one another and with bright stars, eclipses of the sun and the moon and their risings and settings and the lunar crescent. In Chapter 7 of that text appears three verses giving the approximation formula for the sine function by means of a rational fraction which is amazingly accurate — a maximum error of less than one percent! In other words, if we plot the actual sine value in the range 0–90 degrees, as well as Bhaskara's approximation to it (as we do in Chapter 11), they are so close that we cannot tell them apart! Indeed, rational approximations such as this are often more accurate than power series in representing non-algebraic functions. A discussion of this approximation formula that Bhaskara attributed to Aryabhata I will be postponed to the chapter on Indian trigonometry (Chapter 11).

In 629 CE, Bhaskara I wrote his commentary, the *Aryabhatiyab-hasya*, on the *Aryabhatiya*. The commentary contains a discussion with illustrations of the thirty three verses on mathematics in the original text. It provides a lucid commentary on all the mathematical results contained in the *Aryabhatiya*, notably the solution of the indeterminate equations of the first degree and derivation of the trigonometric formulae and tables of which more will be heard of in a later chapter.[3]

Other subjects discussed in the *Aryabhatiyabhasya* include numbers and symbolism, the classification of different branches of mathematics, the names and solution methods of equations of the first degree, quadratic equations, cubic equations and equations with more than one unknown. Bhaskara I introduced the 'Euclidean algorithm' (a method for calculating the Greatest Common Divisor of two numbers or the largest number that divides both of them without leaving a remainder) for solving linear indeterminate equations. The text contains detailed commentaries and examples elucidating Aryabhata's short terse rules. It is a remarkable work justifying later commentators to describe Bhaskara I as an 'all-knowing' scholar. References were also made by Bhaskara I to the works of Indian mathematicians/astronomers earlier than Aryabhata. Subsequent notable commentators on Aryabhata and Bhaskara I arose mainly in Kerala. Their work will be detailed in a later chapter.

6.2   The Mathematics in *Aryabhatiya*

6.2.1   The style and notation

Irrespective of whether Aryabhata was a native of Kerala or not, there is no doubt that his text, *Aryabhatiya*, had a profound influence on the later mathematicians/astronomers of Kerala. Within India, it may be said to occupy the status akin to Euclid's *Elements* in Western mathematics or to *Jiu Zhang Suan Shu* ('Nine Chapters on the Mathematical Arts') in Chinese mathematics, although the contexts and contents are very different in all three. A knowledge of the contents of *Aryabhatiya* is essential for understanding the subsequent development of Indian mathematics.

The *Aryabhatiya* is the earliest clearly datable extant Indian text of mathematics and astronomy. It has 121 stanzas in all, written in a style marked by brevity and conciseness. There are four 'chapters' or sections (*padas*), namely *Gititkapada, Ganitapada, Kalakriyapada* and *Golapada*. The last two concentrates on astronomy and hence we will not discuss them here.[4] The main discussion will be the second *pada* which is concerned with *ganita* (or computational techniques).

The first chapter of the book, *Gitikapada*, consisting of 13 stanzas, is in two parts setting out basic definitions and important astronomical parameters and tables:

(i) **Dasagitka** describes an alphabetic scheme in Sanskrit for representing numerals based on distinguishing between classified

(*varga*), unclassified (*avarga*) consonants and vowels. The *vargas* fall
into five phonetic groups: *ka-varga* (guttural), *ca-varga* (palatal),
ta-varga (lingual), *ta-varga* (dental) and *pa-varga* (labial). Each
group has five letters associated to it where the letters run from
k (क) to m (म) in the Sanskrit alphabet. Thus, it is possible to
represent numbers from 1 to 25 by $5 \times 5 = 25$ letters. The Sanskrit
letters y (य) to h (ह) consist of seven *avargas* of semi-vowels and
sibilants representing numerical values $30, 40, 50, \ldots, 90$; and the
eighth *avarga* used as a means of extending to the next place value.
The 10 vowels denoted successive integral powers of 10 from 100
onwards.[5]

As mentioned in an earlier chapter, this form of representation,
compared to other schemes of earlier notation, such as *bhuta-
samkhya*, and later ones such as *katapayadi*, popular with the
Kerala School, has the advantage of brevity and conciseness but the
disadvantage of having limited potential for formations of words that
are pronounceable and meaningful, both necessary attributes for easy
memorisation.[6]

(ii) **Aryasta-satam** is a system of dividing time into intervals.
According to Vedic cosmology, the basic unit of time is the *kalpa*.
A *kalpa* in turn is divided into 14 *manus* and each *manu* into 71
maha yugas. The *maha yuga* is in turn sub-divided into four smaller
yugas: krta, treta, dvapara and *kali* in order of descending magnitudes
(expressed in years) spanning 17,28,000, 12,96,000, 8,64,000 and
4,32,000 respectively *or* their relative magnitudes in the ratio 4:3:2:1.
The use of the factors 14 and 71 in devising this scheme of time-
interval remains a puzzle. Aryabhata simplified this system by
redefining 1 *manu* = 72 *maha yugas* and thereby ensured that
1 *kalpa* = 1008 *maha yugas* = 0 (mod 7) that each kalpa began
on the same day of the week. He also introduced a division of *maha
yuga* into four smaller *yugas* of equal duration where each of the
smaller *yuga* is 10,80,000 years. This equal division of time became
astronomically acceptable, since at the beginning of each 'yuga',
all planets would be in conjunction at the commencement of the
zodiac.

6.2.2 Mensuration

The next chapter, *ganitapada* (or mathematical section), consisting of merely 33 numbered stanzas, gives rules and results in arithmetic, mensuration and geometry, trigonometry, algebra and mechanics.[7] In the arithmetic section appears methods of finding the square and cube of any number[8]; of extracting square and cube roots[9]; operations with fractions[10]; the Rule of Three and the method of inversion.[11] The rules contained in this section are not new; they were part of the tools available at that time. Aryabhata expressed them in a concise but highly cryptic form. The commentaries, notably the influential one by Bhaskara's *Aryabhatiyabhasya* (629 CE), are essential in understanding Aryabhata's work and will be incorporated with the main text where the exposition requires it to be done.

6.2.3 Geometry

The section on geometry and mensuration **(Verses 6–12)** deals with the square, the cube, the triangle, the trapezium, the pyramid, the circle, the sphere and gnomonic shadow. The latter part of **Verse 6** of this section gives the volume of a right pyramid (i.e. one with a square base) that has aroused some controversy and in some quarters has been perceived as a good example of the Indian weakness in geometry. It may be translated thus:

> The product of the perpendicular [dropped from the vertex to the base] and half the base gives the measure of the area of a triangle. Half the product of that area [i.e. of the triangular base] and the height is the volume of a solid called "six-edged".

The first sentence of this quotation gives a result that has been widely known to most ancient cultures. It does not assume any 'sophisticated' knowledge of geometry (or algebra). It is a good illustration of the principle of 'dissection and reassembly' used in both Chinese and Indian mathematics.[12] Consider the following ('Not to Scale') diagrams [Figure 6.1].

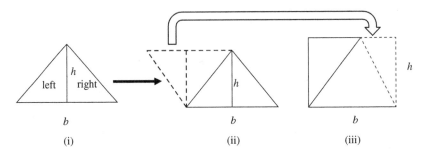

Figure 6.1. Dissection and reassembly.

The 'left' triangle in Figure 6.1(i) is rotated and inserted in the space shown as a dotted-line triangle in Figure 6.1(ii). Then do the same to the 'right' one. A section of it is dissected and moved to the slot for a triangle shown in Figure 6.1(iii) as dotted lines. Half the area of the resulting rectangle [base (b) × height (h)] in Figure 6.1(iii) gives the area of the triangle we started within Figure 6.1(i). Note that the triangle in question can be any one of the three: equilateral, isosceles or scalene. The second sentence in **Verse 6** above gives the rule for the volume of a "six-edged" solid.

This is clearly wrong being based on the analogy of the area of a triangle.[13] The reasoning behind this false rule may be as follows. Consider a large equilateral triangle which is divided into four equal equilateral triangles. Fold up the peripheral equilateral triangles over the central one to form a pyramid. It is obvious that the area of this pyramid is equal to the area of the large equilateral triangle. Now, the base area of this pyramid is one-quarter the area of the large triangle. Six such pyramids (which properly reassembled will constitute a rectangular solid) together will have volume equal to half the product of the area of the original triangle multiplied by the height of the pyramid.

Bhaskara I in his *Aryabhatiyabhasa* extends the discussion on finding the area of a triangle in case where the lengths of the three sides are given but not the altitude.

Figure 6.2 shows a scalene triangle whose sides (a, b and c) are known.

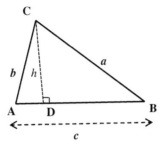

Figure 6.2. Area of a triangle.

Applying the Pythagorean theorem to triangles BDC and ADC gives:

$$(BC^2 - CD^2) - (AC^2 - CD^2) = (a^2 - h^2) - (b^2 - h^2) = a^2 - b^2,$$

or

$$BD^2 - AD^2 = a^2 - b^2$$

i.e.

$$(BD + AD)(BD - AD) = a^2 - b^2.$$

Therefore

$$c(BD - AD) = a^2 - b^2.$$

So

$$BD - AD = \frac{a^2 - b^2}{c} \quad \text{and} \quad BD + AD = c.$$

The length of segments of the base (BD and AD) can be calculated and then the altitude (h) and finally the area of the triangle.

Bhaskara gives the following example to illustrate this calculation:

> In a scalene triangle one of the 'flanks' (or sides) is 13 and the other is 15. The base is two times seven only. O my friend what would be the measure of the area of the triangle?

The calculation is as follows:

$$(BD - AD) = \frac{(15^2 - 13^2)}{14} = 4 \quad \text{and} \quad BD + AD = 14.$$

So

$$2 \times BD = 18 \quad \text{or} \quad BD = 9.$$

Applying the Pythagorean theorem to triangle BDC, we get the altitude (h)

$$h = \sqrt{15^2 - 9^2} = 12.$$

Therefore, the area of the triangle[14] is half the base multiplied by height, i.e.

$$= \frac{1}{2}(14 \times 12) = 84.$$

The controversy regarding the accuracy of Indian geometry continues into the next verse dealing with circles. **Verse 7** is cryptic almost to the point of incomprehension.

Half of the circumference multiplied by the semi-diameter is the area of a circle. That multiplied by its own root is the volume of the circular without remainder...

Two interpretations of this rule have been offered: it could either contain the rule for finding the volume of a sphere *or* the rule for obtaining the area of the surface of a hemisphere. The first was favoured by commentators, such as Bhaskara I and the Kerala mathematician/astronomer Paramesvara. The second is of a more recent vintage of which Elfering's interpretation (1977) is a good example. A modified **Verse 7** reads:

(i) Half the circumference multiplied by half the diameter that gives the area of the circle. The volume of the sphere is the product of the area of the circle and it's (area's) square root.

(ii) Half the circumference multiplied by half the diameter, that is the area of the circle. This one (i.e. the circumference) multiplied by its defining base (i.e. the radius) is exactly the surface (area) of the hemisphere.

Symbolically,

$$A = \frac{1}{4}Cd \equiv \frac{1}{2}Cr,$$

where A is the Area, C the Circumference, d the diameter and r the radius of the circle.

Then

(i) $V = A \cdot \sqrt{A} \equiv \pi r^2 \cdot \sqrt{\pi r^2} \equiv \sqrt{\pi^3} \cdot r^3$, where $V = $ volume,
(ii) Area of the hemisphere $= Cr = 2\pi r^2$.

Here, (i) is wrong while (ii) is correct.[15]

Verse 8 gives the rule for calculating the area of a trapezium.[16] It also gives a method for calculating the lengths of the segments of the diagonals of the trapezium, translated as the 'lines on their own fallings'.

Bhaskara I's solution and rationale are as follows (See Figure 6.3)[17]:

If h_1 and h_2 are perpendiculars from the intersection of the diagonals, it would follow that

$$h_1 = \frac{b \times h}{a + b}; \quad h_2 = \frac{a \times h}{a + b}, \quad \text{where } h = h_1 + h_2.$$

Also,

$$\text{Area of trapezium} = \frac{1}{2}(b + a) \times h.$$

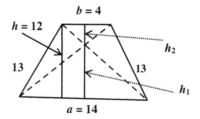

Figure 6.3. Area of a trapezium.

Bhaskara I illustrates with the following example which the reader is invited to work out:

> Let the earth (base) be fourteen units and the 'face' (top) four units. The two 'ears' (non-parallel sides) each measures thirteen and the height is 12 units. Find the lines whose top is the intersection of the diagonals and also find the area.
> [Answers: 48/18; 168/18; 108]

Bhaskara I has yet another example involving isosceles trapeziums:

> The two faces of a drum (*panava*) are eight and eight, the separation is two and the length of drum is said to be sixteen. What is the area of that which is shaped in the form of the drum.

Figure 6.4 (not to scale) shows the drum-shaped musical instrument known as a *Panava*. The two parallel faces of the figure are 8 units in length and the central unit is 2 units in width. The length between the faces is 16 units. The area of the *Panava* is the combined area of the two identical 'trapeziums' AFED and BFEC as follows[18]:

$$\text{Area} = \frac{1}{2}\left[\frac{\text{AD} + \text{BC}}{2} + \text{EF}\right]\text{AB}$$

$$= \frac{1}{2}\left[\frac{8+8}{2} + 2\right]16 = \mathbf{80} \text{ square units.}$$

Bhaskara I provides a generalised approach which can be used to obtain the area of a triangle, rectangle and trapezium using the same formula which can be adapted depending on the figure:

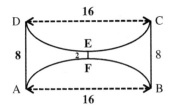

Figure 6.4. Area of a *Panava*.

Thus,

$$\text{Area} = \frac{1}{2}(\text{base} + \text{face}) \times \text{height} = \frac{1}{2}(b + f) \times h$$

(i) If $f = 0$, the formula reduces to the appropriate one to find the area of a triangle.
(ii) If $f = b$, the formula reduces to the appropriate one to find the area of a rectangle.
(iii) The original formula is the one for finding the area of a trapezium.

6.2.4 'Squaring' a circle

Verse 10 contains the following statements:

(i) ... the chord of a sixth part of the circumference is equal to the semi-diameter[19];
(ii) Add 4 to 100, multiply by 8 and add 62,000; the result is *approximately* the circumference of a circle whose diameter is 20,000.

These two verses are critical for the subsequent development of Indian mathematics.

The first provides the basis for the construction of trigonometric tables. Its proximity to the next would indicate that Aryabhata was aware of the common approach to "squaring the circle", i.e. find a regular polygon of sufficiently large number of sides so that its perimeter is approximately equal to a circle of the same area.[20] This approach begins with a polygon whose side was a "chord of the sixth part of the circumference" (or of a regular hexagon). From **Verse 10**, it could be inferred that Aryabhata used a regular inscribed polygon of 348 sides, to arrive at an implicit value for π as $62832/20000 = 3.1416$. And it was the word "approximately"[21] that gave food for further thought.

Bhaskara I is one of the first commentators to raise the important question as to why only an approximate value for the circumference is available, and not the true circumference. He answers it himself by stating that there is no such method by which the accurate circumference is computed. The *karani* (i.e. surd or square root) would lead us to believe that the circumference is the square root of 10 times the diameter. But this is only a traditional presumption and not a proof.

An important motivation for the Kerala work on infinite series arose from a recognition of the impossibility of arriving at an exact value for the circumference of a circle, given the diameter (due to the incommensurability of π).[22] To quote Nilakantha's explanation from his commentary on *Aryabhatiya* (c. 1500 CE):

> Why is only the approximate value (of circumference) given here? Let me explain. Because the real value cannot be obtained. If the diameter can be measured without a remainder, the circumference measured by the same unit (of measurement) will leave a remainder. Similarly, the unit which measures the circumference without a remainder will leave a remainder when used for measuring the diameter. Hence, the two measured by the same unit will never be without a remainder. Though we try very hard we can reduce the remainder to a small quantity but never achieve the state of 'remainderlessness'. This is the problem.

Verse 11 shows the computation of the Indian sine (Rsine) table. It states:

> Divide a quadrant of the circumference of a circle (into as many parts as desired). Then from (right-handed) triangles and quadrilaterals, one can find as many Rsines of equal arcs as one likes, for any given radius.

Bhaskara I's commentary is particularly useful in interpreting the above cryptic or even incomprehensible instruction. We will take this up in Chapter 11, but it would be useful to consider an important relationship between the chord and circumference of a circle.

A Digression: Relationship between the Chord and the Circumference of a Circle

"The chord of a sixth of the circumference [of a circle] is equal to the semi-diameter."

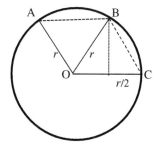

Figure 6.5. The chord and circumference of a circle.

Arc AB is one-sixth of the circumference of a circle. AOB and BOC are equilateral triangles (Figure 6.5). Therefore, AB = BC = chord 60° = semi-diameter = r.

6.2.5 Sundials and shadows

The next few verses introduce problems of sundials and shadows leading to practical methods of constructing geometrical shapes such as circles, triangles and quadrilaterals. The solution to shadow problems mainly uses the properties of similar triangles. Bhaskara I presents an interesting rule to find the distance of separation between the Sun and the Earth.

Verse 16 states:

> The upright side is the distance between the tips of two shadows multiplied by a shadow divided by the decrease. The upright side multiplied by the gnomon,[23] divided by its shadow becomes the arm (base).
>
> (Note: The terms 'upright' and arm refers to the vertical and horizontal sides of a right-angled triangle.)

The diagram in Figure 6.6 constructed on the basis of Bhaskara I's commentary hopefully clarifies this rule:

The shadows of two equal gnomons (length g) are 10 (s_1) and 16 (s_2) *angulas* respectively and the distance between the tips of the shadows is 30 (f) *angulas*. Find the length of the upright side (h) and the base (u).

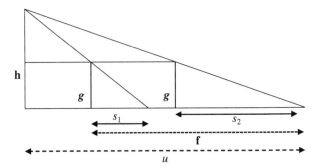

Figure 6.6. A shadow problem.

The solution clarified by Bhaskara I involves the rule:

The length of the shadow × the distance between the ends of the shadows divided by the 'difference'.

An ambiguity arose from the denominator labelled as 'difference'. Its meaning was interpreted as the distance between the ends of two shadows $(s_2 - s_1)$.

$$\frac{u}{s_1} = \frac{f}{s_2 - s_1} \rightarrow u = \frac{f s_1}{s_2 - s_1} \quad \text{or} \quad u = \frac{300}{6} = 50.$$

Further,

$$\frac{h}{g} = \frac{u}{s_1} = \frac{g u}{s_1} \quad \text{or} \quad h = \frac{600}{10} = 60$$

when the height of the gnomon is 12 *angulas*.

Verse 17 is a statement of the Pythagorean theorem known in India at least from the time of the *Baudhyana's Sulbasutra* (*c*. 800 BCE). This is presented together with results relating to the inner segments of a circle, notably the result that in a circle of diameter AB and a chord CD with which the diameter intersect at right angles at E, then AE · EB = $(\frac{1}{2}CD)^2$. This follows from the properties of similar triangles BCE and AEC (Figure 6.7).

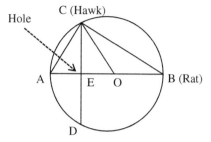

Figure 6.7. A hawk and a rat.

Bhaskara I has a well-known problem involving a hawk and a rat to illustrate the use of this rule:

A hawk, sitting on a wall of height 12 *hastas* sees a rat 24 *hastas* away at the foot of the wall. As the rat runs towards a hole in the wall where it lives, the hawk dives down and kills the rat. How far is the rat away from its hole when it is killed and what is the distance travelled by the hawk before the kill? (Assume that the hawk and the rat travel at the same speed)

AB represents roaming ground of the rat whose death occurs at the spot marking the centre of the circle (point O). The hypotenuse OC is the hawk's path. EB is the distance that the rat has to travel to reach the safety of its home. The square of the height of the position of the hawk is $(\frac{1}{2}CD)^2 = 12^2 = 144$. The rat's roaming ground is AB = AE + EB = 24. Estimate the quotient $144/24 = 6$. Add and then subtract this quotient from the roaming ground of the rat to get 30 and 18 *hastas* respectively. Their respective halves (15 and 9) give the path of the hawk (OC) and the remaining distance that the rat has to cover to reach the safety of its hole.

Another interesting application of the Pythagorean theorem is also found in Bhaskara I's commentary.

There is a rectangular tank of water of dimension 6 × 12. At the north-east of the tank there is a fish; and at the north-west there is a crane. Frightened by the crane, the fish crosses the tank hurriedly

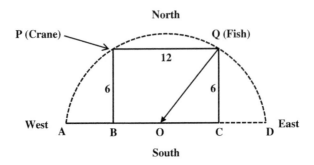

Figure 6.8. A crane and a fish.

going towards the south but was killed by the crane who came along the sides of the tank. Find the distances travelled by them, assuming that their speeds were the same.

In Figure 6.8, BCQP is a rectangular reservoir of dimensions 12 by 6 units where QC = 6 is on the eastern side and PB = 6 is on the western side. The fish is at Q and the crane is at P. The fish swims towards the southern side of the reservoir and reaches O. At the same time, the crane walks southwards along the edges PB and then BO, where it catches the fish at O. Since it is assumed that the speeds are the same, the fish travels the same distance QO as the crane PB + BO (i.e. QO = PB + BO).

Extend OB to A such that OA = OQ

Now

$$AC = AO + OC$$
$$= OQ + OC$$
$$= PB + BO + OC$$
$$= 12 + 6 = 18.$$

Let AC be extended so that it intersects the circle at D. Then we get

$$CD \cdot AC = (OD - OC)(AO + OC)$$
$$= (OD - OC)(OD + OC) = OD^2 - OC^2$$
$$= OQ^2 - OC^2 = QC^2 = 36 \quad \text{(Pythagorean Theorem)}.$$

So

$$CD = \frac{QC^2}{AC} = \frac{36}{18} = 2.$$

Therefore,

$$AD = AC + CD = 18 + 2 = 20$$

so that

$$AO = \frac{1}{2}AD = 10 = OQ.$$

This means that the crane and the fish travel **10** units each.

There are other interesting problems involving the application of the Pythagorean theorem not dissimilar to the ones found in other mathematical traditions, notably in Chinese and early European mathematics. The reader may wish to solve the following problem which have an international provenance. The answers are given. [Note: 1 '*hasta*' is equal to 24 *angulas* or about 18 inches]

The Broken Bamboo Problem

A bamboo of 16 *hastas* was broken by the wind. It fell on to the ground, with its tip hitting the ground at 8 *hastas* from its root. Where was it broken by the Lord of the wind? (*Answer:* 10 *hastas* from the top and 6 *hastas* from the ground)

The Lotus Problem

A lotus in bloom is of height eight *angulas* above [the surface of] of the water. Blown by the wind, it sinks 1 hasta, Quickly tell me the height of the lotus and the depth of the water. (*Answer:* height 40 and water 32)

In the calculations of eclipses the following geometrical result, expressed in **Verse 18**, is important.[24] In Figure 6.9, two circles of diameter d_1 and d_2 intersect (at two points). The "arrows" of the intercepted arcs of the two circles (a_1 and a_2) constitute the obscured part of an eclipsed body where the obscuration a is the sum of the corresponding arrows a_1 and a_2.

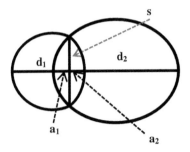

Figure 6.9. Calculation of eclipses.

It can be shown that the total curve obscured by both circles is
$a = a_1 + a_2$
where,

$$a_1 = \frac{(d_1 - a) \cdot a}{(d_1 - a) + (d_2 - a)} \quad \text{and} \quad a_2 = \frac{(d_2 - a) \cdot a}{(d_1 - a) + (d_2 - a)}.$$

Consider the following illustrative example from Bhaskara I.

If 8 minutes out of 32 minutes of the angular diameter of the
moon is eclipsed (at a lunar eclipse) by the shadow of diameter
80 minutes, I want to know what are the 'arrows' (*sara*) of the
intercepted shadow and the full moon.

Here we have $d_1 = 32$ minutes, $d_2 = 80$ minutes, $a = 8$ minutes.
Substituting the values in the above expressions gives $a_1 = \mathbf{2}$ and
$a_2 = \mathbf{6}$.

6.2.6   Series

In **Verse 19**, Aryabhata (according to Bhaskara I) states alternative
ways of summing an arithmetical progression:

(i) The desired number of terms minus 1, halved, plus the number
of terms which precedes, multiplied by the common difference
between the terms, plus the first term, is the mean value. This
multiplied by the number of terms desired is the sum of the
desired number of terms.

(ii) Or the sum of the first and last terms is multiplied by half the number of terms.

In modern notation, if n is the number of terms in an arithmetical progression, d is the common difference, a and z are the first and last terms of the progression respectively, then we get the familiar formulae:

(i) Mean $= a + \frac{n-1}{2}d$;

(ii) Sum $(S) = n\left(a + \frac{n-1}{2}d\right)$;

(iii) Sum $(S) = \frac{n}{2}(a + z)$.

These results may have already been known in earlier work, notably in Jain mathematics.

To show how the rules are applied, Bhaskara I provides both simple and complex illustrations. Two examples are given below:

(1) The first term is 2, the common difference is given to be 3. The number of terms is given as 5. State the values of the mean and of the whole [number of terms].

Here, the first term $a = 2$, the common difference $d = 3$, the number of terms $n = 5$. The procedure (ii) outlined for solution is as follows:

> The desired number of terms is 5, decreased by one, 4, halved 2, having the common difference for multiplier, 6, increased by the first term 8; this is the mean value. Just this multiplied by the desired number of terms produces the whole value, 40.
> Also if z, the last term, is known to be 14, then from (iii), $S = \mathbf{40}$.

Bhaskara I also provides a more complicated illustration:

(2) A king makes offerings in the month of Kartika of two units on the first (day), increasing by 3 units day after day. When the fifteenth day has gone, a learned Brahmin arrives and the amount collected is given to the Brahmin during the next ten days.

During the next five days, it is given to another Brahmin. Under this arrangement what is the amount given to both?

The reader may wish to check the solutions given by Bhaskara I that works out to a total offering of **605** units to the first Brahmin and **415** units to the second.

Verse 20, as elaborated in Bhaskara's commentary, contains a method of finding the number of terms (n) of an arithmetic progression:

Multiply the sum of the progression by eight times the common difference, add the square of the difference between twice the first term and the common difference, take the square root of this, subtract twice the first term, divide by the common difference, add one, divide by two. The result will be the number of terms.

Expressed in modern notation where S = sum of the progression:

$$n = \frac{1}{2}\left[\frac{\sqrt{8dS + (2a - d)^2} - 2a}{d} + 1\right].$$

For this formula to be known, it may be conjectured that the method of 'completing the square' was already known, although the more likely approach today would involve the solution of quadratic equations of the form $ax^2 + bx + c = 0$.

The solution offered by Bhaskara where the first term (a) of the series is 5, the common difference (d) is 7 and the sum (S) is 95, is merely a restatement in words of the formula for n given above

That is,

$$n = \frac{1}{2}\left[\frac{\sqrt{(8 \times 7 \times 95) + (2 \times 5 - 7)^2} - (2 \times 5)}{7} + 1\right] = 5$$

is the number of terms.

Verse 21 relates to the 'sum of the sum' series of the kind: $1 + (1 + 2) + (1 + 2 + 3) + \cdots + n$ terms:

Of the series which has one for the first term and one for the common difference, take three terms in continuation, of which the first is equal to the given number of terms and find their continued product. That (product), or the number of terms plus one subtracted from the cube of that, divided by 6, gives the sum of the series to any number of terms shown as a *citighana*.[25]

Expressed in modern notation, the rules for finding the sums of the following series are given as:

(i) Sum of (cubes) $= 1^3 + 2^3 + 3^3 + \cdots + n^3 = (1 + 2 + 3 \cdots + n)^2$ which corresponds to summing the series (1), $(3 + 5)$, $(7 + 9 + 11)$, ..., etc.

(ii) Sum of the 'piled solid of squares'

$$= \sum_{j=1}^{n} \sum_{i=1}^{j} \frac{n(n+1)(n+2)}{6} \quad \text{or} \quad \sum_{j=1}^{n} \sum_{i=1}^{j} \frac{(n+1)^3 - (n+1)}{6}.$$

Verse 22 contains an important result useful to the Kerala School whose work is discussed in Chapter 10:

The sixth part of the product of three quantities consisting of the number of terms, the number of terms plus one, and twice the sum of terms plus 1 is the sum of the squares. The square of the sum of the (original) series is the sum of the cubes.

Symbolically,

$$S_1 = \sum \frac{n(n+1)}{2},$$

where S_1 is the sum of the first n natural numbers.

$$S_2 = \sum \frac{n(n+1)(2n+1)}{6},$$

where S_2 is the sum of squares of first n natural numbers.

$$S_3 = (S_1)^2 = \sum \frac{n^2(n+1)^2}{4},$$

where S_3 is the sum of cubes of first n natural numbers.

The work on series reported in **Verses 19–22** in *Aryabhatiya* has a long history possibly going back to Vedic times. An interest in the sum of a finite geometric progression is found in the work of Pingala (*c.* 200 BCE),[26] in Jain texts and the Bakhshali Manuscript. And the work would continue after Aryabhata, by his commentators and successors: Bhaskara I, Mahavira, Bhaskara II, Narayana Pandita and finally the Kerala mathematicians, notably Nilakantha. Their contributions will be discussed in chapters later in the book.

6.2.7   Algebra manipulations, 'rule of three' and method of inversion

Verses 23–26 relate to rules of algebraic manipulations, and in (iii) below the famous 'Rule of Three', that we came across in an earlier chapter.

(i) Subtract the sum of the squares of two factors (given as a and b) from the square of their sum. Half the result is the product of the two factors.

In present-day notation:

$$\frac{(a+b)^2 - (a^2 + b^2)}{2} = ab.$$

(ii) Multiply the product of two [positive]^{27} factors by the square of two (or 4); add the square of differences between the two factors, take the square root, add and subtract the difference between the two factors, and divide the result by two. The result will be either of the two factors.

$$\frac{\sqrt{4ab + (a-b)^2} + (a-b)}{2} = a;$$

$$\frac{\sqrt{4ab + (a-b)^2} - (a-b)}{2} = b.$$

(iii) In the 'Rule of Three' (*trairasika*), multiply the *phala-rasi* (fruit) by the *iccha-rasi* (desire or requisition) and divide by the *pramana* (measure or argument). The required result *iccha-phala* (or fruit corresponding to desire or requisition) will be thus obtained.

Symbolically (as we saw in the last chapter), these rules can be expressed given p, i, m and f are *phala-rasi, iccha-rasi, pramana* and *iccha-phala*, respectively:

$$f = \frac{p \times i}{m}.$$

Rules (i) and (ii) are rules of two operations known as *Samkramana* involving two quantities when their sum and difference are given. In (i), the product of two quantities (ab) is calculated, given their sum $(a + b)$ and the sum of their squares $(a^2 + b^2)$. In (ii), the rule is given to find the two quantities a and b given their product and difference $[ab$ and $(a - b)]$. The subject remained of abiding interest to commentators and others who came after Aryabhata I.[28]

The statement of Rule (iii) occurs here for the first time in Indian mathematics and given the technical name of the 'Rule of Three' and involves four numerical quantities.[29] However, the concise manner in which the rule is given would indicate that it was already well known and that Aryabhata was merely restating it as a prelude to its use in astronomical computations. The antecedents of this rule have been traced back about a 1000 years to a verse in the *Vedanga Jyotisa* (Kuppana Sastry, 1985) and discussed S.R. Sarma (2002, 135–136).[30]

Bhaskara I's commentary on Aryabhata's work also contains a detailed discussion of this Rule and how it can be extended to encompass Rules of Five, Seven . . . Bhaskara also introduces the question of the logical sequence in which the three numerical quantities should be set down and the order in which the multiplication $(p \times i)$ and division by m should be carried out. Brahmagupta's formulation of the Rule became a model for subsequent writers to bring out to explore the fact that the three quantities should be set down in such a way that the first and last would be of like denomination and the middle one of a different denomination. The Rule is mentioned by Sridhara (*c.* 750), Mahavira (*c.* 850) and Aryabhata II (*c.* 950) without adding much to it. However, Bhaskara II in his *Lilavati* states the important point

that nearly the entire arithmetic is based on the 'Rule of Three' and
that most topics require variants of this 'Rule'.

> Just as the universe is pervaded by Hari and his manifestations,
> even so all that has been taught [in arithmetic] is pervaded by the
> Rule of Three with its variations.

Again, in the *Siddhantasiromani*, Bhaskara II reiterates this
point:

> (Apart) from squaring, square-root, cubing and cube-root, what-
> ever (needs to be) calculated (involves) a variation of the Rule
> of Three, (and) nothing else. For increasing the comprehension of
> duller intellects like ours, what has been written in various ways
> by the learned sages... has become arithmetic."

It is at this point that Aryabhata introduces a rule for finding
the monthly interest payable (r), given the principal (p), the 'interest
on interest' (a) and the time in months (t). The rule is stated in
Aryabhatiya as:

> The interest on the capital, together with the interest [on the
> interest], with the time and capital for multiplier, increased by the
> square of half the capital. The square root of that, decreased by
> half the capital and divided by the time, is the interest on one's
> own capital.

Expressed in symbolic terms:

$$r = \frac{\sqrt{pta + \left(\frac{p}{2}\right)^2} - \left(\frac{p}{2}\right)}{t},$$

Bhaskara I illustrates this rule with an example:

> The [monthly] interest on 100 is not known. However, this interest
> increased by the interest [on the interest] for four months is 6. Find
> the monthly interest on 100.

The solution is given thus:

> The interest on capital increased by the interest (on the interest),
> 6, is multiplied by the time and the capital to obtain $6 \times 4 \times 100 =$
> 2400. Now square half the capital, 2500. and add to 2400 to obtain

4900. Find the square root of the result, 70, and decrease this by half the capital, 50, to get the result 20. Divide this result by the time, four months to get the monthly interest on capital which is five.[31]

or

$$r = \frac{\sqrt{100 \times 4 \times 6 + \left(\frac{100}{2}\right)^2} - \left(\frac{100}{2}\right)}{4} = 5.$$

Verse 27 is concerned with the rules relating to the manipulation of fractions where multiplication, division and simplification by reduction to a common denominator would be easily recognised as the ones that we use today.

Verse 28 discusses a procedure known as the reversed (or inverse) procedure which was first stated in *Aryabhatiya*:

In a reversed [operation], multipliers become divisors and divisors, multipliers; additive [quantity] is a subtractive [quantity], a subtractive an additive [quantity].

An example from a commentator on *Aryabhatiya*, Paramesvara (*c.* 15th century CE) reads:

Example: What is the number which, multiplied by 3, divided by 5, the quotient increased by 5, the square-root of the sum diminished by 1, and again squared yields the result 4?

The result obtained by reversing the operations can be expressed as:

$$\{[(\sqrt{4} + 1)^2 - 6]5 \div 3\} = 5.$$

The method reappears in the work of Bhaskara I and other later mathematicians.

It was also at this juncture in **Verses 29–30** that Aryabhata introduced the methods of calculating quantities from their sums. In modern notation:

Find n unknown quantities $x_1, x_2, x_3, \ldots, x_n$ and their unknown sum $S = \sum_1^n x_i$, given that we know the sums $S_i = S - x_i$ for $1 \leq i \leq n$.

The solution is obtained by calculating $S = \frac{1}{n-1} \sum_1^n S_i$ and then each unknown x_i which is equal to $S - S_i$.

An example from Bhaskara I's *Aryabhatiyabhasya*, modified for the sake of clarity, would help to illustrate the rule.

> In a forest, there are four herds of elephants made up of one group of males in heat, one not in heat, one females and one consisting of the young. Leaving out one group at a time in order in which the groups are listed (i.e. males in heat, males not in heat, females and the young), gives the following sums 30, 36, 49 and 50, respectively. Find the number of elephants in each of the group.

The suggested solution is expressed in modern notation:

In this case, $n - 1 = 3$, $S_1 = 30$, $S_2 = 36$, $S_3 = 49$, $S_4 = 50$, then $S = \frac{1}{3}[30 + 36 + 49 + 50] = 55$.

Therefore,

Number of elephants in heat $= S - S_1 = 55 - 30 = 25$,

Number of elephants not in heat $= S - S_2 = 55 - 36 = 19$,

Number of female elephants $= S - S_3 = 55 - 49 = 6$,

Number of young elephants $= S - S_4 = 55 - 50 = 5$.

Aryabhata also introduced an important algebraic identity frequently used by the Kerala School whose work is discussed in Chapter 10. The statement is cryptic and difficult to express verbally. In modern notation, it is

$$\frac{1}{n-1} \sum_i \left(\sum_1^n S_r^i - S_r \right) \equiv \sum_1^n S_r.$$

6.2.8   Indeterminate equations[32]

At this point in *Aryabhatiya* appears one of the most important contributions of Aryabhata, namely, his solution of linear indeterminate equations[33] by the 'pulveriser' calculation (*kuttakaraganita*). The term 'pulveriser' is to be understood as a means of reducing something to finer sizes by repeated operation. In mathematics, it refers to repeated division by which the given numbers are made smaller. It arose in the field of astronomy with the problem of solving first-order indeterminate equations, notably where there was a need to determine the orbits of planets.[34] While there is some fragmentary evidence of earlier work in India on this subject, a systematic treatment of it only came with Aryabhata. This method was developed by Aryabhata in the 5th century to solve indeterminate equations of the form

$$ax - by = c.$$

The problem that he addresses may be expressed in modern terms as follows:

Find an integer (N) which when divided by another integer (a) leaves the remainder (r_1) and when divided by another integer (b) leaves the remainder (r_2).

Symbolically, the two problems may be expressed thus:

$$N = ax + r_1 = by + r_2. \tag{1}$$

(1) may be rewritten as:

$$by - c = ax, \quad \text{where } c = r_1 - r_2. \tag{2}$$

It is suggested that c be kept always positive by labelling r_1 and r_2 such that $r_1 > r_2$.

The (*Kuttaka*) problem may be stated thus: Given (a, b, c) are integers, solve for integers (x, y). Expressed in modern language, this is the familiar problem of solving for two unknowns (x, y) given only one equation (2). There can be any number of integral solution sets (x_i, y_i).

The solution[35] is offered in **Verses 32 and 33**. There is some controversy as to how they are to be interpreted.[36] Faced with the ambiguities and obscurities, we are once again dependent on Bhaskara I. In his commentary on the *Aryabhatiya*, the following numerical example is given to illustrate the method:

A quantity divided by 12 leaves a remainder of 5. Furthermore, if such a quantity divided by 31 leaves a remainder of 7, what should one such quantity be?

Bhaskara I's explanation of the solution procedure of Aryabhata is clear and concise when translated into modern notation.[37] In Bhaskara I's example, integers N, x, and y are sought such that $N = 31x + 7 = 12y + 5$ or $N = 31x + 2 = 12y$. The 'mutual division' is understood to mean continued divisions recasting the original equation between two unknown quantities with smaller and smaller coefficients until it is reduced or 'pulverised' into a form that can be solved by inspection.[38] Thus

$$y = \frac{31x + 2}{12} = 2x + w,$$

$$x = \frac{12w - 2}{7} = 1w + v,$$

$$w = \frac{7v + 2}{5} = 1v + u,$$

$$v = \frac{5u - 2}{2}.$$

At this point, a 'clever' integer solution is found by inspection showing that $u = 2$ and $v = 4$. By working our way to the top through a chain of substitutions, we find the minimum solution set is ($x = 10$, $y = 26$ and $N = 317$).

The method of solution came to be known as *kuttaka*. This word is derived from *kutt* meaning 'to crush', 'grind' or 'pulverise', and describes a method which is a successive process of breaking down into smaller and smaller pieces: making the values of the coefficients [a and b in (1)] smaller and smaller. All the great mathematicians of the subsequent period dealt with the *kuttaka*, and it is one of the very few topics in Indian mathematics to be made the subject of a special monograph, entitled *Kuttakara Siromani*, written by a commentator on Aryabhata I named Devaraja. In later chapters of this book, notably in the discussion of the works of Brahmagupta and Bhaskara II (Bhaskaracharya), we take up the subject again.

Endnotes

[1] Indeed, Bhaskara I's definition of the term '*ganita*' and his discussion of its scope both in mathematics and astronomy remain one of the most comprehensive expressions in early Indian mathematics.

[2] Bhaskara's major contribution was to elaborate on Aryabhata's stray remarks about the technical terms, allowing himself considerable latitude in interpreting *Aryabhatiya*. For example, in Verse 6, Aryabhata introduces rules regarding triangles. Bhaskara distinguishes between 'equal' (equilateral), 'bi-equal' (isosceles) and 'unequal' (scalene) triangles. Further, a 'square' is specified as not only a quadrilateral having equal sides but also equal diagonals.

[3] One of the approximations for π used for many centuries was $\sqrt{10}$. Bhaskara I criticised this approximation. He regretted that an exact measure of the circumference of a circle in terms of diameter was not available and he believed that such a measure was not possible.

[4] The *Kalakriyapada* is concerned with methods of computing planetary positions as well as presenting a geometrical model implied by the computational methods. The last section, *Golapada*, discusses a medley of subjects, including earth's shape, origins of lighting on different planets and calculation of eclipses.

[5] The following example from *Aryabhatiya* illustrates the representation of numbers by alphabets:

$$khya = 2 + 30 = 32; \ khyu = (2 + 30) \times 10^3 = 32{,}000.$$

Bhaskara I in his commentary, after listing the names of notational places from 1 (10^0) to billion (10^9), asks why there is need for these notational places. He answers that it is easy for traders to label prices of their various commodities.

[6] For a fuller summary of different systems of number representation in India, see Joseph (2011: 339–344).

[7] Aryabhata distinguished between the mathematics of 'fields' (containing subjects such as series, shadows, etc.) and of 'quantities' (containing subjects such as proportions, pulverisers, etc.). By doing so, he was able to include a discussion on astronomical techniques in the same book containing computational methods in mathematics.

[8] Both a geometrical representation and an arithmetical operation are implied in the terms 'square and squaring' (*varga*) and 'cube and cubing' (*ghana*). Thus, Verses 3(a–b) and 3(c–d) may be translated respectively as:

> "An equilateral quadrilateral with equal diagonals is a 'square' (*samacaturasra*). The product of two equal quantities is also called '*varga*'."
>
> "The continued product of three equal quantities gives (rectangular) solid having twelve (equal) edges and is called a '*ghana*'."

Bhaskara I points out that an equi-quadrilateral with unequal diagonals (a 'diamond' shape) cannot be a square. Therefore, a square is an 'equi-diagonal-equi-quadrilateral'. It is at this point that a diagram is introduced 'to convince the dull-minded of the truth of the statement'. For details, see Keller (2006, Vol. 2, pp. 2–6).

[9] The actual algorithms for calculating square and cube roots are explained in **Verses 4** and **5** respectively. Bhaskara I's commentary is particular useful here.

Verse 4 states cryptically

> One should divide repeatedly the non-square [place] by twice the square root. When the square has been subtracted from the square [place], the quotient is the root in a different place.
> (Note: The 'square' and 'non-square' places refer to even and odd powers of 10 respectively)

Verse 5 states cryptically

> One should divide the second non-cube place by three times the square of the root of the cube. The square [of the quotient] multiplied by three and the former [quantity] should be subtracted from the first [non-cube place] and the cube from the cube [place]...

As far as we know, this is the first time that the subject of extraction of cube roots appears in Indian mathematics. The rationale behind this cube root algorithm is found in understanding that, for example, a three digit number can be expressed in the form $ax^2 + bx + c$ where a, b and c are integers and $x = 10$. These methods are the precursors of the ones taught in schools before the advent of the calculator and computer. For complete details of these verses, see Shukla's critical edition and translation of *Aryabhatiya*. (1976, pp. 36–38). For a clear exposition through an example, see Plofker (2009, p. 126).

[10] **Verse 27(a–b)** dealing with the simplification of the quotient of fractions ('the numerators and denominators of the multipliers and divisors should be multiplied by one another') follows from the discussion of the 'Rule of Three' in the previous verse. *Verse 27(c–d)* discusses the rule for the reduction of two fractions to a common denominator: 'Multiply the numerator as also the denominator of each fraction by the denominator of the other fraction'.

[11] The important 'Rule of Three' is stated clearly in **Verse 26**; and the method of inversion in **Verse 28**:

> "In the method inversion multipliers become divisors and divisors become multipliers, additions become subtractions and subtractions become additions."

The Rule of Three and its development has already been discussed in some detail in the previous chapter in relation to the Bakhshali Manuscript.

[12] For a statement of this ancient method of proof, see endnote 27, Chapter 3 and Joseph (2003).

[13] A 'six-edged' solid may be interpreted as a pyramid with a triangular base. The rule given here may be expressed as Volume = $\frac{1}{2}$ (Area of triangular base) × height. The correct rule is found in **Verse 44** of the *Brahmasphutasiddhanta* of Brahmagupta (b. 598). It states: "The volume of a uniform excavation divided by three is the volume of the needle-shaped solid."

This is equivalent to: Volume of a pyramid

$$= \frac{1}{3}[(\text{area of base}) \times \text{height}].$$

[14] An alternative approach, given by Brahmagupta, more in tune with the present-day method follows from:

Area of the triangle

$$= \sqrt{S(S-a)(S-b)(S-c)}, \quad \text{where } S = \frac{1}{2}(a+b+c).$$

This is derived from the application of the Pythagorean theorem.

[15] This controversy is strangely reminiscent of the one that occurred in the interpretation of a result in Egyptian mathematics relating to the area of a curved surface. Was the result applicable to the surface area of a semi-cylinder or the surface area of a hemisphere? If it was the latter, the Egyptian geometry was remarkably advanced for it would antedate the work of Archimedes by 1,500 years. For details, see Joseph (2011, 116–118).

[16] The product of two sides and the height divided by their sum are the 'two lines of their own falling' (i.e. the two unequal segments produced by the intersection of the diagonals of the trapezium). When the height is multiplied by half the sum of both widths, one will discover the area.

[17] There was a particular interest in isosceles trapezium which, as we saw in the chapter on *Sulbasutras*, was the shape of the altars in which *soma* sacrifices occurred.

[18] Note that because these 'trapeziums' have curved surface, this is only an approximate estimate of the area of the *panava*.

[19] In modern notation, this means a chord subtending 60° at the centre = Radius (R) or $R \sin 30° = R/2$. Further discussion is found in the chapter on Indian trigonometry.

[20] One particular case of inscribing a regular polygon that would have been known to Aryabhata is the case of a hexagon inscribing a circle which would have led to the result that the side of an inscribed regular hexagon is equal to the radius of the circle. This is a result fundamental to the derivation of the Indian sine.

[21] The word 'approximately' may be a mistranslation of the Sanskrit word *asanna* since the latter also conveys the notion of 'unattainability' or in this context 'irrationality'. There is an interesting discussion beginning with Bhaskara I about why the approximate circumference can never be the true circumference, or why the true value of the circumference can never be attained. The example taken is one we came across in our discussion of Jain mathematics in Chapter 4 and asks whether the square root of 10 times the square of the diameter can ever be equal to the circumference of the circle. Bhaskara states that if such an equality is stated it is merely a tradition and not a proof. And this would include Aryabhata's estimate of the circumference. By implying that 'Circumference' is a *karani* or a number whose exact value cannot be stated, it is clear that Bhaskara and the Indian mathematicians who followed him were aware of the 'irrational' nature of certain numbers, although the discussion never took the Greek form of thinking in terms of ratios of integers.

[22] From ancient times, Indian mathematics has been interested in evaluating irrationals through iterative approximations. Examples from the *Sulbasutras*, as we saw in Chapter 3 included an approximation for the square root of 2 and a crude one for what we would now identify as π. In *Aryabhatiya*, a search through iterative methods of algorithms for the square and cube roots of any number is yet another example. This interest in iterative computational methods served both as an inspiration and engine for the growth of Kerala mathematics as we will see in a later chapter.

[23] A gnomon may be defined as the projecting piece on a sundial that shows the time by the position of its shadow.

[24] An explanation of this rule is found in Plofker (2009, pp. 130–131).

[25] The term *citighana* literally means 'the solid contents of a pile (of balls) in the shape of a pyramid on a triangular base'. The pyramid is constructed such that there is 1 ball at the top, 1+2 balls in the next lower layer, $1 + 2 + 3$ in the next lower layer and so on. In the nth layer which forms the base, there are $1 + 2 + 3 + \cdots + n$ balls.

[26] In Chapter 8 of Pingala's *Chandahsastra*, there appears a method of summing all the *Sankhyas* which involves the sum of a geometrical series: $1 + 2 + 2^2 + 2^3 + \cdots + 2^n$ which is $2^{n+1} - 1$.

[27] Notice these have to be 'positive factors' otherwise there might be an issue as to whether

$$\sqrt{4ab + (a - b)^2} = a + b \quad \text{or} \quad -a - b.$$

[28] These include contributions of Brahmagupta, Sridhara and Mahavira, Aryabhata II, Sripati, Bhaskara II, Narayana Pandita and the later Kerala mathematicians such as Chitrabhanu and Jyesthadeva. Some of these works will be considered in later chapters.

[29] The relevant passage reads: "The known result is to be multiplied by the quantity for which the result is wanted, and divided by the quantity for which the known result is given." This concise instruction may be expressed in terms introduced by Aryabhata as: "Take the known result *phala-rasi*, multiply it with *iccha-rasi* and divide it by *pramana* to get the result to be known as *iccha-phala*".

[30] In a treatise composed by al-Biruni entitled *Maqala fi rashikat al-hind* ('Treatise on the proportion rules of the Indians'), there is an interesting comparison between the notion of the Indian 'Rule of Three' and the Euclidean notion of a ratio. Al-Biruni implies that the difference between the two notions ultimately rests on formal proof versus computational demonstration.

[31] A verification of this result uses the 'Rule of 5'. In modern terms, the result is obtained from solving the quadratic equation for r:

$$(pt^2)r^2 + (pt)r = a, \quad \text{where } r \text{ is the interest rate.}$$

[32] A distinction should be made between 'Diophantine Equation' (named after the Hellenistic mathematician Diophantus who live in the 3rd century CE) which involves finding rational solutions and 'Indeterminate Equation' concerned with finding integer solutions. Diophantus *Arithmetica* is a curious synthesis of Greek, Egyptian and Mesopotamian mathematics. It was not only one of the first purely number-theoretic and algebraic texts, but the first to use the blend of rhetorical and symbolic exposition known as syncopated algebra. For further details, see Hettle (2015).

[33] An example of an indeterminate equation in two unknowns (x and y) is $3x + 4y = 50$, which has a number of positive whole-number (or integer) solutions for (x, y). For example, $x = 14$, $y = 2$ satisfies the equation, as do the solution sets $(10, 5)$, $(6, 8)$ and $(2, 11)$.

[34] Bhaskara I presents in his *Mahabhaskraiya* a number of astronomical problems in which *kuttaka* is used. For details, see Keller (2006, Vol. 2, pp. 160–185).

[35] Verses 32 and 33, as translated by Keller (2006, Vol. 1, p. 128) remains the authoritative text.

[36] There is a long history of interpretations which include those of Rodet (1879), Kaye (1908), Heath (1910), Mazmudar (1912), Sen Gupta (1927), Ganguly (1928), Clark (1930) and Datta (1932). The interpretations of Rodet and Kaye are now accepted as arising from faulty translations. But the damage has persisted with the adoption of Kaye's translation and interpretation by some Western and Indian historians of mathematics, notably Heath and Mazmudar. Sen Gupta's is based on Brahmagupta's; Clark's on that of Paramesvara. Datta's and Ganguly's which refer to Bhaskara I's interpretation are now acknowledged to be the more satisfactory ones.

[37] A translation of the full verbal explanation of the procedure is given in Keller (2006, Vol. 1, p. 131). The explanation of this procedure in modern notation is based on the exposition by Plofker (2007, pp. 416–417).

[38] To illustrate the process of mutual division: Taking $y = \frac{31x+2}{12}$, we divide to get $2x + [(7/12)x + 1/6]$. Setting $w = (7/12)x + 1/6$ which reduces to $7x = 12w - 2$, we continue with the process of mutual division. Bhaskara I, while commenting on this solution procedure, called it *kuttaka*.

Chapter 7

Riding the Crest of a Wave: From Brahmagupta to Mahavira

7.1 Brahmagupta

The revival in mathematical activity, spearheaded by Aryabhata and Bhaskara I, came in the middle of the first millennium CE when established channels of communication within India were concentrated in three centres of learning — Kusum Pura and Ujjain in the North, Mysore in the south and outside, with other cultures, notably first Persia, and later the Islamic world and China. The scene was set for the transmission of Indian mathematical ideas to the West and the incorporation of important Babylonian and Hellenistic ideas, mainly from Alexandria, into Indian astronomy. In contributing to this circulation of innovative ideas, Brahmagupta holds a special place in the history of mathematics.[1]

Brahmagupta was born in 598 in Bhillamala (the present-day Bhinmal in Rajasthan) during the reign of Vyaghramukh of the Chapa dynasty. In his youth, he moved to Ujjain to become eventually the head of the astronomical observatory there. He followed in the footsteps of Varahamihira (505–587) who had worked there and built up a strong school of astronomy.[2] In addition to the *Brahmasphutasiddhanta* ('Corrected Doctrine of Brahma') which he composed in 628 probably under the patronage of Vyaghramukh of the Chapa dynasty. He composed a second work at the age of 67 and called it *Khandakhadyaka* (meaning 'an edible morsel' because it claimed to be a more palatable version of *Aryabhatiya*). It was primarily an astronomical handbook that employed the Aryabhata's system of starting each day at midnight. There was also a couple of his texts entitled *Durkeamynarda* and *Cadamakela* of which we know very little.

7.1.1   The content and scope of *Brahmasphutasiddhanta*

The *Brahmasphutasiddhanta* contains 24 chapters consisting of 1008 verses of which the first 10 may have formed an earlier abridged version of the same text, given the existence of a number of manuscripts containing only these chapters.[3] In the chapters on astronomy, Brahmagupta discussed the average and real motions of the planets, the problems of place-time-distance concerning the earth, sun, and planets, planetary conjunctions and the rising and setting of celestial objects. He correctly described the phenomena of solar and lunar eclipses as being caused by the moon and earth casting shadows. A whole chapter is devoted to the description and use of various astronomical instruments.

In addition to astronomy there are chapters and sections dealing with mathematics. In the two chapters (Chapters 12 and 18 containing 66 and 101 verses respectively), Brahmagupta laid the foundation of two major fields of Indian mathematics, *patiganita* ('mathematics of procedures') and *bijaganita* ('mathematics of seeds'), which roughly correspond to arithmetic (including mensuration) and algebra respectively according to the subject divisions today. Chapter 12 is simply titled as *"ganita"* and covers arithmetical operations and certain practical topics which were prerequisites for a qualified calculator (*ganaka*). Chapter 18 is named *Kuttaka* and called as such probably because there were no specific name for this area of algebra then.

The *Brahmasphutasiddhanta* and *Khandakhadyaka* were both composed in concise verses, more comprehensible than *Aryabhatiya*, but requiring at places explanations, especially since no proofs or demonstrations were offered. As customary then, the task of rendering and interpreting difficult verses in simple language was left to the commentators of whom the best known in this case was Prthudaksvami (*fl.* 864).[4]

In the chapters on mathematics, Brahmagupta established rules and procedures for various operations. He discussed certain results already known to his predecessors and played a pioneering role in systematically framing the direct and inverse 'Rule of Three' which was introduced in the last chapter. Among his major accomplishments, Brahmagupta initiated a discussion on the four fundamental operations with zero, gave rules for operations of negative numbers ("debts") as well as surds. He also obtained partial solutions for certain indeterminate equations of the first and second degree with two unknown variables.[5] Perhaps his best known discovery followed from an investigation of the properties of a cyclic quadrilateral (a four-sided polygon whose vertices all reside on a circle). He also gave a valuable interpolation formula for computing sines which will be examined in a separate chapter on Indian trigonometry.

Although Brahmagupta was familiar with the works of astronomers following in the tradition of *Aryabhatiya*, it is not known if he knew Bhaskara I's *Aryabhatiyabhasa*. In any case, Brahmagupta directed a string of criticism of certain astronomers, notably Aryabhata I and Latadeva,[6] initiating one of the well-known schisms in Indian astronomy. The difference of opinion was about the application of mathematics to the physical world, notably the choice of astronomical parameters and theories. Critiques of rival theories are scattered throughout the first 10 chapters while the eleventh chapter is wholly devoted to criticism of these theories.

7.2 The Mathematics in the *Brahmasphutasiddhanta*[†]

7.2.1 Arithmetic of calculation

In the beginning of chapter twelve [12.1] of his *Brahmasphutasid-dhanta*, entitled *Calculation*, Brahmagupta details 20 operations

---

[†]The specific references to *Brahmasphutasiddhanta* in this chapter are given as [Chapter and Verse]

beginning with addition and concluding with eight procedures or practices relating to operations with fractions.[7] A student then was expected to know basic arithmetical operations covered in the *Aryabhatiya* as far as taking the square and square root and finding the cube and cube-root of an integer. However, the rules that Brahmagupta gave for operating with fractions was a relatively new subject in Indian mathematics and as relevant today in the classroom as they were when first stated.[8]

Once Brahmagupta had discussed the 20 operations and the eight procedures, his facility with computations becomes clearly evident from the shortcuts and tricks that he adopts in calculations. The following is a couple of his 'tricks' that give us a flavour of his approach to computational arithmetic.[9]

[12.58] (i) Evaluation of a Quotient (m/n) using Euclidean Division[10]

$$\frac{m}{n} = \frac{m}{n+h} \pm \frac{m}{n+h} \cdot \frac{h}{n}.$$

Example 1: If $m = 9999$, $n = 95$ and $h = 4$, then

$$\frac{9999}{95} = \frac{9999}{95+4} \pm \frac{9999}{95+4} \cdot \frac{4}{95} = 101 + \frac{404}{95}.$$

Apply the rule again taking $h = 6$

$$\frac{404}{95} = \frac{404}{95+6} \pm \frac{404}{95+6} \cdot \frac{6}{95} = 4 + \frac{24}{95}.$$

Therefore,

$$\frac{9999}{95} = 101 + 4 + \frac{24}{95} = 105 + \frac{24}{95} = 105\frac{24}{95}.$$

(ii) *The Samkrama Technique*: Find two quantities (x and y) given their sums and differences:

[12.63] (1) $x^2 = (x-y)(x+y)+y^2$: $[137^2 = (137-3)(137+3)+3^2 = 18769$ taking $y = 3]$,

[**18.36**] (2) $\frac{1}{2}[(x+y)+(x-y)] = x$; $[\frac{1}{2}[(x+y)-(x-y)] = y$ given $(x+y)$ and $(x-y)]$,

[**18.36**] (3) $\frac{1}{2}\left(\frac{x^2-y^2}{x-y} + (x-y)\right) = x$; $\frac{1}{2}\left(\frac{x^2-y^2}{x-y} - (x-y)\right) = y$ given $(x-y)$ and (x^2-y^2).

An extension of (3) above led Brahmagupta to derive the *samkrama* and *vismakrama* respectively for astronomical purposes.[11]

[**18.98**] $\frac{1}{2}(x+y) \pm \frac{1}{2}\sqrt{2(x^2+y^2)+(x+y)^2} = x, y$ given $(x+y)$ and (x^2+y^2),

[**18.99**] $\frac{1}{2}\sqrt{4xy+(x-y)^2} \pm (x-y) = x, y$ given (xy) and $(x-y)$.

In the eight 'procedures' mentioned earlier, there is a considerable overlap between Brahmagupta's text and the topics covered in *Aryabhatiya* and Bhaskara I's commentary. It is not the intention in this book to reproduce earlier work. Instead, we will pick out piecemeal those subjects in which Brahmagupta made significant contributions.

In the section on procedures, one topic relates to interest computations. The problems considered are more complicated than the ones in *Aryabhatiya*, such as calculating the Principal (P) from the final amount (A) and the length of the loan period (t), **or** the time taken for the original amount to become a multiple of itself given the interest charged. Brahmagupta provides a generalised version of the quadratic rule for computing interest discussed in in *Aryabhatiya*.[12] There is little that is new in the section on 'Series' in Brahmagupta's work. The rules for the sum and number of terms in an arithmetic progression is as stated in *Aryabhatiya* and elaborated by Bhaskara I.

One of the great contributions of Brahmagupta was his treatment of crucial mathematical objects, such as operations with the number zero and with negative numbers. The *Brahmasphutasiddhanta* is the earliest known text to treat zero as a number in its own right, rather than as simply a placeholder digit in representing another number as done by the ancient Babylonians, **or** as a symbol for a lack of quantity, **or** as a direction separator as found in ancient Egyptians. In chapter 18 of his *Brahmasphutasiddhanta*, Brahmagupta sets a trend to be followed by others of describing the arithmetic operations with positive and negative numbers and zero (*sunyaganita*) almost in the same way as we would do today. In a masterly summary which has strong resonance even today, Brahmagupta's discussion of the operations is clear and so we will leave him to state them in his own words, with modifications and additions included in the square brackets for sake of clarity. In five verses, beginning first with a description of addition and subtraction, he then proceeds to other operations.

[**18.30**] [The sum] of two positives is positive, of two negatives negative; of a positive and a negative [the sum] is their difference; if they are equal it is zero. The sum of a negative and zero is negative, [that] of a positive and zero [is] positive, [and that] of two zeros [is] zero.

[**18.31**] If a smaller [positive] is to be subtracted from a larger positive, [the result] is positive; [if] a smaller negative from a larger negative, [the result] is negative; [if] a larger [negative or positive is to be subtracted] from a smaller [positive or negative, the algebraic sign of] their difference is reversed — i.e. negative [becomes] positive and positive [becomes] negative

[**18.32**] A negative minus zero is negative, a positive [minus zero] positive; zero [minus zero] is zero. When a positive is to be subtracted from a negative or a negative from a positive, then it is to be added.

[**18.33**] The product of a negative and a positive is negative, of two negatives positive, and of positives positive; the product of zero and a negative, of zero and a positive, or of two zeros [are all] zero.

[**18.34**] A positive divided by a positive or a negative divided by a negative is positive; a zero divided by a zero is zero; a positive divided by a negative is negative; a negative divided by a positive is [also] negative.

[**18.35**] A negative or a positive divided by zero has that [zero] as its divisor, or zero divided by a negative or a positive [has that negative or positive as its divisor]. The square of a negative or of a positive is positive; [the square] of zero is zero. That of which [the square] is the square of [its] square-root

Brahmagupta's rules for arithmetic of negative numbers and zero are quite close to our present understanding, except that in modern mathematics division by zero is often left undefined or equal to infinity.[13] Mahavira's (*fl.* 850 CE) rules are identical except his incorrect inference that a number remains unchanged when divided by zero. It is only with Bhaskaracharya (b. 1114) that this subject was advanced further.

Brahmagupta concludes this section with a short discussion of various operations with surds (*karani*),[14] of which a notable result relates to the addition/subtraction of two surds. The following instruction is given:

> The surds divided by a suitable optional number, and the square of the sum of the square roots of the quotients should be multiplied by that optional number (in the case of addition), and the square of the difference of the square roots of the quotients being so treated will give the difference of the surds.

Expressing this cryptic rule in modern symbolic notation, we get

$$\sqrt{a} \pm \sqrt{b} = \sqrt{c\left\{\sqrt{\left(\frac{a}{c}\right)} \pm \sqrt{\left(\frac{b}{c}\right)}\right\}^2},$$

where c is a suitably chosen optional number such that $(a/c, b/c)$ are perfect squares. For example, if we wish to evaluate $\sqrt{8} + \sqrt{2}$ and $\sqrt{8} - \sqrt{2}$, we may choose $c = 8$ and arrive at the answers

$$\sqrt{8} + \sqrt{2} = \sqrt{18} \quad \text{and} \quad \sqrt{8} - \sqrt{2} = \sqrt{2}.$$

Operations with surds has a long history. It is first found in Bhaskara I's commentary on *Aryabhatiya*. The clearest statement of addition and subtraction rules are found in Brahamagupta; and he devotes a few other verses on multiplication and division of surds. Mahavira and Bhaskaracharya restate these rules without any significant innovations. There are some interesting extensions in Narayana Pandita whose work is discussed in Chapter 9. However, the whole subject of arithmetical operations with surds that interested Indian mathematicians over a long time fell into disuse. Datta and Singh (revisions by Shukla) [1993] provides a useful survey which discusses multiplication and division of surds. It remains one of the best illustrations of the Indian fascination with computational mathematics.

7.2.2   Triangles and quadrilaterals

It is when we come to the "Figures" (or geometry) section that Brahmagupta's contribution becomes very important. His work on geometry may be classified as finding the areas, sides and diagonals of triangles and quadrilaterals. The rules that he suggests generate both 'approximate' and 'accurate' results. It is intriguing that Brahmagupta views a triangle as a quadrilateral with one side of length zero. The answer may lie in the fact that he never refers to a 'quadrilateral' as such but to the neologism *tricaturbhuja* (i.e. 'tri-quadrilateral'), a term that he coined and used in only two verses [**12.21** and **12.27**] with it occurring in no subsequent mathematical text in Sanskrit. The general criticism that Brahmagupta was not able to distinguish between properties of a triangle and that of a quadrilateral is based on an implicit assumption that his rules applied to both a triangle and an (unrelated) quadrilateral at the same time. Further, it is necessary to re-examine the oft-stated assertion that Brahmagupta did not grasp the necessity for his 'quadrilaterals' to be inscribed if some of his rules were to be valid. As Kichenassamy (2010, pp. 15–28) and Fillozat (2002, p. 28) have argued that by submitting the relevant sections of his mathematical texts to literary analysis, several supposedly obscure, inaccurate or objectionable passages can be clarified and explained, including the vexed issue of Brahamagupta's failure to distinguish between a cyclic and non-cyclic quadrilateral.

Among Brahmagupta's most famous results in geometry are his rules for all quadrilaterals that can be inscribed within circles (i.e. cyclic quadrilaterals) where the sides, diagonals, perpendiculars, segments areas and the diameters of circles are integers.[15] Given the lengths of the sides of any cyclic quadrilateral, Brahmagupta gave an approximate and an exact formula for the figure's area.[16]

In [12.21], he states: "The 'gross' (i.e. approximate) area of a triangle or a quadrilateral is equal to the product of the half the sum of the opposite sides. The 'accurate' (area) is the square root from the product of the halves of the sum of the sides, being reduced by (each) of the four sides side."

In modern notation if a, b, c, d are the four sides of the Figure 7.1. Then

$$\text{'Gross' area} = \left[\frac{(a + c)}{2}\right] \cdot \left[\frac{b + d}{2}\right]$$

and

$$\text{Exact' area} = \sqrt{(s - a)(s - b)(s - c)(s - d)},$$

where $s = \frac{1}{2}(a + b + c + d)$.

Note that the formula for finding the accurate area of a triangle (i.e. the quadrilateral with side $d = 0$ and hence a become $a*$) is

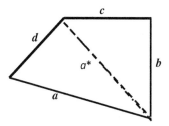

Figure 7.1. Area of an inscribed triangle.

usually attributed after Heron of Alexandria who lived around the beginning of the Common Era. In Figure 7.1, the area of triangle of sides a^*, b and c is given by[17]:

$$A^* = \sqrt{(s^* - a*)(s^* - b)(s^* - c)},$$

where $s^* = \frac{1}{2}(a^* + b + c)$.

Brahmagupta's other significant contribution to the geometry was the theorem on the diagonals of a cyclic quadrilateral.

The theorem states: [12.28].

Divide mutually the sums of the products of the sides attached to both the diagonals and then multiply both the quotients by the sum of the products of the opposite sides; the square roots of the results are the diagonals of the quadrilateral. The 'gross' area is the product of the halves of the sums of the sides and opposite sides of a triangle and a quadrilateral. The accurate [area] is the square root from the product of the halves of the sums of the sides diminished by [each] side of the quadrilateral.

In Figure 7.2, let a, b, d and c (the circle is not shown) be the sides of a convex cyclical quadrilateral (ABCD) and p and q are the

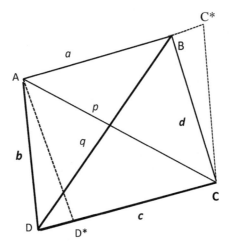

Figure 7.2. A cyclical quadrilateral.

diagonals. Then

$$p = \sqrt{\frac{(ab+cd)(ac+bd)}{(ad+bc)}}, \quad q = \sqrt{\frac{(ad+bc)(ac+bd)}{(ab+cd)}}$$

and

$$pq = ad + bc.$$

It is not known how Brahmagupta arrived at this formula. However, the Kerala mathematicians Jyesthadeva (*fl.* 1500 CE) gave a rigorous proof in his *Yuktibhasa* and left the impression that the proof was known well before his time. The proof involves using the properties of the similar triangles DAD* and BCC* and the right-angled triangle theorem. For details of the proof, see Gupta (1977).

In the next few verses Brahmagupta gives methods of constructing various figures including triangles (isosceles, right, scalene), different types of trapezia. In [**12.38**], Brahmagupta shows a way of constructing a cyclical trapezium using properties that he explored earlier. The processes involved in constructing it may be gathered from the Figures 7.3(i)–(iii) and expressed succinctly as follows:

> The uprights and sides of two rectangular triangles reciprocally multiplied by the diagonals are the four dissimilar sides of a trapezium. The greatest is the base; the least is the top; and the other two are the flanks.

The four triangles (ii), (iii), (v) and (vi) in Figure 7.4 are slotted in to get the quadrilateral shown in Figure 7.5. Please note that the final diagram is not according to any scale.

This construction (Figure 7.5) is based on a more detailed extension by Bhaskaracharya in his *Lilavati* (1150) and by Ganesha Daivajana's *Buddhivilasini* (*c.* 1500) on Brahmagupta's construction. The sides of the quadrilateral in Figure 7.5 are $a = c_2b_1$, $b = c_1b_2$, $d = c_2a_1$ and $c = c_1a_2$ and the diagonals are $AC = (a_1b_2 + a_2b_1)$ and $BD = (a_1a_2 + b_1b_2)$. The circumradius of both triangle ABC and ACD can be shown to be $\frac{1}{2}(c_1c_2)$. And the expression for diagonals

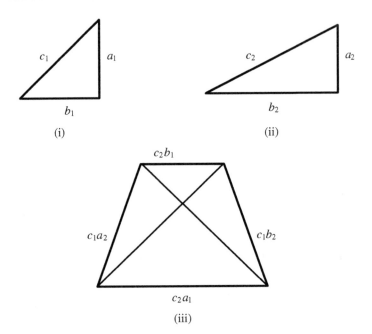

Figure 7.3. Construction of a cyclic trapezium.

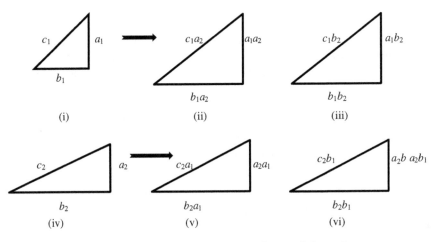

Figure 7.4. Construction of a cyclic quadrilateral.

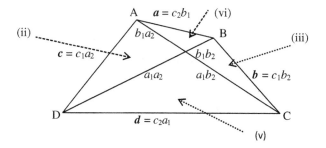

Figure 7.5.

can be shown to be:

$$AC = (a_1 b_2 + a_2 b_1) = \sqrt{\frac{(ab + cd)(ac + bd)}{(ad + bc)}};$$

$$BD = (a_1 a_2 + b_1 b_2) = \sqrt{\frac{(ad + bc)(ac + bd)}{(ab + cd)}}.$$

This is identical to the rule for the determination of the diagonals of a cyclic quadrilateral established earlier.

In [12.21], Brahmagupta shows a way of finding both the *abadhas* (i.e. segments of the base) and the altitude. The proof is simple and is largely derivative of the work of the early Aryabhatan School discussed in the last chapter.

$$AB^2 = BD^2 + AD^2$$

$$= (BC - DC)^2 + AD^2 = BC^2 - 2BC \cdot DC + (DC^2 + AD^2)$$

or

$$c^2 = a^2 - 2aa_2 + b^2$$

Therefore,

$$a_2 = \frac{a^2 + b^2 - c^2}{2a} = DC.$$

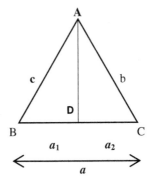

Figure 7.6. Finding the *Abadhas* and the altitude.

Similarly, it can be shown that the second segment of the base is BD (or a_1) = $\frac{a^2-b^2-c^2}{2a}$ and the altitude

$$AD = \sqrt{c^2 - a_1^2} = \sqrt{b^2 - a_2^2}.$$

We have seen in a previous chapter that from the time of the *Sulbasutras*, the properties of right-angled triangles, commonly known as the Pythagorean theorem, were well-known in Indian mathematics. Brahmagupta was the first Indian mathematician to study properties of *rational* right-angled triangles.[18] His solution for a right-angled triangle (Figure 7.7(a)) with rational sides a, b, c (and rational area) is in the form: $a = 2mn$, $b = m^2 - n^2$ and $c = m^2 + n^2$ where m and n are two unequal rational numbers. If $m = 2$ and $n = 1$, we get the Pythagorean triple ($a = 4$, $b = 3$, $c = 5$). Other combinations of (m, n) such as $(4, 3)$ and $(6, 3)$ yield triples $(24, 7, 25)$ and $(36, 27, 45)$.

For a rational isosceles triangle, Brahmagupta gives the following solution in terms of Figure 7.7(b). If m and n are two unequal rational numbers a rational isosceles triangle is the result of merging two identical right triangles each of hypotenuse $m^2 + n^2$, one side $m^2 - n^2$, with a third common side of $\frac{1}{2}mn$.[19]

Following from that, Brahmagupta gives a further result for constructing a scalene triangle made up of two triangles with two rational sides, one of them being the common altitude (m).

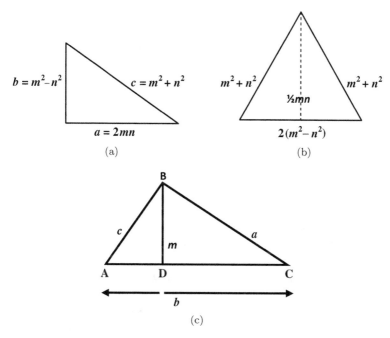

Figure 7.7. (a–b) A rational isosceles triangle. (c) A scalene triangle.

(See Figure 7.7(c).) To find the sides of the two triangles, Brahmagupta gives a rule. Expressed in modern notation, the sides are[20]:

$$\text{AB} = \frac{1}{2}\left[\frac{m^2}{b} + b\right], \quad \text{AD} = \frac{1}{2}\left[\frac{m^2}{b} - b\right],$$

$$\text{BC} = \frac{1}{2}\left[\frac{m^2}{a} + a\right], \quad \text{DC} = \frac{1}{2}\left[\frac{m^2}{a} - a\right],$$

$$\text{AC} = \frac{1}{2}\left[\frac{m^2}{a} + \frac{m^2}{b} - a - b\right].$$

[12.39] An interesting application of the right-angled triangle theorem that benefits from a clear commentary by Prthudakasvami is as follows:

On the top of a hill there are two holy men. One of them possesses magic powers which allows him to travel through the air. Springing from the top of the hill he ascends a certain height and then

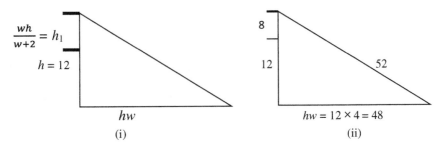

Figure 7.8. Two holy men.

proceeds to descend diagonally to a neighbouring town. The other man walks down the hill and then to the town. If the distance of the two journeys are equal, find the distance of the town from the hill and the height attained by the magician before his descent.

In Figure 7.8(i) we have a right-angled triangle whose height is $h + h_1 = h + \frac{wh}{w+2}$, the base is hw and w is a multiplier. So the 'descent diagonal' (d) will be $d = \sqrt{\left(h + \frac{wh}{w+2}\right)^2 + (wh)^2}$. The distance travelled by the 'jumping' holy man is $\frac{wh}{w+2} + d$ while the other person travels $h + wh$.

Since distances travelled by the two holy men are the same,

$$h + wh = \frac{wh}{w+2} + \sqrt{\left(h + \frac{wh}{w+2}\right)^2 + (wh)^2}.$$

Note: Since w is any chosen arbitrary number unrelated to h, the above relationship holds for all w. Figure 7.8(ii) is the numerical solution provided by Prthudakasvami. Here $h = 12$ and $w = 4$. The vertical jump by one of the holy men is $\frac{wh}{w+2} = 8$. The height of the right angle triangle is $h + \frac{wh}{w+2} = 12 + 8 = 20$ and the base of the triangle is $wh = (12 \times 4) = 48$. Applying the right-angled theorem, we calculate $d = 52$. So the distance travelled by each holy men is **60**.

[12.23] A further contribution of Brahmagupta was the study of a quadrilateral in which the square root of the sum of the products of the opposite sides is the diagonal and the square root of square of the diagonal minus the square of half the sum of the base and the face is the altitude. Such a quadrilateral could be a square, a rectangle or an isosceles trapezium. A proof of this result is also given by Prthudakasvami,

Brahmagupta's work is rich in formulae for the lengths and areas of geometric figures such as the circumradius of an isosceles trapezium and a scalene quadrilateral, and the lengths of diagonals in a scalene cyclic quadrilateral. This leads up to Brahmagupta's famous theorem, given in **[12.30–12.31]**, and loosely translated as:

> In Figure 7.9, there are two triangles within [a cyclic quadrilateral] with unequal sides, the two diagonals are the two bases. Their two segments are separately the upper and lower segments [formed] at the intersection of the diagonals. The two [lower segments] of the two diagonals are two sides in a triangle; the base [of the quadrilateral is the base of the triangle]. Its perpendicular is the lower portion of the [central] perpendicular; the upper portion of the [central] perpendicular is half of the sum of the [sides] perpendiculars diminished by the lower [portion of the central perpendicular].

In other words, the theorem states that if a cyclic quadrilateral has *perpendicular diagonals*, then the perpendicular to a side from

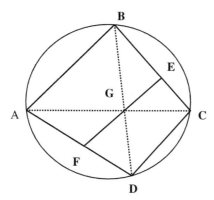

Figure 7.9. A cyclic quadrilateral.

the point of intersection of the diagonals always bisects the opposite side.

In Figure 7.9, let A, B, C and D be four points on the circle such that the lines AC and BD are perpendicular to one another. Denote the intersection of AC and BD by G. Drop a perpendicular from G to the line AD. Let F be the intersection of the lines EG and AD. The theorem states that F is the midpoint of AD or AF = FD.

To show that AF = FD = FG we note that the angles FAG and CBG are equal, because they are inscribed angles on the same arc of the circle and that angles EBG and CGE are both complementary to angle BCG (i.e. they add up to 90°), and are therefore equal. Finally, the angles CGE and FGA are equal. Hence, AFG is an isosceles triangle, and thus the sides AF and FG are equal. Similarly, it can be shown that FD = FG. Hence, the angles FDG, BCG, BGE and DGF are all equal, so DFG is an isosceles triangle and so FD = FG. It follows that AF = FD, as the theorem claims.

Brahmagupta also gives constructions of various figures with arbitrary sides. He provides the formulae for the lengths and areas of geometric figures, such the lengths of diagonals in a scalene cyclic quadrilateral, and of the 'needle figure' (i.e. a quadrilateral with two opposite sides extended until they intersect).

7.2.3   Geometry of circles

In the section on the circle Brahmagupta gives both 'approximate' and 'accurate' rules for finding the circumference and area of the circle.

> [**12.40**] The diameter and the square of the radius [each] multiplied by 3 are [respectively] the practical (i.e. approximate) circumference and the area [of a circle]. The accurate [values] are the square-roots from the squares of those two multiplied by ten.

If C, A and r are the circumference, area and radius respectively, then Brahmagupta's rules may be stated as:

$$C \approx 3 \times 2r, \quad A \approx 3r^2, \qquad \text{approximately,}$$
$$C = 2\sqrt{10r^2}, \quad A = \sqrt{10} \times r^2, \quad \text{accurately.}$$

It is interesting that although Brahmagupta should have known Aryabhata's more accurate value of $62832/20000$ for π, he continued to use the Jain value of $\sqrt{10}$.

Brahmagupta's discussion of chords and 'arrows'[21] does not go much beyond the work of Aryabhata and Bhaskara I. However, a couple of interesting results should be noted: [**12.40–12.41**]

> In a circle, the chord is the square root of the diameter less the arrow taken into the arrow and multiplied by four; The square of the chord divided by four times the arrow and then added to the arrow, is the diameter. Half the difference of the diameter and the root extracted from the difference of the square of the diameter and the chord is the smaller arrow.

In Figure 7.10, O is the centre of the circle, $BC = d$ is the diameter, $CD = a$ is the *sara* (or 'arrow') and Chord $AF = 2AD$. Now it can be shown that chord $2AD = \sqrt{4a(d-a)}$ and arrow $(sara) = \frac{1}{2}(d - \sqrt{d^2 - chord^2})$.

This follows from the fact that angle $B\hat{A}C = 90°$, $\triangle s\,ABD$ and ACD are similar and so $\frac{AD}{DB} = \frac{CD}{AD}$

Therefore

$$AD^2 = CD \cdot DB = a(d-a)$$

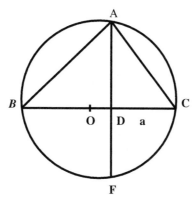

Figure 7.10.

So,

$$\text{Chord} = AF = 2AD = \sqrt{4 \times a(d-a)}.$$

Also

$$\sqrt{d^2 - Chord^2} = \sqrt{d^2 - [4 \times (d-a)]} = \sqrt{(d-2a)^2} = d - 2a.$$

Therefore, the 'arrow'

$$(sara) = a = \frac{1}{2}[d - (d-2a)] = \frac{1}{2}[d - \sqrt{d^2 - Chord^2}].$$

7.2.4   Intersecting circles: erosion

In Figure 7.11, the two intersecting circles (centre O_1 and centre O_2) has diameters d_1 and d_2 respectively. Their common chord is AF and their arrows (*saras*) are GD $= a_1$ and HD $= a_2$. We need to find the length of 'arrows' GD (a_1) and HD (a_2), given diameters d_1 and d_2 and erosion GH $= e = (a_1 + a_2)$.

[**12.42**] states:

> The erosion subtracted from both diameters, the remainders, multiplied by the erosion and divided by the sum of the remainders will give the arrows.

$$a_1 = \frac{e(d_2 - e)}{d_1 + d_2 - 2e}, \quad a_2 = \frac{e(d_1 - e)}{d_1 + d_2 - 2e}.$$

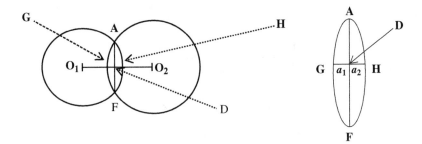

Figure 7.11. Erosion.

The proof follows from expressing the common chord AF of the two circles in terms of the two diameters and 'arrows' of the circles, substituting for $e = a_1 + a_2$ and using the identity for the chord AF: $(d_1 - a_1)a_1 = (d_2 - a_2)a_2$

A related result is given in Verse 43 as:

The square of half the chord (AF) being divided severally by the given arrows, the quotients added to the arrows respectively, are the diameters. The sum of the arrows is the erosion: and that of the quotients is the residue of subtracting the erosion.

Or expressed in modern notation:

$$d_1 = \frac{Chord^2}{4a_1} + a_1; \quad d_2 = \frac{Chord^2}{4a_2} + a_2; \quad e = a_1 + a_2.$$

At the conclusion of his discussion of the measurement of circles, Brahmagupta moves on to the volume of solids and surface areas (in particular, empty spaces excavated out of solids). This includes the volume of rectangular prisms, pyramids and the frustum of a square pyramid. For the volume of a frustum[22] of a pyramid [12.44–12.46], he gives two estimates: the approximate and accurate.

The area, calculated from the square of half of the sums of the sides at top and at bottom is multiplied by the depth to give the practical (or approximate) value of the volume. Subtracting the practical content from the other, divide the difference by three and add the quotient to the practical content. The sum is the exact value of the volume.[23]

If the depth is h units and the side of its square face is s_1 units and s_2 units and volume is V.
Then

$$V = h\left(\frac{s_1 + s_2}{2}\right)^2 \qquad \text{Approximate,}$$

$$V = h\left(\frac{s_1 + s_2}{2}\right)^2 + \frac{h}{3}\left[\frac{s_1^2 + s_2^2}{2} - \left(\frac{s_1 + s_2}{2}\right)^2\right] \qquad \text{Accurate.}$$

An Example: A square well, measures 10 cubits at the top and six at the bottom is 30 cubits deep. What is the approximate and accurate contents of the well? (*Answer*: Approximate Volume = 1920 cubits³; Accurate Volume = 1920 + 40 = 1960 cubits³).

This section concludes with a discussion of how to find the volume of a stack and in particular the number of bricks in a trapezoidal stack.

7.2.5 Shadow problems

In Chapter 19, involving problems on gnomons and shadows, Brahmagupta has a section on determining heights and distance of the objects by observing their reflection in water. Shadow measurements and ensuing calculations have formed an important part of astronomy from early times. Brahmagupta's contributions have been significant in this area.

> [**19.17**] When the distance between a man and his house is divided by the sum of the heights of the house and the man's eyes and when this quotient is multiplied by the height of his eyes, the top of the image of the house as seen [in the] reflecting water is at a distance equal to the above product.

In Figure 7.12 (Not to Scale), let AB be the height of the house and CD the height of the man's eyes. Let E be the reflecting point. Then, the man will see the tip of the shadow of the house when

$$\mathbf{BE} = \frac{\mathrm{BD} \times \mathrm{CD}}{\mathrm{AB} + \mathrm{CD}} \quad \text{and the height of the house is } \frac{\mathrm{CD} \times \mathrm{BE}}{\mathrm{DE}}.$$

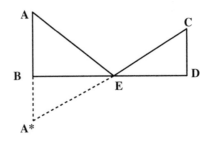

Figure 7.12. Shadow problem.

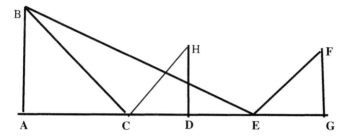

Figure 7.13. Heights from shadows.

The proof follows from showing that the two triangles A*BE and CDE are similar.

Yet another problem involving shadows is illustrated with Figure 7.13 below

[**19.19**] The distance between the first and second positions of water is divided by the difference between the distance of the man from water. When the resulting quotient is multiplied by the height of the man's eyes it gives the height of the house. If however the quotient is multiplied by the distance between the man and the water it gives the distance between the water and the house.

In Figure 7.13, AB is the height of the house. HD and GF are the two positions of the man. C and E are the points of reflection in the water. Then

$$\text{The height of the house (AB)} = \frac{CE \cdot HD}{EG - CD},$$

$$\text{The distance of the house (AC)} = \frac{CE \cdot CD}{EG - CD}.$$

Use the previous result with Figure 7.13 and the properties of similar triangles (namely, Δs ABC, CHD, EFG, AEB) to prove the above results. Additionally, a problem which combines shadow and reflection is used to work out the height at which the light from a given source can be seen.

Another topic relates to the shadow of the gnomon at two different positions (Figure 7.14)

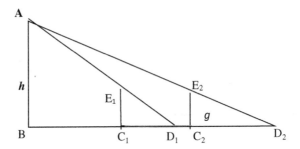

Figure 7.14. Shadow from a gnomon.

[12.54] The shadow multiplied by the distance between the tips of the shadows and divided by the difference of the shadows is the base. The base multiplied by the gnomon and divided by the shadow is the height of the flame of the light.

Let the distance between the two position of the gnomon (C_1 and C_2) be d. Let $C_1D_1 = s_1$ and $C_2D_2 = s_2$ be the length of the shadows of the gnomon at two different position. Let $D_1 \, D_2$ be the distance between the tips of the shadows. Let $AB = h$ be the altitude and $BD_1 = b$ be the base and $C_1E_1 = C_2E_2 = g$. Using the properties of two pairs of similar triangles (ABD_1 and $E_1C_1D_1$) and (ABD_2 and $E_2C_2D_2$) we can show that

$$\frac{h}{b} = \frac{g}{s_1} \quad \text{and} \quad \frac{h}{b + D_1D_2} = \frac{g}{s_2}.$$

Further manipulation and substitution will give the results expressed in words as:

$$\text{Base} = \text{Shadow length} \times \frac{\text{Distance between tips of shadows}}{\text{Difference of Shadows}},$$

$$\text{Height} = \frac{\text{Base} \times \text{Gnomon}}{\text{Shadow}}.$$

This is the rule stated in **[12.54]**.

7.2.6   Brahmagupta's contribution to indeterminate analysis

In a list of topics that mathematicians of his day needed to master before being awarded the coveted title of '*Acharya*', Brahmagupta pointed to the solution of the *kuttaka* problem and then devoted a hundred verses to this subject. It was Aryabhata I who provided a highly terse description of a systematic procedure for solving indeterminate equations of the first order, supplemented by the useful commentary by Bhaskara I. Given three unknowns $(N, x$ and $y)$ and two equations:

$$N = ax + r_1,$$
$$N = by + r_2,$$

where (r_1, r_2) are remainders such that $r_1 > r_2$; and (a, b) are 'divisors' in the sense that when they divide N, they leave remainders r_1 and r_2.

The problem is to find integer solutions for the unknowns N, x and y given that a, b, r_1 and r_2 are also integers. And that was the problem posed by Aryabhata and elucidated by Bhaskara I. Chapter 6 contains an explanation of that solution.[24]

Brahmagupta's main contribution to the subject was to devise a method for solving second-order indeterminate equations (*vargaprakriti*) of the form:

$$x^2 - Ny^2 = K \quad (N > 0, \text{ a non-square integer}) \tag{1a}$$

or

$$\left[\frac{(x^2 + Ny^2)}{K}\right]^2 - N\left[\frac{(2xy)}{K}\right]^2 = 1.$$

One reason for solving this problem was to find a rational approximation to the square root of N.

A special case of (1a) is where $K = 1$, i.e.

$$x^2 - Ny^2 = 1.$$

Then

$$\left|\sqrt{N} - \frac{x}{y}\right| \leq \frac{1}{2xy} < \frac{1}{2y^2}.$$

More specifically, given $x^2 - Ny^2 = K$ we have the rational solution

$$\left[\frac{(x^2 + Ny^2)}{K}\right]^2 - N\left[\frac{(2xy)}{K}\right]^2 = 1.$$

Also if one trial solution of the equation $x^2 - Ny^2 = 1$ is obtained, an infinite number of solutions can be found from the solution set:

$$(x, y) \rightarrow (x^2 + Ny^2, 2xy). \tag{1b}$$

Further, Brahmagupta was aware that an integral solution to $x^2 - Ny^2 = K$ if $K = \pm 1, \pm 2$ or ± 4 exists. This is known as the *Bhavana* principle. And the principle may be stated thus:

For an equation of the type $x^2 - Ny^2 = 1$, it is only necessary to obtain one solution set to start with before infinitely many solution sets are found. However, in the Brahmagupta approach there is need to start with a trial solution (x, y). In Bhaskaracharya's *Chakravala's* method there is even no need for a trial solution as we will find in Chapter 8.

An illustrative example from *Brahmasphutasiddhanta:*

$$x^2 - 92y^2 = 1.$$

Start with $(x, y) = (10, 1)$ which gives

$$10^2 - 92.1^2 = 8. \tag{2}$$

Applying the *Bhavana* principle to $(x, y) \rightarrow (x^2 + Ny^2, 2xy)$, we get

$$(10, 1) \rightarrow (10^2 + 92 \times 1^2, 2 \times 10 \times 1).$$

Therefore we get,

$$192^2 - [92 \times 20^2] = 64. \tag{3}$$

Dividing throughout by 64 to get the solution set $(x, y) = [\mathbf{24}, \mathbf{2.5}]$

$$24^2 - 92 \times \left(\frac{5}{2}\right)^2 = 1. \tag{4}$$

Applying the Bhavana principle again gives

$$\left[24, \left(\frac{5}{2}\right)\right] \rightarrow \left[24^2 + 92\left(\frac{5}{2}\right)^2, 2 \times 24 \times \left(\frac{5}{2}\right)\right].$$

So that,

$$1151^2 - 92 \times 120^2 = 1 \quad \text{or a solution set } (x, y) = (\mathbf{1151}, \mathbf{120}). \quad (5)$$

Further solution sets can be obtained by applying the *bhavana* (composition) principle repeatedly.

It was Brahmagupta who first gave the method of solving this form of indeterminate equation and he did so without the use of continuous fractions. About a 1,000 years later, the French mathematician, Pierre Fermat (1601–1655) issued a challenge to British mathematicians of his time to find the integral solution of the equation $Nx^2 + 1 = y^2$ where N is a positive integer and not a perfect square. Such an equation is incorrectly called even today the Pell equation after John Pell (1610–1685) when Leonhard Euler (1707–1783) referred to that equation by that name without any historical justification. It is more fitting that this '*vargaprakriti*' equation be renamed as Brahmagupta–Bhaskara II equation after the two Indian mathematicians who contributed to its solution.[25]

7.3 An Early Commentator on the Works of Brahmagupta: Prthudakaswamin

Known as *Caturveda* or 'one who knows the four Vedas', Prthudakas-vami wrote influential commentaries on Brahmagupta's *Brahmas-phutasiddhanta* and *Khandakhadyaka*. The date of publication of the former has been estimated as 864 CE. It contains good exemplars of Brahmagupta's rules. We will select those involving simple, quadratic and simultaneous equations to illustrate his versatility, and particularly his contribution to the emergence of Indian algebra. The reader is invited to solve these problems.

Example 1: Nine traders shared the cost of purchasing a horse with each contributing one share increasing by one up to nine shares. The horse was then sold for five less than 500. What are the shares of each trader?

(*Answer*: The share of the sale proceeds was: 11, 22, 33, 44, 55, 66, 77, 88 and 99).

Example 2: Take the square root of what remains of the revolution around the sun less 2, further reduced by 1, multiplied by 10 and then increased by 2. When will this be equal to what remains of the revolution of the sun less 1?

Hint: Let the remains of the revolution of the sun be $(x^2 + 2)$. Solve for x the following:

$$10[\sqrt{(x^2 + 2) - 2} - 1] + 2 = (x^2 + 2) - 1$$

i.e.

$$10(x - 1) + 2 = x^2 + 1$$

or

$$x^2 - 10x + 9 = 0.$$

(*Answer*: $x = 9$ or 1 of which **9** was chosen).

Example 3: Find the smallest integer that leaves remainders 5, 4, 3 and 2 when divided by 6, 5, 4, 3 respectively

Answer: Smallest number which satisfies the above conditions is $N = 59$. This is a known as a 'remainder' problem involving simultaneous indeterminate equations. An explanation of the solution is found in Puttaswamy (2012, pp. 220–221).

7.4 Sridhara

There remains some controversy over Sridhara's time and place of birth; some scholars suggest he came from Bengal, and others from southern India. What is definitely known is that he wrote the partly extant *Patiganita* and the highly influential *Trisatika*.[26] The latter proved to be one of the most popular textbooks on arithmetic before *Lilavati* of Bhaskaracharya about three centuries later. In it he deals with elementary operations, including extracting square and cube roots, and fractions. Eight rules are given for operations involving zero (but not division). His methods of summation of different arithmetic and geometric series were to become standard references in later works. He is one of the only two known mathematicians of that period (the other being Mahavira) who concentrated on mathematics only and excluded astronomy.

7.4.1 Solution of quadratic equations

There has been a tendency to neglect the contributions of Sridhara in the histories of Indian mathematics. Sridhara was one of the first mathematicians to give a general rule for the solution of quadratic equations, However, the original contained in his *Bijayganita* is no longer extant and we have to rely on a quotation of the rule from Bhaskaracharya.[27] The rule states:

> Multiply both sides of the equation by a four times the coefficient of the square of the unknown; add to both sides the square of the coefficient of the unknown. Then take the square root.

Expressed in modern notation:
Given

$$ax^2 + bx = c.$$

Multiply by $4a$ to get

$$4a^2x^2 + 4abx = 4ac.$$

Add b^2 to both sides to get

$$4a^2x^2 + 4abx + b^2 = 4ac + b^2.$$

So we have

$$(2ax + b)^2 = 4ac + b^2.$$

So

$$2ax + b = \sqrt{4ac + b^2}.$$

Therefore

$$x = \frac{\sqrt{4ac + b^2}}{2a} - \frac{b}{2a}.$$

Sridhara gives an illustration in *Trisatika* of this quadratic formula to find the number of terms of an arithmetic progression [*Trisatika*, R. 14].

Let **S** be the sum of the first n terms of the arithmetic progression: $a + (a + b) + (a + 2b) + \cdots$.

Then

$$S = \frac{n}{2}[2a + (n - 1)b]$$

or

$$2S = bn^2 + n(2a - b).$$

Applying Sridhara's rule given above we get:

$$n = \frac{\sqrt{4b(2S)ac + (2a - b)^2}}{2b} - \frac{(2a - b)}{2b},$$

or

$$n = \frac{\sqrt{8bS + (2a - b)^2}}{2b} - \frac{(2a - b)}{2b}.$$

To illustrate with an example adapted from *Trisatika:* the first term $a = 20$, Sum $S = 245$, difference $b = 5$, find the number of terms n:

$$n = \frac{\sqrt{8bS + (2a - b)^2}}{2b} - \frac{(2a - b)}{2b} = \frac{\sqrt{9800 + 1225}}{10} - \frac{35}{10} = 7.$$

In the works of Sridhara, Mahavira, Bhaskara II and others are found a number of fascinating problems, clearly devised to stimulate the interest of the reader. Let us consider a couple the first from *Trisatika* and the next from Mahavira's *Ganitasarasangraha*. The first contains a favourite theme of a thrilling amorous moment enriched by vivid poetic imagery, taken up later by Mahavira who adds further fantasy and more complicated mathematics to an essentially straightforward problem of Sridhara involving a linear equation[28]:

> One-third of the pearls of a necklace broken during a 'love quarrel' got scattered on the floor. One-fifth fell on couch, one-sixth was collected by a servant maid nearby and one-tenth was taken by the lover. Six pearls remained on the necklace. How many pearls were there on the necklace?

Solution: Let x be the number of pearls in the necklace. Then form the equation:

$$\frac{1}{3}x + \frac{1}{5}x + \frac{1}{6}x + \frac{1}{10} + 6 = x \quad \text{and} \quad \text{solve for } x \text{ to get } x = \mathbf{30}.$$

Mahavira's Version

> One night in spring, a certain young lady was lovingly happy with her husband on the floor of a big mansion, white like the moon, situated in a pleasure garden full of trees heavy with flowers and fruits. The whole place was resonant with the sweet sounds of parrots, cuckoos and bees intoxicated with the honey from the flowers in the garden. In the course of a "love quarrel" between the couple, the lady's necklace came undone and the pearls got scattered all around. One-third of the pearls reached the maid-servant who was sitting nearby; one-sixth fell on the mattress; one half of what remained (and one half of what remained thereafter and again one half of what remained thereafter and so on, counting six times in all) were scattered everywhere. On the broken necklace, there remained 1161 pearls. Oh my love, tell me quickly the total number of pearls on the necklace.

The solution is tedious (but not difficult today) and takes us into the realms of fantasy. We will not attempt it here except to point out that if x is the total number of pearls on the necklace. Then the answer is obtained from the solution of the following equation:

$$x = \frac{1}{3}x + \frac{1}{6}x + \left(\frac{1}{2}\right)^2 x + \left(\frac{1}{2}\right)^3 x + \left(\frac{1}{2}\right)^4 x$$
$$+ \left(\frac{1}{2}\right)^5 x + \left(\frac{1}{2}\right)^6 x + \left(\frac{1}{2}\right)^7 x + 1161.$$

With the answer of **1,48,608** pearls, this is truly a fantasy necklace to counteract the stern and cold logic of mathematics. It reflects a fascination with large numbers alluded to previously.

Yet another example found in both *Trisatki* and *Ganitasarasan-graha* requires the solution of a quadratic equation:

> One-twelfth of a pillar, multiplied by one-thirty part thereof was found submerged under the water. One-twentieth of the rest of the pillar, multiplied by three-sixteenth was buried in the mud below. Twenty *hasta* of the pillar were found above the water. What is the length of the pillar which is an integer? (1 *hasta* = 18 inches).

Solution in Modern Notation: Let the length of the pillar in *hastas* be in x.

Then

$$\left(x - \frac{x}{12} \cdot \frac{x}{30}\right) - \frac{1}{20}\left(x - \frac{x}{12} \cdot \frac{x}{30}\right)\left(x - \frac{x}{12} \cdot \frac{x}{30}\right)\frac{3}{16} = 20,$$

or

$$\left(x - \frac{x^2}{360}\right) - \frac{3}{320}\left(x - \frac{x^2}{360}\right)^2 = 20.$$

By assuming that $y = x - \frac{x^2}{360}$, we obtain the equation $3y^2 - 320y + 6400 = 0$ giving us $y = 80$ or $80/3$. Taking the integer value $y = 80$ and resubstituting to get the equation in x as $x^2 - 360x + 28800 = 0$. There are two integral solutions: $x = $ **240** or **120**. It is amazing that without symbolic algebra, Indian mathematicians were able to solve problems of such complexity.

7.4.2   Volume of a sphere

We discussed earlier the Brahmagupta's formula for the volume of the frustum of a square pyramid. The method may be used to find the volume of any regular frustum. Sridhara extended this work to find a rule for the volume of the frustum of a cone. In *Trisatki*, he writes:

> The square root of ten times the square of the sum of the squares of the diameter at the top, the diameter at the bottom and the

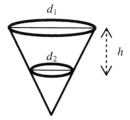

Figure 7.15. Volume of a cone.

and the sum of these diameters, when multiplied by the depth and divided by 24 gives the volume of a circular well.

In Figure 7.15, let the diameters of the bottom and top of the frustum be d_1 and d_2 respectively and h be the depth. Then, according to above rule, the volume of a cone may be obtained from:

$$V = \frac{h}{24} \sqrt{10 \left\{ (d_1^2 + d_2^2) + (d_1 + d_2)^2 \right\}^2},$$

follows from Sridhara's approximate rule for finding the volume of a heap with circular base calculated taking $\pi = 3$. For further details of the derivation of this rule, see Puttaswamy (2012, pp. 215–217).[29]

7.5 Mahavira

Mahavira was the best-known Indian mathematician of the ninth century. A Jain by religion, he was familiar with Jain mathematics, which may have inspired his owns book, the *Ganitasarasangraha*. It is highly likely that Mahavira knew the works of Aryabhata and Brahmagupta. But unlike his predecessors, Mahavira was not an astronomer — his work was confined only to mathematics. He was a member of the court of the ninth century king Amoghavarsa of the Rashtrakuta dynasty, judging from high praise that he heaped on the king in his book. It was a time of political expansion, architectural achievements and memorable literary creations. The dynasty provided great encouragement to mathematical and scientific endeavours, attracting Jain mathematicians and other scholars to settle. Some of the architectural wonders of the period are to be found in two of the popular tourist spots and World Heritage sites in India: the *Kailasanath Temple* at *Ellora* and the sculptures of Elephanta Caves.

7.5.1   The *Ganitasarasangraha*

The *Ganitasarasangraha* ('A Summary of the Essence of Mathematics') is the first work in Sanskrit to survive in its entirety. The book was widely used in southern India and translated into Telugu, a regional language, during the eleventh century. It is 1100 verses long and gives a good idea of mathematical knowledge of that time, although the arrangement of the topics is unusual compared to the classifications adopted by earlier writers. It contains a lucid classification of arithmetical operations with a number of examples to illustrate the rules.[30] His treatment do not correspond to the eight procedures laid out in the *Brahmasphutasiddhanta*, although it was supposedly a commentary on Brahmagupta's work. It includes:

1. A detailed examination of operations with fractions, with some ingenious methods for decomposing integers and fractions into unit fractions (a subject of practical utility for the ancient Egyptians);
2. A restatement of general rules of operations with zero and positive and negative quantities;
3. An extension and systematisation of the Jain work on permutations and combinations, for both of which he provides the well-known general formulae illustrated with examples involving combination of flavours, of a necklace made out of different precious stones and a garland containing different flowers[31];
4. Solutions of different types of quadratic equations, as well as an extension of his predecessors' work on indeterminate equations; and
5. Geometric work on right-angled triangles whose sides are rational and, something unusual in Indian mathematics, attempts (albeit unsuccessful) to derive the formulae for the area and perimeter of an ellipse.

Mahavira's contribution can be looked at in two ways. *Ganitasarasangraha* [GSS] may be seen as the culmination of Jain work on mathematics (indeed it is the only substantial treatise on Jaina mathematics that we have). Alternatively, Mahavira can be seen as summarising and extending the mathematical content of the works of his predecessors such as Aryabhata, Bhaskara I and Brahmagupta which we discussed in the last and present chapters. He was very conscious of the debt he owed those who came before him. In the introductory chapter of his book, he wrote:

> With the help of the accomplished holy sages, who are worthy to be worshipped by the lords of the world ... I glean from the great ocean of the knowledge of numbers a little of its essence, in the manner in which gems are [picked] from the sea, gold from the stony rock and the pearl from the oyster shell; and I give out according to the power of my intelligence, the *Sarasangraha*, a small work on arithmetic, which is [however] not small in importance.

7.5.2   Contents of *Ganitasarasanghraha*

The book begins with an invocation of his name-sake, the founder of Jainism to whom he ascribes 'the shining lamp of knowledge of numbers by whom the whole universe is made comprehensible'. This is followed by verses praising his patron Amoghavarsa and then one of most famous eulogy on mathematics (*ganita*), already quoted in the beginning of Chapter 2.

In the next section [**GSS Chapter 1**], Mahavira names the eight fundamental arithmetic operations which does not include addition and multiplication. Instead, after listing the first six operations from multiplication to extraction of cube roots, he concludes with addition and subtracting quantities in progressions and series.[32] His section on operations with zero are similar to the ones discussed by Brahmagupta, except the statement that division by zero results in zero (wrong) and negative numbers can have no square root (correct).[33] In the subsequent chapters he discusses rules relating to fractions, solution to equations in one unknown, rules of three quantities, mixtures (problems on interest and investment etc.), geometry, excavations and shadows.

7.5.3   Unit fractions [GSS, Chapter 2, Verses 75–80]

Mahavira was the first Indian mathematician to express a fraction as the sum of several fractions which have unity in the numerators (i.e. unit fractions) His methods were as follows:

7.5.3.1   *Expressing 1 as the sum of n unit fractions*

Let the number of unit fractions be n. Excluding the first ($\frac{1}{2}$) and last fractions, let the rest be in a geometric progression (GP) whose first term is $a = \frac{1}{3}$ and the common ratio is $r = \frac{1}{3}$. Then the sum of the GP with $(n-2)$ terms is:

$$1 = \frac{1}{2} + \frac{1}{3} + \frac{1}{3^2} + \frac{1}{3^3} + \cdots + \frac{1}{3^{n-2}} + \frac{1}{2 \times 3^{n-2}}.$$

If $n = 5$, we get

$$1 = \frac{1}{2} + \frac{1}{3} + \frac{1}{3^2} + \frac{1}{3^3} + \frac{1}{2 \times 3^3}.$$

7.5.3.2   *Expressing any fraction as the sum of unit fractions*

Let p/q be a given fraction such that $p < q$. Find c such that $(q+c)/p$ is an integer $= r$
    Then

$$\frac{p}{q} = \frac{rp}{rq} = \frac{q+c}{rq} = \frac{1}{r} + \frac{c}{rq}.$$

Repeat the process for $\frac{c}{rq}$ and so on. If $c < p$, the process ends after a finite number of steps.

Consider the following example from Egyptian mathematics (Joseph, 2011, p. 92) that involves the decomposition of $13/89$:

$$\frac{13}{89} = \frac{(1+4+8)}{89} = \frac{1}{89} + \left(\frac{1}{30} + \frac{1}{178} + \frac{1}{267} + \frac{1}{445} \right)$$

$$+ \left(\frac{1}{15} + \frac{1}{89} + \frac{2}{267} + \frac{2}{445} \right)$$

$$= \frac{1}{15} + \frac{1}{30} + \frac{1}{60} + \frac{1}{178} + \frac{1}{267} + \frac{1}{356} + \frac{1}{445}$$
$$+ \frac{1}{534} + \frac{1}{890} + \frac{2}{267} + \frac{2}{445}.$$

And then a further decomposition of the last two terms of the above expansion. This is extremely tedious, particularly where arithmetic operations of multiplication and division are dependent on the decomposition into unit fractions as in Egyptian mathematics.[34]

7.5.3.3 *The 'mother of all fractions'*

The problems here involve the combinations of additions, subtractions, increasing or decreasing an integer or fraction by a fraction of itself. The example given (GSS, Chapter 2, Verses 139–140):

> Sum a third, a fourth, a half of a half, a sixth of a fifth, one divided by three fourths, one divided by five halves, one and one sixth, one and one fifth, a half and its own third, two sevenths plus its own sixth, one min a ninth, one minus a tenth, an eighth minus its own ninth, a fourth minus its own fifth. Add them as if they constitute a lotus garland of flowers and give the answer.

The answer is computed as **121/15**.

7.5.4 A medley of problems [GSS, Chapter 4]

As the title suggests there is a collections of problems and methods for solving equations in one unknown. Mahavira divides them into 10 categories depending on the nature of the given quantities, whether fractions or combination of integer and fractions, etc. A few examples will be given below to illustrate his methods.

Example 1. One-third of a herd of elephants and three times the square root of the remaining part of the herd were seen on the mountain slopes. In a lake, a male elephant was seen along with three female elephants. How many elephants were there in the herd?

Modern Solution: Let x be the number of elephants in the herd. Then we have:

$$\frac{x}{3} + 3\sqrt{\frac{2x}{3}} + 4 = x \Rightarrow \frac{2x}{3} - 3\sqrt{\frac{2x}{3}} - 4 = 0. \tag{1}$$

Let $y = \sqrt{\frac{2x}{3}}$, so (1) becomes $y^2 - 3y - 4$. Then

$$y = \frac{3 \pm \sqrt{9 + 16}}{2} = \frac{3 \pm 5}{2} \Rightarrow y = 4 \quad \text{or} \quad y = -1.$$

Since $y > 0$, discard the value of $y = -1$, and take $y = 4$.

If $y = 4$, squaring we get $\frac{2x}{3} = y^2 = 16$. Hence

$$2x = 48 \Rightarrow x = \mathbf{24} \text{ elephants.}$$

Example 2. [**GSS Chapter 4, Verse 41**]: Out of a certain number of Sarasa birds [cranes], one-fourth the number are moving about among the lotus plants; one-ninth together with one-fourth as well as 7 times the square root of the total number of birds are found on a hill nearby; 56 birds remain on the Vakula (popularly known as a cherry tree). What is the total number of birds?

Solution: In modern notation the solution follows:

If x is the total number of birds, this gives the equation

$$x = \frac{x}{4} + \frac{x}{9} + \frac{x}{4} + 7\sqrt{x} + 56 \text{ birds}$$

or

$$\left(x - \frac{x}{4} - \frac{x}{9} - \frac{x}{4} - 7\sqrt{x} \right) = 56 \Rightarrow \frac{x}{18} + \sqrt{x} = 56$$

which solved for x gives **576** birds.

Example 3. A Problem in two unknowns: A wizard possessing magical powers observes a cock-fight and speaks to the two owners of the fighting cocks. To one he declares that he is prepared to give the owner two-thirds of the stake money if his cock loses, but expects the owner to give him (the wizard) all the stake money. To the other owner the wizard promises a similar arrangement except that he will pay three-quarters of the stake-money to the owner if his cock loses. Tell me O mathematician, the stake money of each of the owners if from either of them he earns a profit of 12 gold coins.

Solution: Let the owners be labelled as A and B and x and y be their respective stakes. If A wins and B loses, the wizard gets x gold coins from A and gives $(3/4)y$ to B. If B wins and A loses, the wizard get y coins from B and gives A $(2/3)x$. The profit earned by the wizard in each case is 12 gold coins.

So we get two equations:

$$x - \left(\frac{3}{4}\right) y = 12,$$

$$y - \left(\frac{2}{3}\right) x = 12.$$

Solve the two simultaneous equations for $x = $ **42** and $y = $ **40**, the stakes by A and B respectively. Mahavira also gave solutions to two other types of simultaneous equations involving quadratic Forms. For further details, see Puttaswamy (2012).

7.5.5 'Procedure of figures'

In the chapter on geometry, Mahavira makes a distinction similar to Brahmagupta between approximate and accurate results. Approximate rules are first given for computations involving figures such as triangular, quadrilateral and circular, followed by accurate ones. Hardly any difference is found between Brahmagupta and Mahavira in their treatment of many subjects and we will not repeat them. Instead, we concentrate on those topics in which Mahavira made a mark, notably in the area of non-regular "circular" figures in geometry. These included mensuration of a 'long-circle' (i.e. an ellipse) and an 'outer [or] wheel-circle'.

It is interesting that Mahavira was the only known mathematician from ancient and medieval India who discussed the mensuration of an ellipse. Mahavira's initial classification of regular closed curves into equal and elongated circled may be traced back to early Jain cosmology which we mentioned in a previous chapter.

7.5.5.1   The 'long circle': circumference of an ellipse major and minor D_2 and D_1

In GSS, Chapter 7, Mahavira states the rules for calculating the circumference and area of an ellipse:

> The *ayama* [i.e. half the longer diameter D_2 in Figure 7.16(a)] combined with *vyasa* [i.e. shorter diameter D_1 in Figure 7.16(a)] and doubled gives the circumference. One fourth the shorter diameter multiplied by the circumference gives the area [GSS, Chapter 7, 21]

$$\text{Circumference of Ellipse } (C) = 2\left(D_1 + \frac{1}{2}D_2\right); \quad \text{Area} = C \cdot \frac{D_1}{4}.$$

It would seem likely that Mahavira obtained his incorrect formulae from an arbitrary extension of the rules for the circle. An 'approximate' value for the circumference of a circle is $\pi d = 3d$. Of the three units of diameter, two units are arbitrarily allocated to the major axis and one to the minor axis. The values given as exact for the circumference and area of the ellipse are calculated by taking $\pi = \sqrt{10}$ and distributing 10 between $(D_1)^2$ and $(D_2)^2$.

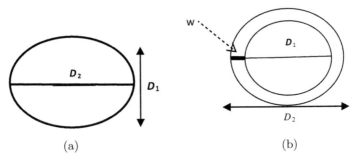

Figure 7.16. (a) An ellipse, (b) An annulus.

Thus Mahavira's exact value of the circumference of an ellipse is $\sqrt{6D_1^2 + 4D_2^2}$ and the area of the ellipse is $C \cdot \frac{D_1}{4}$. Although these formulae are clearly wrong, it can be shown that the circumference of an ellipse reduces to $\pi D_2 \sqrt{1 - \frac{3}{5}e^2}$ where e is the 'eccentricity' such that $e^2 = \frac{(D_1 - D_2)^2}{D_2}$ This is a remarkable approximation to the true circumference.

7.5.5.2 The 'wheel circle': an annulus

Figure 7.16(b) represents an annulus (a ring-shaped object containing a region bounded by two concentric circles). Let D_1 be the diameter of the inner circle, D_2 be the diameter of outer circle and w be the width of the annulus. Then $D_2 = D_1 + 2w$.

Now, the Area of the annulus (A) = Area of the outer circle − Area of the inner circle.

or,

$$A = \frac{\pi}{4}(D_2^2 - D_1^2) = \frac{\pi}{4}(D_2 + D_1)(D_2 - D_1)$$

$$= \frac{\pi}{4}(D_1 + D_1 + 2w)(2w) = \pi w(D_1 + w).$$

Mahavira gave both an 'approximate' and an 'accurate' value of the area of annulus in terms of the circumferences (C_1 and C_2) and the diameters of the two circles:

$$A = \frac{C_1 + C_2}{2} \cdot \frac{D_2 - D_1}{2} \qquad \text{('approximate' where } \pi = 3),$$

$$A = \frac{C_1 + C_2}{6} \cdot \frac{D_2 - D_1}{2} \cdot \sqrt{10} \quad \text{('accurate' where } \pi = \sqrt{10}).$$

7.5.5.3 Areas bounded by circles

In one of the largest sections on geometry in *Ganitasarasangraha* Mahavira discusses areas in a number of shapes. Here we have two examples of areas of regions in circles.

(i) [**GSS, Chapter 7, 82**] In Figure 7.17(a), three equal circles of diameter d touch each other. The area marked as **A** is the region

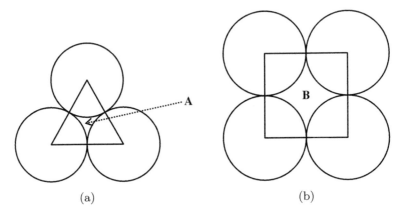

Figure 7.17. Regions bounded by circles.

bounded by the arc of these circles. Then

A = Area of the equilateral triangle $-\dfrac{1}{2}$ (Area of one of the circles).

Note that the area of an equilateral triangle with side

$$d = \frac{1}{2}d \times \sqrt{d^2 - \left(\frac{1}{2}d\right)^2} = \frac{\sqrt{3}}{4}d^2.$$

Therefore, the region bounded by the arc of the circles (**A**)

$$= \frac{\sqrt{3}}{4}d^2 - \frac{1}{2}\pi\left[\frac{d^2}{4}\right] = \frac{d^2}{4}\left(\sqrt{3} - \frac{1}{2}\pi\right).$$

[**GSS, Chapter 7, 82**] In Figure 7.14(b), four equal circles of diameter d touch one another. A square is formed where the centres of each circle provided its vertices. Then the area of the region bounded by the arcs o the four circles, marked as **B**, is given by:

B = Area of the square − Area of one of the circles = $d^2 - \dfrac{\pi(d^2)}{4}$.

The chapter on 'procedure of figures' (geometry) includes not only rules for computing other figures, but also operations with *karanis* (surds) and series computations. The penultimate chapter is on the related topic of 'excavations' and the final chapter is on 'shadows'. There is little that is new from what we found in the

Aryabhatiya and *Brahmasphutasiddhanta's ganita* sections. However, Mahavira introduced certain equations of higher degree in dealing with problems involving geometric series. An example would suffice.

Example 1: The first term of a geometric progression is 5, the number of items in the series is 4 and the sum is 200. Find the common ratio

Applying the rule embodied in the formula: $a \cdot \frac{x^n - 1}{x - 1} = s$ where a is the first term, $\mathbf{n}$ = number of terms and s = sum, the common ratio $\mathbf{x}$ is found as follows:

$$5 \left(\frac{x^4 - 1}{x - 1} \right) = 200 \Rightarrow \left(\frac{x^4 - 1}{x - 1} \right) = 40.$$

Four successive divisions taking the divisors $x = 3$ would exhaust 40.

7.6   Conclusion

In this brief survey, Mahavira's contribution has been highlighted. It is noticeable that over a period of less than 400 years there were a few seminal texts which worked on a number of similar problems sharing results and methods although diverging in certain important respects when it came to the fundamental organisation and presentation of mathematics. This is a striking difference from Greek mathematics occurring over a similar time scale which had a uniform methodology and proof.

A comparison between the work of Mahavira and Sridhara is a useful indicator of the nature and scope of transmission of ideas between two mathematicians from about the same time. The range of topics discussed in the *Ganitasarasangraha* and the *Trisatika* are more or less similar and the arrangement of topics almost parallel each other. They chose similar illustrative examples and there are certain topics which are only discussed in the two texts.

It would appear, despite the past controversies, that Mahavira not only knew the writings of Sridhara but was also influenced by them. References were made by later mathematicians notably Sripati (*c.* 1050 CE) and Bhaskara II who gave the same example to illustrate the 'inverse rule of 3'.[35] Neither of the earlier mathematicians, Bhaskara I or Brahmagupta, mentioned the example. Another rule

which was first introduced by Sridhara and Mahavira, but not found
in the works of Aryabhata I, Brahmagupta and Bhaskara, relates to
finding the area of a segment of a circle. If a denotes the chord of a
circle and h is the height of the segment, the following 'approximate'
rules were stated for calculating the area of the segment (A).[36]

$$1.\ \text{A} = \sqrt{\frac{10}{9}}h\left(\frac{a+h}{2}\right) \quad \text{Sridhara,}$$

$$2.\ \text{A} = h\left(\frac{a+h}{2}\right) \quad \text{Mahavira.}$$

The equivalence of the two rules is easily established if $\sqrt{\frac{10}{9}}$ is taken
to equal 1.

There is another similarity in the manner in which Sridhara and
Mahavira treat fractions. Both place great emphasis on this topic.
In the *Ganitasarasangraha*, out of nine chapters two are given to
the subject of fractions, both the eight operations with fractions in
one and followed by miscellaneous examples involving fractions. In
Trisatika, 21 examples out of 107 are concerned with fractions. Such
preferential treatment of fractions is found in only these two books.
However, the importance placed on unit fractions in Mahavira's work
is not found in Sridhara's.

There is yet another example which was alluded to earlier. It
relates to a pearl necklace, Sridhara's illustration is a straightforward one where with the broken necklace and the scatter of the
pearls expressed as fractions, the answer sought is the total number of
pearls, which gives a realistic number of pearls. Mahavira's example
is stated in a manner that is more romantic and takes considerable
poetic licence and leads to an answer in which the number of pearls
is only found in a fantasy necklace. The same problem also appears
in the *Lilavati* of Bhaskara II.

There are some notable differences between the two texts. There
is the difference in the arrangement of fundamental arithmetic operations. Mahavira strangely only discusses the operations of additions
and subtractions after the other operations. Certain subjects such

as treatment of geometric series, the ellipse, permutations and combinations and indeterminate equations are not found in Sridhara's *Trisatika* ('Three hundred Verses'). But this could be explained by the abbreviated nature of this text. In any case these differences are not very significant compared to their great similarities. Sridhara has remained a shadowy figures in recent histories of Indian mathematics, although in earlier times Sridhara enjoyed a high reputation almost paralleling that of Bhaskaracharya, being held up as the ideal of an accomplished mathematician at that time. It should be remembered that while Mahavira described his text as a collection (*sangraha*) of presumably works of previous writers, Sridhara had made some original contributions which now tend to be forgotten.

Mahavira's *Ganitasarasangraha* is the first comprehensive text on mathematics only. It deals with a wide range of topics from arithmetic operations, including the rules of signs and zero operations, pulveriser and other subjects treated today as algebra. The strength of his book is not only its clarity in both the organisation and presentation of mathematics, but its accessibility to a wider audience. In his treatment, he was true to the expression in his preface of the versatility of '*ganita*' (computation). After Mahavira, there were two scholars of significance, both notable for their work in astronomy rather than mathematics. Aryabhata II is believed to have lived after Sridhara and before Bhaskara II which meant between 950–1100 CE. He is known for his major work *Mahasiddhanta*. His contribution to mathematics was quite modest, mainly on geometry, on the areas of a rhombus, triangle, square, rectangle and isosceles trapezium all of which had been investigated by mathematicians who had come earlier. Sripati (*fl.* 1025 CE) wrote *Siddhantasekhara* and *Ganitatilaka* ('A Mark on the Head of Calculation') was also primarily an astronomer and astrologer although, as noted earlier in this chapter, he devised an alternative method of solution from that of Sridhara to the quadratic equation of the form $ax^2 + bx + c = 0$. Both these scholars were precursors to the emergence of the dominant mathematician of the late classical period, Bhaskaracharya, whom we turn to in the next chapter.

Endnotes

[1] It was through Brahmagupta's *Brahmasphutasiddhantha* that the Islamic world became familiar with Indian mathematics and astronomy.

[2] There is a likelihood that Brahmagupta saw himself as a defender of orthodoxy since he was a severe critic of both the Jain cosmological view and of Aryabhata I's view of the earth as a spinning sphere, a view that gained widespread acceptance on account of his contemporary, Bhaskara I's espousal of that viewpoint.

[3] There is a view that these chapters were a corrected and modified version of an earlier text called *Brahma Siddhanta* written by a Sakalya. Little is known of the author and the antecedents of this text.

[4] The others worthy of note were Sridhara (*fl.* 750), Lalla (*fl.* 750) and Bhattotpala (*fl.* 950). We will refer to the work of Sridhara later in this chapter.

[5] This involves solution of equations of the first order $(ax + by = c)$ and of the second order $(Nx^2 + 1 = y^2)$.

[6] *Latadeva* (ca. 500 CE) helped to revise the astronomical text, *Suryasiddhanta*.

[7] With the help of Prthudaksvamin, his notable commentator, it is possible to list the 20 operations mentioned by Brahmagupta: addition, subtraction, multiplication, division, squaring and square root, cubing and cube root and five methods of operating with fractions. The list is completed with 'procedures' which consist of 'mixtures', series, 'figures' (geometry), 'excavations' (volume), 'piles', 'heaps' and a procedure involving 'shadows'. Some of these operations have already been discussed in earlier chapters and will therefore be ignored here.

[8] Given the fractions m/n and p/q and the integer r, the following results apply:

$$\frac{m}{n} \pm \frac{p}{q} = \frac{mq \mp pn}{nq}; \quad r + \frac{m}{n} = \frac{rn + m}{n}; \quad \frac{m}{n} \cdot \frac{p}{q} = \frac{mp}{nq};$$

$$\frac{m}{n} \div \frac{p}{q} = \frac{mq}{np}; \quad \left(\frac{m}{n}\right)^2 = \frac{m^2}{n^2}; \quad \sqrt{\frac{m}{n}} = \frac{\sqrt{m}}{\sqrt{n}}.$$

[9] The cultivation of 'shortcuts' in Indian mathematics has its resonance in the recent popularity of the so-called 'Vedic Mathematics' of Swami Tirthaji which contains ingenious methods of shortcut calculations.

[10] Euclidean division is the conventional process of division of two integers producing a quotient and a remainder. This result is usually attributed to Nichomachus of Gerasa who lived around 100 CE.

[11] A topic commonly discussed by early Indian mathematicians was *samkrama* (meaning concurrence). From Brahmagupta onwards, the

topic was included in subsequent texts in what we would now describe as algebra, while in others it was part of arithmetic. It usually involved seeking solutions to problems of two unknown quantities which were changing together in their sum and difference. This included simultaneous equations of the form $x + y = a$ and $x - y = b$. Where it involved cases of simultaneous quadratic equations of the forms:

$$x^2 - y^2 = p, \quad x - y = q,$$
$$x^2 - y^2 = p, \quad x + y = q$$

a special name *vismakrama* was used. An indeterminate equation of the first order involving two unknowns quantities x and y in the form $x + y = c$ was, as mentioned earlier, known as *Kuttakara* (*or Kuttaka*).

[12] Brahmagupta gave two rules for the solution of the quadratic equation $ax^2 + bx - c = 0$, expressed in modern notation:

$$x = \frac{\sqrt{b^2 + 4ac}}{2a} - \frac{b}{2a}; \quad x = \frac{\sqrt{ac + (b/2)^2} - (b/2)}{a}.$$

Brahmagupta applies the second rule in computing interest. For a discussion of Brahmagupta's additions to the subject of 'Mixtures', see Datta and Singh (1938/62, Vol. 1 (pp. 219–233) and Vol. 2).

[13] In [**18.41–18.42**] Brahmagupta, in discussing the solution of equations, makes the important point that 'like' power indices (such as $x^2, x^3, \ldots$) can be added or subtracted while unlike terms should be treated separately. An analogy is drawn between the product of 'like' powers and 'unlike' powers and different types of 'colour-products'.

[14] One of the clearest definition of the term *karani* is given by Sripati (1039 CE) as: "A number whose square root cannot be obtained exactly" and is therefore an irrational number. It is expressed by the abbreviation *ka* so that ka 8 is square root of 8.

[15] At no point does Brahmagupta state that he is referring to a cyclic quadrilateral, i.e. a quadrilateral that is inscribed in a circle. This has raised the question whether Brahmagupta was aware of the fact that some of the properties that he referred to only applied to a cyclic quadrilateral. Also until recently there was little recognition of the fact that a 'gross' and 'accurate' formula was provided in a number of cases.

[16] An interesting question is why Brahmagupta gave both a 'gross' and 'accurate' formulae in a number of instances. In some cases, such as the one discussed here, the product of half-sums of opposite sides is grossly inaccurate, unless one is considering the trivial case of a square or a rectangle or an equilateral triangle.

[17] It is worth restating the point that this formula is 'accurate' for quadrilaterals only if they can be inscribed in a circle or are cyclic. Although Brahmagupta did not say so, it is likely that he was aware of this cyclic conditions when he was stating his rules for quadrilaterals.

[18] A *rational triangle* is a triangle such that all three sides measured relative to each other are rational. A rational number is any number that can be expressed as the quotient or fraction p/q of two integers, p and q.

[19] Brahmagupta offers various suggestions for the construction of a rational scalene triangle [**12.34**], a cyclic quadrilateral with rational sides and for a rational isosceles trapezium [**12.36**].

[20] For the derivation of this rule, see (Sarasvati Amma, 1979, pp. 136–138).

[21] An arc of a circle is like a bow and so is called a bow. A versine is the 'arrow' (in the bow) and known in Sanskrit as *sagitta* in Latin. For further explanation, see Chapter 11.2.

[22] A frustum is that part of a solid (normally a cone or pyramid) that lies between two parallel planes cutting it.

[23] It is remarkable in this context that almost 2,500 years the Egyptians may have found the correct formula for the volume of a truncated square pyramid as: $V = \frac{h(s_1^2 + s_1 s_2 + s_2^2)}{3}$. For further details, see Joseph (2011, pp. 114–116).

[24] However, Brahmagupta's explanation of the method is clearer and easier to comprehend than *Aryabhatiya*. He also introduced some refinements. For example, he pointed out if the divisors (a, b) are not co-prime, their common factor should be a factor of the difference between the remainders. Otherwise, the equation would have no solution.

[25] There is a problem of considerable historical interest for which Bhaskara II offers the first complete solution. The problem is to solve $61x^2 + 1 = y^2$ for minimum x and y. The solution is the incredible $x = 226,153,980$ and $y = 1,766,319,049$. This was the problem set by Fermat as a challenge in 1657 but solved by Lagrange a hundred years later.

[26] He was believed to be the author of a third book, *Bijaganita* which has not survived and only known by repute, having been mentioned by notable figures who came after him, including Bhaskara II.

[27] In his book on *Siddhantasiromani*, an astronomer Sripati (c. 1025 CE) gave another method of solving the same quadratic equation, $ax^2 + bx = c$. The solution given, expressed in modern notation:

$$x = \frac{\sqrt{ac + \left(\frac{b^2}{4}\right)} - \frac{b}{2}}{a}.$$

[28] There is a similar problem with a similar theme in Bhaskaracharya's *Lilavati*. The solutions are identical.

[29] A mathematician by the name of Virasena (*fl.* 750 CE) gave another method in his book *Dhavala* to find the frustum of a cone. The method involves removing a cylindrical core from the centre of the frustum. For further details, see Sarasvati Amma (1979, pp. 203–204).

[30] Mahavira is conscious of the aesthetic element in mathematics. For example, he points to 'palindromes' in numbers which he labels as 'garland numbers'. These are those numbers that remain the same from right to left or vice versa. Some examples:

$$139 \times 109 = 15151 \quad \text{and} \quad 27994681 \times 441 = 12345654321.$$

[31] The following problem in Chapter 6 of Mahavira's book is quite deep and throws an interesting light on the society of his time.

> Five men are enamoured of a courtesan, of whom only three she finds attractive. However, to each one of them separately, she says : 'You are my only beloved'. How many of her statements are true?

The solution offered by Mahavira: Multiply the total number of men (5) by the number of those found attractive plus one ($3 + 1$). Diminish this product by twice the number found attractive (2×3) and you will get the number of false statements. The square of the number of men (5^2) minus the number of false statements gives the number of true statements. Or

$$\text{Number of false statements} = (5 \times 4) - (2 \times 3) = 14;$$

$$\text{Number of true statements} = 25 - 14 = 11.$$

It is important to recognise here that the rule concerns the truth values of a set of n *explicit* statements and ($n^2 - n$) *implicit* statements. For further details, see Plofker (2007, 446–447).

[32] In *Aryabhatiya*, as we have seen in the last chapter, the rules for sums of an arithmetic progression can be applied to either an entire series starting with first term or to the final part of it ending with its last term. Mahavira distinguishes between these two operations the first identified as addition and the other as subtraction.

[33] The reason given is that the square of any number whether positive or negative is always positive. Only in 1797 did Caspar Wessel (1745–1818) incorporate the square root of a negative number into an extended system of complex numbers.

[34] Operating with unit fractions is a singular feature of Egyptian mathematics, and is absent from almost every other mathematical tradition. A substantial proportion of surviving ancient Egyptian calculations makes use of such operations. Of the 87 problems in the Ahmes Papyrus

(*c.* 1650 BCE), only six do not. Two reasons may be suggested for this great emphasis on fractions. In a society that did not use money, where transactions were carried out in kind, there was a need for accurate calculations with fractions, particularly in practical problems such as division of food, parcelling out land and mixing different ingredients for beer or bread. The dependence on unit fractions in arithmetical operations, together with the peculiar system of multiplication led to a third aspect of Egyptian computation. Every multiplication and division involving unit fractions would invariably lead to the problem of how to double unit fractions. Now, doubling a unit fraction with an even denominator is a simple matter of halving the denominator. Thus doubling 1/2, 1/4, 1/6 and 1/8 yields 1, 1/2, 1/3 and 1/4. But it was in doubling unit fractions with other odd denominators that difficulties arose. For some reason unknown to us, it was not permissible in Egyptian computation to write two times $1/n$ as $2/n$. Thus the need arose for some form of ready reckoner which would provide the appropriate unit fractions that summed to $2/n$, where $n = 5, 7, 9, \ldots$. At the beginning of the Ahmes Papyrus there is a table of decomposition of $2/n$ into unit fractions for all odd values of n from 3 to 101.

[35] An intriguing example of the use of the inverse rule of 3 is found in the *Brahmasphutasiddhanta*: "If a 16-year old girl, her voice as sweet as that of a cuckoo and *Saurus* crane and can dance like a peacock receives 600 coins, what will one of 25 years receive?" The solution is left to the reader!

[36] The treatment of this topic appears in Chinese mathematics in the premier text *Chiu Chang Suan Shu* ('Nine Chapters') almost nine centuries before Mahavira which has led to a suggestion of transmission of this rule from China to India.

Chapter 8

The 500 Year Climax: Bhaskaracharya and His Legacy

8.1 Introduction

Bhaskara II or Bhaskaracharya (Bhaskara the Teacher) as he is still popularly known in India, lived in Vijjadavida (i.e. modern Bijapur) in the Sahyadri region in Karnataka. His father, Mahesvera, came from a family of scholars enjoying a long tradition of royal patronage. Little else is known about him, except that his grandson Changadeva helped to set up a school for the study of his writings. His fame rests on *Sidhantasiromani* (Crown of treatises) a work in four parts: *Lilavati* (arithmetic), *Bijaganita* (algebra), *Goladhyaya* (a chapter on the celestial globe) and *Grahaganita* (*the* mathematics of planets).[1] The first two parts are essentially texts on mathematics and the last two relate to astronomy. It was a highly influential treatise written in 1150 when the author was 36 years old.

Lilavati (literally meaning 'beautiful' or 'playful'), based mainly on the works of Brahmagupta, Sridhara and Aryabhata II, shows a profound understanding of arithmetic. Bhaskaracharya's work included topics on fundamental operations, rules of three, five, seven, nine and eleven, work on permutations and combinations, and rules of operations with zero.[2] They speak of a maturity, a culmination of 500 years of mathematical development.

In 1587, on the instructions of the Mughal emperor Akbar, the court scholar Fyzi translated *Lilavati* into Persian. Fyzi tells a charming story of the book's origin. *Lilavati* was the name of Bhaskaracharya's daughter. From casting her horoscope, he discovered that the auspicious time for her wedding would be a particular hour on a certain day. He placed a cup with a small hole at the bottom in a vessel filled with water, arranged so that the cup would sink at the beginning of the propitious hour. When everything was ready and the cup was placed in the vessel, *Lilavati* suddenly out of curiosity bent over the vessel and a pearl from her hair fell into the cup and blocked the hole in it. The lucky hour passed without the cup sinking. Bhaskaracharya believed that one way to console his dejected daughter, who now would never get married, was to write her a manual of mathematics!

Bhaskaracharya's *Bijaganita* contains problems on determining unknown quantities, evaluating surds (i.e. square roots that cannot be reduced to whole numbers), solving simple and quadratic equations, and some general rules which went beyond Sridhara in dealing with the solution of indeterminate equations of the second degree and even equations of the third and fourth degree.[3] Bhaskaracharya's 'cyclic' (*Chakravala*) method for solving indeterminate equations of the form $ax^2 + bx + c = y$ was rediscovered in the West by William Brouncker in 1657. He also developed the subject of permutations and combinations which he named *Anka Pasha* (in contrast to the Jains who named the subject as *Vikalpa*). Apart from repeating the rule for obtaining the value of $_nP_r$ and $_nC_r$ which had been given by Mahavira, if not earlier, Bhaskaracharya stated that the number of permutations of r things split into $k_1, k_2, \ldots$ classes is, expressed in our notation, $\frac{r!}{k_1!k_2!\cdots}$. Bhaskaracharya was the first mathematician to record this result.

In *Siddhantasiromani*, Bhaskaracharya demonstrates his knowledge of trigonometry, including the sine table and relationships between different trigonometric functions. Certain preliminary concepts of the infinitesimal calculus and analysis may be traced to this work, concepts which would be taken up the Kerala School of mathematics in their work on infinite series some 200 years later.

He won such a great reputation in different parts of India that his manuscripts were still being copied and commented upon as late as the beginning of the 19th century. A medieval monastery, built in his honour, has a stone inscription that refers to him in the following terms:

> Triumphant is the illustrious Bhaskaracharya whose feats are revered by the wise and the learned. A poet endowed with fame and religious merit, he is like the crest on a peacock.

It was generally believed until recently that mathematical developments in India came to a virtual halt after Bhaskaracharya. This opinion has had to be revised in the light of recent research on what one could describe as medieval Indian mathematics. A number of the manuscripts of this period are yet to be published, or even subjected to critical scrutiny. We will begin with an examination of the mathematical contents of the two seminal texts of Bhaskaracharya, the *Lilavati* (an arithmetical manual) and *Bijaganita* (an algebra text).

8.2   The Contents of Lilavati

8.2.1   Arithmetical operations

The contents of *Lilavati* is mainly arithmetic although it contains some geometry, with a preponderance of problems on right-angled triangles, and a chapter on solving *kuttaka* (i.e. indeterminate equations). Starting with an invocation to the deity of Ganesa,[4] [**Verses 2–11**] the text continues, without any detailed explanations, metrology of weights and measures prevalent at that times, and names of decimal places up to 10^{14}. This is followed in [**Verses 12–47**] by a discussion of the eight arithmetic operations (addition, subtraction, multiplication, division, methods of finding squares, cubes, square roots and cube roots) and eight operations on fractions. They are all found in previous texts such as the *Brahmasphutasiddantha* and the *Ganitasarasangraha*. There is little that is new so far from previous work. This section ends with [**Verses 45–47**] listing the eight operations with zero: addition, subtraction, multiplication, division, square, square root, cube and cube root. These rules had become standard by then, and like Brahmagupta, suggests any quantity divided by zero remained 'zero-divided'. Bhaskaracharya returns to this topic in the first chapter of *Bijaganita*, when he infers that division by zero gives *khahara* (infinity), the first clear statement of this result in the history of mathematics.[5]

8.2.2   Method of inversion

[**Verses 48–55**] describes the method of inversion, i.e. the reverse process of getting an unknown quantity from known quantities. The method is explained thus: 'A divisor should be taken as multiplier and vice versa, the square as a square root and vice versa, addition as a subtraction and vice versa'. An example given in *Lilavati* will illustrate the method:

> O maiden with playful eyes, tell me, since you understand the method of inversion, what number multiplied by 3, then increased by three-quarters of the product, then divided by 7, then diminished by one-third of the result, then multiplied by itself, then diminished by 52, whose square root is then extracted before 8 is added and then divided by 10, gives the final result of 2?

Solution: The solution offered is elegant and simple. We start with the answer 2 and work backwards. When the problem says divide by 10, we multiply by that number; when told to add 8, we subtract 8; when told to extract the square root, we take the square, and so on. It is precisely the replacement of the original operation by the inverse that gives the method its name of 'inversion'.

Therefore the original number is obtained thus:

$$[(2 \times 10) - 8]^2 + 52 = 196$$

$$\Rightarrow \sqrt{196} = 14$$

$$\Rightarrow \frac{(14)\left(\frac{3}{2}\right)(7)\left(\frac{4}{7}\right)}{3} = 28.$$

8.2.3   To find an unknown quantity

There was a range of problems that involves a form of pseudo-algebra for finding unknowns before the methods of algebra came into existence. These would include finding quantities that satisfy certain restrictions or conditions. For example, to discover an unknown number x, choose any convenient number and carry out the operations such as multiplication, division etc. Next multiply x by the result given in the problem and divide this product by the number obtained. Here we find the unknown x by the method of false assumption.[6] An example from Bhaskaracharya's text would help to clarify this method.

Example 1:

> A pilgrim sets out on a journey with a certain amount of gold coins. He gave away half the amount at Prayaga [in modern Allahabad]. He spent two-ninths of the remaining amount at Kashi [Varanasi]. One-fourth of what remained was paid as a fee. He then spent six-tenth of the remainder at Gaya [in modern Patna]. He returned home with the 63 gold coins. Find the initial amount he had [before his journey] using the method of fractional residues.

Solution: Assume that he had 1 gold coin to begin with.

He spent $\frac{1}{2}$ in Prayaga and so he had $\frac{1}{2}$ left.

In Kashi he spent $\frac{1}{2} \times \frac{2}{9} = \frac{1}{9}$ and was left with $\frac{1}{2} - \frac{1}{9} = \frac{7}{18}$.

The fee paid was $\frac{7}{18} \times \frac{1}{4} = \frac{7}{72}$ and was left $\frac{7}{18} - \frac{7}{72} = \frac{21}{72} = \frac{7}{24}$.

In Gaya he spent $\frac{7}{24} \times \frac{6}{10} = \frac{7}{40}$ and returned home with: $\frac{7}{24} - \frac{7}{40} = \frac{7}{60}$.

Applying the method of fractional residues, the remaining amount was:

$$\frac{63}{\frac{7}{60}} = 9 \times 60 = 540 \text{ gold coins.}$$

Example 2: The Pearl Necklace

A problem in different style and detail that keeps recurring in the texts of the period is the problem of the pearl necklace. It appears, as we have seen in the last chapter, not only in the works of Brahmagupta and Mahavira but does so also in *Lilavati*. The example here does not attempt to capture the poetry and atmospherics of the one in Mahavira's *Ganitasarasamgraha*, although it also involve an amorous couple in search of pearls. The problem is given below and the reader is invited to solve it using the method of false assumption without any fractional residues. The answer is supplied.

> In a love play of a couple, the lady's pearl necklace was broken. One-third of the pearls fell on the ground, one-fifth went under the bed. The lady collected one-sixth and her lover collected one-tenth. Six pearls remained on the necklace. Find the total number of pearls in the necklace. (*Answer*: **30** pearls)

8.2.4 A digression: the method of false assumption in Egyptian mathematics

In the Ahmes Papyrus (dated 1650 BCE) appears the earliest example of the method of false assumption:

> A quantity, its $\frac{1}{4}$ added to it so that 15 results (i.e. A quantity and its quarter added becomes 15. What is the quantity?)

Solution: In terms of modern algebra, the solution is straightforward and involves finding the value of x, the unknown quantity, from the following equation:

$$x + \frac{1}{4}x = 15, \quad \text{so} \quad x = 12.$$

The scribe (Ahmes), however, may have reasoned as follows: if 4 is the answer, then $4 + (1/4 \text{ of } 4)$ would be 5. The number that 5 must be multiplied by to get 15 is 3. If 3 is now multiplied by the assumed answer (which is clearly false), the correct answer is $4 \times 3 = 12$.

The scribe was using the oldest and probably the most popular way of solving linear equations before the emergence of symbolic algebra, the method of false assumption (or false position). Variants of 'quantity' problems of this kind included adding a multiple of an unknown quantity instead of a fraction of the unknown quantity. For example, Problem 25 of the Moscow Papyrus (another source of ancient Egyptian mathematics dating back to 1850 BCE) asks for a method of calculating an unknown quantity such that twice that quantity together with the quantity itself adds up to 9. The instruction for its solution suggests assuming the quantity as 1 and that together with twice the assumed quantity gives 3. The number that 3 must be multiplied by to get 9 is 3. So the unknown quantity is 3. It is interesting to reflect that such an approach was still in common use in Europe and elsewhere until about a 100 years ago. The method was brought into Europe by Fibonacci in the thirteenth century in his book *Liber Abaci* under the name of method of false position.[7]

8.2.5   Finding unknowns from sums and differences

Suppose we need to find two numbers given their sum and difference. [Verses 61–62] give us the method:

i. The sum decreased [or] increased by the difference and being halved gives two quantities. This is called *sankramana* (or method of elimination). [Verse 61]
ii. The difference of the squares divided by the difference of two quantities will be the sum. From it two quantities will be obtained as mentioned before [Verse 62]

In modern notation, let the two numbers be a and b. Given $c = (a + b)$ and $d = (a - b)$

Then

$$a = \frac{c+d}{2} \quad \text{and} \quad b = \frac{c-d}{2}. \quad \text{[Verse 61]}$$

Since

$$a^2 - b^2 = (a + b)(a - b).$$

Therefore

$$\frac{a^2 - b^2}{a - b} = a + b. \quad \text{[Verse 62]}$$

Consider two examples to illustrate the methods:

Example 1: O my dear child! If you know the method of transition, find two numbers whose sum is 101 and whose difference is 25.

By the above method

$$a = \frac{101 + 25}{2} = \textbf{63}, \quad b = \frac{101 - 25}{2} = \textbf{38}.$$

Example 2: O mathematician! Tell me two numbers whose difference is 8 and the difference of whose squares is 400.

By the above method:

$$\frac{a^2 - b^2}{a - b} = \frac{400}{8} = 50$$

$$a + b = 50, \quad (a - b) = 8$$

$$a = \frac{50 + 8}{2} = \textbf{29}, \quad b = \frac{50 - 8}{2} = \textbf{21}.$$

8.2.6   Quadratic equations and right-angled triangles [Verses 72–78]

The *Lilavati* is famous for adding substantially to a store of fascinating recreation problems which are there to ignite the interests of students, especially those who are not well-disposed to the subject. At the same time they illustrate or explain certain rules in a painless fashion. Bhaskaracharya carried on this tradition which started with Bhaskara I and continued with Sridhara and Brahmagupta (and his commentator, Prthudakasvami), all of whom in a number of cases used the same or similar examples. We will consider a few of these problems, confining ourselves here to those dealing with quadratic equations[8] and right-angled triangles.

Example 1: *The Flight of the Swans*

> O girl, at the beginning of the rainy season, ten times the square root of a number of swans who were swimming in a lake flew away to the bank of Manasa Sarovar [a sacred lake on the Himalayas] and one-eight flew to a forest called Sthala Padmini. Three pairs remained in the lake disporting themselves swimming from one end to the other. Tell me, how may swans are there in the lake.

Solution: In modern notation, let x be the number of swans swimming in the lake. [The 'x' in the original solution correspond to an appropriately assumed (false) value.]

Then

$$10\sqrt{x} + \frac{x}{8} + 6 = x \quad \text{or} \quad 10\sqrt{x} = \frac{7x}{8} - 6.$$

Multiplying throughout by 8, we get

$$80\sqrt{x} = 7x - 48.$$

Squaring both sides, we get

$$6400x = (7x - 48)^2 = 49x^2 - 672x + 2304$$

or

$$(49x - 16)(x - 144) = 0.$$

Therefore $x = \mathbf{144}$ and $x = \mathbf{16/49}$.

Since the answer has to be a whole number, the number of swans swimming the lake is **144**.

Example 2: Wandering Bees

From a swarm of black bees, square root of the half went to the *malati* (jasmine) tree; 8/9th of the total bees eventually went to the *malati* tree to suck honey. Of the remaining two, one got caught in a lotus whose fragrance captivated him. He started wailing and his beloved responded. How many bees were there in the swarm?

Solution: Let x be the total number of bees in the swarm
We have the equation:

$$\sqrt{\frac{x}{2}} + \frac{8}{9}x + 2 = x$$

$$\sqrt{\frac{x}{2}} = \frac{1}{9}x - 2 = \frac{x - 18}{9}.$$

Squaring both sides, we get

$$\frac{x}{2} = \frac{(x - 18)^2}{81}$$

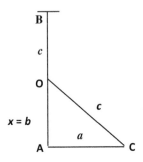

Figure 8.1. The broken bamboo.

or

$$2x^2 - 153x + 648 = 0$$
$$(x - 72)(2x - 9) = 0.$$

The only admissible answer is $x = \mathbf{72}$.

Example 3: *The Broken Bamboo Again*

Oh mathematician! A bamboo 32 cubits long, is standing on a level ground. A blast of wind breaks it and it touches the ground 16 cubits from its foot. Can you tell me at what height the bamboo broke. In Figure 8.1, let O be the breaking point and AB be the bamboo before breaking and let the end of the broken bamboo touch the ground at C.

Given

$$AB = 32 \quad \text{and} \quad AC = 16 \quad \text{and} \quad OC = OB = 32 - x$$

we have

$$OA^2 = OC^2 - AC^2 \quad \text{or} \quad x^2 = (32 - x)^2 - 16^2$$
$$x^2 - 64x + 1024 = x^2 + 256 \Rightarrow 64x = 768,$$

therefore

$$x = \frac{768}{64} = \mathbf{12} \text{ cubits.}$$

A Digression: *The Broken Bamboo Problem Elsewhere*

This is a famous problem in the history of mathematics. It kept reappearing in the works of Indian mathematicians, from Mahavira to Bhaskaracharya in the twelfth century, and eventually in European works, probably thus charting a westward migration of Chinese mathematics from the *Nine Chapters* written around the beginning of the Common Era via India and the Islamic world to Europe where it makes an early appearance in Philippi Calendri's *Arithmetic* of 1491.

The appearance of this problem in China [and discussed by Chinese mathematician, Yang Hui (1261)] may be expressed in the form

Given a and $b + c$, find b (See Figure 8.1).

Suggested solution: Take the square of the distance from the foot of the bamboo to the point at which its top touches the ground, and divide this quantity by the length of the bamboo. Subtract the result from the length of the bamboo, and halve the resulting difference. This gives the height of the break.

Expressed in modern notation, with reference to Figure 8.1
Let

$$a = \text{distance from foot of bamboo,}$$

$$b + c = \text{length of bamboo,}$$

$$b = \text{height of erect section of bamboo.}$$

Then the above rule is equivalent to

$$b = \frac{1}{2}\left(b + c - \frac{a^2}{b+c}\right)$$

which yields $b = \frac{1}{2}\left(32 - \frac{16^2}{32}\right) = 12$ cubits, which is the same answer we got before.

Example 2: Two Bamboos

Two bamboos of heights 15 and 10 feet are standing on a level ground. Tight strings are tied from the tip of either bamboo to the foot of the other. What is the point of intersection of the two strings above the ground?

In Figure 8.2 above, let AB and CD are two bamboo shoots of heights $a = 15$ and $b = 10$ feet respectively. The two strings AD and

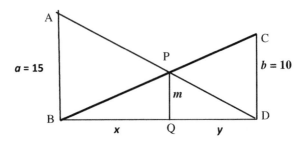

Figure 8.2. Two bamboos.

BC intersect at P. And $PQ = m$, the perpendicular distance of the intersection from the level ground, which needs to be calculated.

Let $BQ = x$ and $QD = y$.

Triangle PQD is similar to triangle ABD and so

$$\frac{y}{x+y} = \frac{m}{a}. \tag{1}$$

Triangle PQB is similar to triangle CDB and so

$$\frac{x}{x+y} = \frac{m}{b}. \tag{2}$$

Add (1) and (2) and simplify further to get:

$$m = \frac{ab}{a+b}.$$

Therefore, if $a = 15$ and $b = 10$, the distance of the intersection point and the ground is:

$$PQ = m = \frac{15 \times 10}{15 + 10} = 6 \text{ feet.}$$

Also, if you are given the distance between the bamboos $(x+y) = d$, you can find the values of x and y as:

$$x = \frac{md}{b} = \frac{ad}{a+b} \quad \text{and} \quad y = \frac{bd}{a+b}.$$

Note that the distance of the intersection (m) does not depend on the distance between the bamboos (d).

8.2.7 Solution of cubic and biquadratic equations by reduction

Bhaskaracharya did not go beyond what we would today describe as solution of cubic and higher order equations, except in a limited way using the method of reducing the original equation to simple and quadratic forms employing useful transformations. Two examples that he offers in *Bijaganita* demonstrate the solution of a cubic equation and a biquadratic equation.

Example 1: *Cubic Equation*

Oh Learned One! Can you show me the number when multiplied by 12 and increased by the cube of that number is equal to the sum of 35 and six times the square of the number.

Solution: Let x be the number
Then we have

$$12x + x^3 = 35 + 6x^2 \quad \text{or} \quad x^3 - 6x^2 + 12x - 8 = 27$$

or

$$(x - 2)^3 = 3^3.$$

So that

$$(x - 2) = 3 \Rightarrow x = \mathbf{5} \quad \text{which is the only root.}$$

Example 2: Biquadratic (fourth degree) equation

Can you tell me the number which when multiplied by 200 and added to the square of the number and then multiplied by 2 and the resulting number is subtracted from fourth power of the number is equal to 10,000 minus unity?

Solution: Let x be the number
Then we have

$$x^4 - 2(200x + x^2) = 10000 - 1$$

$$x^4 - 2x^2 - 400x = 9999.$$

Add $4x^2 + 400x + 1$ to both sides to get

$$x^4 + 2x^2 + 1 = 4x^2 + 400x + 10000$$
$$(x^2 + 1)^2 = (2x + 100)^2.$$

Take the square root

$$(x^2 + 1) = \pm(2x + 100).$$

If we take

$$x^2 + 1 = 2x + 100 \Rightarrow x^2 - 2x + 1 = 100,$$

then

$$(x - 1)^2 = 10^2 \Rightarrow x = 11 \text{ or } -9,$$

where $x = 11$ is the only value shown.[9]

8.3   Rule of Quantities and its Variations

We are back on familiar territory. It may be remembered that Aryabhata signalled the importance of this subject by giving the 'Rule of Three' (*trairasikha*) in terms of *phala-rasi* (fruit or p), *iccha-rasi* (desire or i), *pramana* (measure or m) and *iccha-phala* (resulting fruit or f). The rule can be expressed symbolically as:

$$f = \frac{(p \times i)}{m}.$$

The mathematicians that followed Aryabhata generally devoted some space to this subject. Bhaskaracharya gave an extensive treatment of this subject that included the direct and inverse 'Rule of Three', a general procedure for 'Rules' of other odd numbers $(5, 7, 9)$ and application of these rules in barter transactions. A few applications follow:

Example 1: *Barter* [**Verse 96**]

In a market, 300 mangoes can be purchased for 1 *dramma* (16 *paisa*). However, 30 pomegranates of good quality are available for 1 *paise*. Find quickly how many pomegranates can be exchanged for 10 mangoes.

Solution:

300 mangoes can be exchanged for $30 \times 16 = \mathbf{480}$ pomegranates.

10 mangoes can be exchanged for $\frac{480}{300} \times \frac{10}{1} = \mathbf{16}$ pomegranates.

Example 2: Rule of Five

30 planks of wood, each measuring 14 cubits × 16 fingers × 12 fingers cost 100 *niskas* (a unit of currency). Find Oh friend, the cost of 14 such planks whose dimensions are each 4 measures less than the former.

Solution: An example of the application of Rule of 5

$$14 \text{ cubits} : 10 \text{ cubits}$$
$$16 \text{ fingers} : 12 \text{ fingers}$$
$$12 \text{ fingers} : 8 \text{ fingers}$$
$$30 \text{ planks} : 14 \text{ planks}$$
$$100 \text{ } niskas : x \text{ } niskas$$

Therefore,

$$x = \frac{10 \times 12 \times 8 \times 14 \times 100}{14 \times 16 \times 12 \times 30 \times 1} = \frac{\mathbf{50}}{\mathbf{3}} = 16\frac{2}{3} \text{ } niskas.$$

8.4   A Medley of Mixed Quantities

Following the *Brahmasphutasiddhanta*, there were several discussions of the classic 'eight procedures': namely mixtures, series, figures, excavations, piles, sawing, heaps and shadows. The procedure on mixtures deals with standard topics such as interest and investments and there is little new in the way that Bhaskaracharya treats these subjects. However, his examination of two other procedures, namely exchange of jewels and finding the weight of gold is interesting and may be illustrated by the following examples:

Example 1: Exchange of Jewels: [**Verse 109**]

Four merchants have 8 rubies, 10 sapphires, 100 pearls and 5 diamonds respectively. They become friends during the journey and decided to share their jewels, each one of them giving each of the other one jewel from his lot in such a way that they all end up with value of their shares being the same. O friend, find out the price of the four pieces of jewellery.

Solution: (i) Our present-day algebraic approach is fairly straight-forward. Let a, b, c, d be the prices of rubies (R), sapphires (S), pearls (P) and Diamonds (D) respectively.

Then

$$5a + b + c + d = a + 7b + c + d = a + b + 97c + d = a + b + c + 2d$$

or

$$4a = 6b = 96c = d.$$

If we assume that the price of a pearl is one unit, then price of ruby is $96/4 = 24$, price of a sapphire is $96/6 = 16$ and price of a diamond is $96/1 = 96$.

(ii) *Bhaskaracharya's Approach*: Jewels possessed initially: 8R, 10S, 100P, 5D. Number of Gifts to each person $= 1$ and number of persons $= 4$.

So remainders: $8 + (-1 \times 4) = 4$; $10 + (-1 \times 4) = 6$; $100 + (-1 \times 4) = 96$; $5 + (-1 \times 4) = 1$. If we suppose $x = 96$, the prices of R is $96/4 = 24$, S $= 96/6 = 16$, D $= 96/1 = 96$. Note LCM of the remainders is $4 \times 6 \times 96 \times 1 = 2304$. We get the prices of jewels in the proportion of 576:384:24:2304. Choosing a price of a pearl is 1 unit, the price of ruby is 24, the price of a sapphire is 16 and the price of diamond is 96.

Example 2: Mixing Gold [**Verse 111**][10]

> Four types of gold consisting of 10 *Masas* [a measure of weight] of 13 carats, 4 *Masas* of 13 carats, 2 *Masas* of 11 carats and 4 *Masas* of 10 carats are melted to form a new type of gold. Find its fitness. If this is purified, what is its fineness? If the mixed gold when purified has 16 carats, what is its weight?

[The fineness of the mixture equals the sum of the products of weight and fineness of the constituents, divided by the total weight. If the sum is divided by the weight of pure gold one gets the fineness of pure gold and if it is divided by the fineness the result will be the weight of pure gold.]

Solution:

Sum of the products $= (13\times10)+(12\times4)+(11\times2)+(10\times4) = 240$

Therefore, the fineness of the mixture $= \frac{240}{10+4+2+4} = \frac{240}{20} = 12$ *carats*.

Fineness of the purified gold $\frac{240}{16} = 15$ *carats*.

If the fineness is 16 carats, weight $= \frac{240}{16} = 15$ *masas*.

Amount of impurities burnt out is $15 - 10 = 5$ *masas*.

Example 3: *Interest* [**Verse 91**]

A sum of 94 *niskas* is lent at three different interest rates of 5, 3 and 4 percents respectively. An equal interest was earned on the three portions in 7, 10 and 5 months respectively. Tell me O mathematician, the amount of each portion.

Solution: Bhaskaracharya's reasoning (expressed in modern notation) is as follows:

Let the Principal $= P = 94$; Rates of interests: $r_1 = 5$, $r_2 = 3$, $r_3 = 4$; Time periods: $t_1 = 7$, $t_2 = 10$, $t_3 = 5$.

Therefore,

$$r_1 t_1 = 35, \quad r_2 t_2 = 30 \quad \text{and} \quad r_3 t_3 = 20$$

and

$$\frac{100}{35} + \frac{100}{30} + \frac{100}{20} = \frac{235}{21}.$$

The first portion $= \dfrac{100}{35} \times \dfrac{21}{235} \times 94 = \mathbf{24};$

The second portion $= \dfrac{100}{30} \times \dfrac{21}{235} \times 94 = \mathbf{28};$

The third portion $= \dfrac{100}{20} \times \dfrac{21}{235} \times 94 = \mathbf{42}.$

8.5 Combinations

In three verses [**118–120**] Bhaskaracharya discusses combinations. Starting with the number n, one writes down $n, (n-1), (n-2), \dots$, divided by $1, 2, 3, \dots$ to get $\frac{n}{1}, \frac{n-1}{2}, \frac{n-2}{3}, \dots$. Then the number of combinations of n things taken 1, 2, 3 at a time are:

$$\frac{n}{1}, \frac{n(n-1)}{1 \times 2}, \frac{n(n-1)(n-2)}{1 \times 2 \times 3} \ldots \text{ respectively.}$$

Or the number of combinations of n things taken r at a time are

$$\frac{n(n-1)(n-2) \cdots (n-r+1)}{[1 \times 2 \times 3 \times \cdots \times r]}$$

or in present-day notation: $\binom{n}{r}$

Examples 1(i)–(ii): [**Verse 122**]

> (i) A king has a splendid palace with 8 doors. Skilled engineers had constructed 4 open squares which were highly polished and spacious. To get fresh air, 1 door, 2 doors, 3 doors, . . . , 8 doors are kept open. How many different types of ventilations are possible?

Solution: Applying the above rule, we get the following possibilities:

$$\binom{8}{1} + \binom{8}{2} + \cdots + \binom{8}{8} \text{ giving } 8 + 28 + 56 + 70 + 56 + 28 + 8 + 1 = \mathbf{255}.$$

Alternatively, the total number of different ventilations possible $= 2^8 - 1 = \mathbf{255}$.

How many different kinds of relishes can be made using 1, 2, 3, 4, 5 and 6 types of tastes from sweet, bitter, astringent, sour, salty and hot flavours?

Solution:

$$\binom{6}{1} + \binom{6}{2} + \binom{6}{3} + \binom{6}{4} + \binom{6}{5} + \binom{6}{6}$$

$$= 2 + 12 + 30 + 20 = 2^6 - 1 = 63.$$

This example has a long history. As mentioned in an earlier chapter, Sushruta's great work on medicine (*The Sushruta Samhita* from the 6th century BCE) contained the statement that 63 combinations possible with six different tastes (*rasa*) — bitter, sour, salty, astringent, sweet, hot — by taking the *rasa* one at a time, two at a time, three at a time and so on. No explanation was given then on how the total combinations were arrived at.

8.6 Series

Bhaskaracharya gave the rules known at that time for (i) the sums of the first n integers, (ii) the sums of the squares of these integers, (iii) the sums of the cubes of the integers and (iv) the summation of series in arithmetic and geometric progressions. Expressed in modern notation, they are:

(i) $1 + 2 + 3 + \cdots + n = \frac{n(n+1)}{2}$,

(ii) $1^2 + 2^2 + 3^2 + \cdots + n^2 = \frac{n(n+1)(2n+1)}{6}$,

(iii) $1^3 + 2^3 + 3^3 + \cdots + n^3 = \left[\frac{n(n+1)}{2}\right]^2$,

(iv) $a + (a+d) + (a+2d) + \cdots + [a + (n-1)d] = S = \frac{n}{2}[2a + (n-1)d]$, where $a =$ first term and $d =$ common difference in an Arithmetic Progression.

Make d the subject of the formula to get:

$$d = \frac{\frac{S}{n} - a}{\frac{(n-1)}{2}}.$$

(v) Given a is the first term, r the common ratio, and S the sum of a Geometric Series:

$$S = a + ar + ar^2 + \cdots + ar^{n-1} = \frac{a(r^n - 1)}{r - 1}.$$

Example 1: Sums of Squares and Sums of Cubes [**Verse 127**]
   Tell me the sums of $1^2 + 2^2 + \cdots + 9^2$ and of $1^3 + 2^3 + \cdots + 9^3$.

$$1^2 + 2^2 + 3^2 + \cdots + 9^2 = \frac{n(n+1)(2n+1)}{6} = \frac{9(10)(19)}{6} = 285.$$

$$1^3 + 2^3 + 3^3 + \cdots + 9^3 = \left[\frac{n(n+1)}{2}\right]^2 = \left[\frac{9(10)}{2}\right]^2 = 2025.$$

Example 2: *Capturing Elephants* [**Verse 133**]

> A king in pursuit of elephants belonging to an enemy covers 2
> *yojanas* (about 16 miles) on the first day, an then increases the
> distance travelled according to an Arithmetic Progression (AP).
> If he travels 80 *yojanas* altogether in 7 days, find out the extra
> distance travelled by the king each day.

Solution: Let S = Sum = 80, n = number of days = 7 and a =
distance travelled on 1st day = 2 and d = extra distance travelled.
Applying the formula given above
   Then

$$d = \frac{\frac{S}{n} - a}{\frac{(n-1)}{2}} = \frac{\frac{80}{7} - 2}{\frac{(7-1)}{2}} = 3\frac{1}{7} \; yojanas.$$

Example 2: *Charity Disbursement* [**Verse 137**]

> A man gave 2 cowries to a charity and then he gave successively
> twice the amount each day for a month. How many cowries did he
> give away in 1 month?

Solution: Given $a = 2$, $r = 2$, $n = 30$.

$$\text{Total amount} = \frac{a(r^n - 1)}{r - 1} = \frac{2(2^{30} - 1)}{2 - 1} = 2{,}147{,}483{,}646 \; cowries.$$

8.7   Geometry of Triangles and Quadrilaterals

8.7.1   The Pythagorean theorem [Verses 142–165]

Bhaskaracharya's contributions to this topic is noteworthy for its
originality and comprehensiveness. He begins with the Pythagorean
theorem and other rules for constructing right-angled triangles based
on the Pythagorean Theorem. The following results (Figure 8.3)
apply:

(i) *Pythagorean Theorem*: In a right-angled triangle, square root of
   the sum of squares of the base (b) and the altitude (a) is the
   hypotenuse (c). The square root of the difference between the

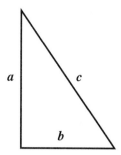

Figure 8.3. A right-angled triangle.

squares of the hypotenuse and base (altitude) is the altitude (base).

$$c = \sqrt{a^2 + b^2}; \quad b = \sqrt{c^2 - a^2}; \quad a = \sqrt{c^2 - b^2}.$$

(ii) If twice the product of two numbers is added to the square of their difference, the result is the sum of their squares. Also if the sum of two numbers is multiplied by their difference, the result is the difference of their squares.

$$a^2 + b^2 = (a + b)^2 - 2ab; \quad a^2 - b^2 = (a + b)(a - b).$$

(iii) Multiply the given hypotenuse by twice a chosen optional number[11] and then divide by the sum of the square of the optional number and unity. This gives the altitude. Multiply this again by the optional number and difference between this product and the given hypotenuse is the base.

In Figure 8.3, if c is the given hypotenuse and m is an optional number, then the altitude (a) and base (b), according to (iii), can be shown to be:

$$a = \frac{2cm}{m^2 + 1};$$

$$b = c - \frac{2m^2c}{m^2 + 1} = \frac{c(m^2 + 1) - 2m^2c}{m^2 + 1} = \frac{c(1 - m^2)}{m^2 + 1},$$

or

$$b = \frac{2m^2c}{m^2 + 1} - c = \frac{2m^2c - c(m^2 + 1)}{m^2 + 1} = \frac{c(m^2 - 1)}{m^2 + 1}.$$

Example 1: [**Verse 151**] If the hypotenuse is 85, find the other sides as integers.

Solution: Choose $m = 2$ so that c is divisible by $m^2 + 1$. Then the other two sides of the right-angled triangle are altitude (a) = 68 and base (b) = 51.

Suryadasa (1538), a commentator on Bhaskaracharya's *Bijaganita*, gave an explanation of how he arrived at the above results and also a method of constructing a rational right-angled triangle, given the base. In one of the notable commentaries on *Lilavati*, entitled *Buddhivilasini*, Ganesh Daivajna (b. 1507) provides a rigorous geometric proof of the rationale provided by Bhaskaracharya for the so-called Pythagorean theorem. However, it is believed that Bhaskaracharya gave the diagram in Figure 8.4 as proof with the exclamation 'Behold'! A more conventional proof is as follows:

Given: (Figure 8.4) ABC is a right-angled triangle where the right-angle is at C. Also, AB = c, AC = b, and BC = a. Complete the square ABED with side = c. Draw EH perpendicular to BC. Draw DF perpendicular to EH. Produce AC to meet FD at G. It can be shown that triangles ABC, AGD, DEF and EBH are congruent. Let A* = area of triangle ABC. Then the sum of the areas of the four triangles = 4A* = $4(\frac{1}{2}ab)$ = 2ab. Now the area of the small square GCHF = $(a - b)(a - b) = (a - b)^2$.

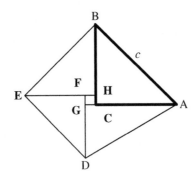

Figure 8.4. Behold!

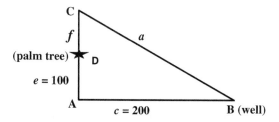

Figure 8.5. Two monkeys.

Therefore, areas of the four triangles plus area of GCHF

$$= 2ab + (a - b)^2 = a^2 + b^2 = c^2 = \text{Area of ABED}.$$

We will consider a few examples from *Lilavati* to illustrate Bhaskaracharya's way of thinking

Example 1: *Jumping Monkeys* [**Verse 163**][12]

Two monkeys were at the top of a palm tree 100 cubits high. At a distance of 200 cubits from the tree was a well. One of the monkeys came down the tree and walked to the well while the other jumped up and slid down along the hypotenuse to the well. If both covered equal distances, find the length of the second monkey's jump. (It is assumed that the jumping monkey travels along a straight line rather than a parabolic route)

The first monkey climbs down DA (i.e. e) and then travels along AB (i.e. c) to the well: $(e + c)$. The second monkey initially jumps from D to C (i.e. f) and then slides down CB (i.e. a) to the well: $(f + a)$.

Given that

$$f + a = e + c,$$

$$a^2 = c^2 + (e + f)^2 = c^2 + e^2 + f^2 + 2ef \quad \text{(Pythagorean Theorem)}$$

and

$$(f + a)^2 = f^2 + a^2 + 2fa = (e + c)^2 = e^2 + c^2 + 2ec.$$

Further simplification will lead to:

$$2ec = 4ef + 2fc \Rightarrow 2f(2e + c) = 2ec$$

$$\Rightarrow f = \frac{ec}{(2e + c)} = \frac{100 \times 200}{2 \times 100 + 200} = \textbf{50 cubits.}$$

Example 2 [**Verse 164**]: Find the sides of a right-angled triangle, given the hypotenuse (c) and the sum of the other two sides ($a + b$).

Solution:

$$c^2 = a^2 + b^2 = (a + b)^2 - 2ab = s^2 - 2ab$$

where $s = (a + b)$

$$(b - a)^2 = (a + b)^2 - 4ab = s^2 + 2(c^2 - s^2) = 2c^2 - s^2.$$

Therefore

$$(b - a) = \sqrt{2c^2 - s^2} \tag{1}$$

also given

$$(b + a) = s. \tag{2}$$

Adding (1) and (2) gives

$$b = \frac{s + \sqrt{2c^2 - s^2}}{2} \quad \text{and} \quad a = \frac{s - \sqrt{2c^2 - s^2}}{2}.$$

Example 3 [**Verse 165**]: In a right-angled triangle, the hypotenuse (c) is 17 and the sum of the other two sides is 23. Find the base (b) and altitude (a).

Solution: Given $c = 17$ and $a + b = 23$, find a and b.

$$a = \frac{23 - \sqrt{2(17)^2 - (23)^2}}{2} = \textbf{8}; \quad b = \frac{23 + \sqrt{2(17)^2 - (23)^2}}{2} = \textbf{15.}$$

Check:

$$8^2 + 15^2 = 17^2.$$

Figure 8.6. A peacock and a snake: an original facsimile.

Example 4: The reader may wish to try out the following problem in *Lilavati* involving the right-angled triangle which is reminiscent of the one we encountered in the discussion of Bhaskara I's 'eagle and rat' example. [**Verses 157–158**] from the *Lilavati* is given in its original in Figure 8.6 below.

[**Verse 158**]: There is a hole at the foot of a pillar nine *hastas* high, and a pet peacock standing on top of it. Seeing a snake returning to the hole at a distance from the pillar equal to three times its height, the peacock descends upon it slantwise. Say quickly, at how many *hastas* from the hole does the meeting of their two paths occur? (It is assumed here that the speed of the peacock and the snake are equal.)

[**Verse 157**]: The square of the pillar is divided by the distance between the snake and its hole; the result is subtracted from the distance between the snake and its hole. The place of meeting of the snake and the peacock is separated from the hole by a number of *hastas* equal to half that difference.

(Answer: The snake and the peacock travel equal distances of 15 *hastas* and the distance of the snake from the hole is 12 hastas.)

8.7.2 An interesting digression: impossible or non-figures (Verses 169–171)

It is impossible that in any polygon (triangle, quadrilateral, ...) the length of one side is greater than the sum of all other sides. If an idiot states that it is possible to construct a quadrilateral of sides 2, 6, 3, 12 or a triangle 3, 6 and 9, provide him with the relevant number of sticks of required lengths and show him that it is impossible to lay down the sticks and form these figures.[13]

However, Bhaskaracharya allows for the possibility that a quantity in a geometric figure can be an absolute negative value. Building on Bhaskara I and Brahmagupta's work on solving a triangle with two base-segments produced by the intersection of the base and altitude, Bhaskaracharya considers the case where the altitude of a triangle falls outside that triangle.

Consider the following example in **Verse 174**: A triangle has sides 10, 17 and base 9, find its area, the two projections (on the base) and altitude.

The rules given to solve this problem may be expressed as follows:

Figure 8.7 shows a triangle ACB, the two sides are $AC = 10$ and $AB = 17$ and $CB = 9$ is the base, and $AD = a$ is the altitude. The base segment DC is negative or in the contrary direction. This is shown in Figure 8.7 as a dotted line.

The projection of $AB = DB = \frac{1}{2}\left(9 + \frac{27 \times 7}{9}\right) = \mathbf{15}$.

And the projection of $AC = DC = \frac{1}{2}\left(\frac{27 \times 7}{9} - 9\right) = \mathbf{6}$.
Applying the Pythagorean theorem:

$$\text{The altitude} = a = \sqrt{10^2 - 6^2} = 8.$$

Therefore, the area of triangle $ACB = \frac{1}{2} \times 9 \times 8 = \mathbf{36}$.

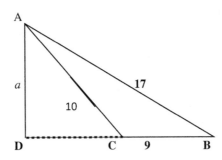

Figure 8.7. Projections.

8.7.3 Triangles and quadrilaterals

Following Brahmagupta's work mentioned in the last chapter, Bhaskaracharya next turned his attention to (cyclic) quadrilateral and triangles, with the latter having already been investigated by the Greek mathematician, Heron (200 BCE). In *Brahmsphutasiddhanta*, Brahmagupta gives the following two results mentioned in the last chapter:

1. The gross area of a quadrilateral is the product of half the sums of the opposite sides while the exact area is given by the square root of the product of four sets of half the sum of the sides (respectively) diminished by the sides.
2. The sums of the products of the sides about the diagonal should be divided by each other and multiplied by the sum of the opposite sides. The square roots of the quotients give the diagonals of a [cyclical] quadrilateral.[14]

These rules, as stated in an earlier chapter, were as follows: Let a, b, c and d be the sides of a cyclic quadrilateral of area A; let $s = \frac{1}{2}(a+b+c+d)$ be the semi-perimeter, and x and y the diagonals. Then the exact area (A) is:

$$A = \sqrt{(s-a)(s-b)(s-c)(s-d)}.$$

And the diagonals are given by:

$$x = \sqrt{\frac{(ab+cd)(ac+bd)}{ad+bc}}, \quad y = \sqrt{\frac{(ad+bc)(ac+bd)}{ab+cd}}.$$

The first statement of the expressions for the diagonals in Western mathematics is found in 1619 in the work of Willebrord Snell, some 1,000 years later.

The derivations of these results are first referred to in a commentary on Brahmagupta's work by Bhattotpala (*c.* 10th century CE),[15] but find their full expression in the sixteenth-century Kerala text *Yuktibhasa*. This contains a detailed discussion of the properties of a cyclic quadrilateral and how it is used to arrive at various trigonometric results. It makes use of Ptolemy's theorem, which states that the product xy of the diagonals of a cyclic quadrilateral is equal to the sum of the products of the two pairs of opposite sides.[16]

A new rule is found in the work of Paramesvara (c. 1430) for obtaining the radius r of the circle in which a cyclic quadrilateral of sides a, b, c and d is inscribed:

$$r = \sqrt{\frac{(ab+cd)(ac+cd)(ad+bc)}{(a+b+c-d)(b+c+d-a)(c+d+a-b)(d+a+b-c)}}.$$

A detailed demonstration of this result is found in a later commentary on Bhaskaracharya's *Lilavati*, from the Kerala School entitled *Kriyakramakari* (c. 1560). This result makes its first appearance in European mathematics in 1782 in the work of l'Huilier.[17]

Bhaskaracharaya's contribution on this subject was limited to repeating the previous work of Brahmagupta but pointing to the "approximate" nature of the rule when applied to non-cyclical quadrilaterals. A loose translation of [**Verses 177–179**] from *Lilavati* states:

> The formula for the area of a quadrilateral is not accurate. This is because the lengths of its diagonals are indeterminate. So can we get an accurate values? ... Even if all four sides of quadrilaterals are equal (e.g. a rhombus), the diagonals can be different and so we get different areas. Unless a perpendicular or a diagonal of a quadrilateral is known the area is indeterminate.

It is clear from the above that Bhaskaracharya and even Aryabhata II (*fl.* 950–1100) before may have recognised the fact that Brahmagupta's rules applied only to cyclical quadrilaterals. The very fact that he is concerned with the mensuration of a rhombus showed that he is confining himself to the geometry of cyclical quadrilaterals.[18]

Example 1: Areas of (cyclical and non-cyclical) Quadrilateral and Triangle [**Verse 176**]

The quadrilateral (Figure 8.8) has base $b = 14$, upper side $d = 9$ and the other two sides ($a = 12$, $c = 13$). Find the area by the

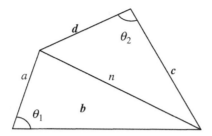

Figure 8.8. Geometry of triangle and quadrilateral.

(imprecise) method. Given $\theta_1 = 40°$ and $\theta_2 = 85°$, find the precise area. Also given the altitude ($n = 12$), find the (imprecise) area

Solutions: Given $s = \frac{1}{2}(a + b + c + d) = 24$

$$A^* = \sqrt{(s-a)(s-b)(s-c)(s-c)}$$

$$= \sqrt{(24-12)(24-14)(24-13)(24-9)}$$

$$= \sqrt{19800} \approx 140.7. \quad \textbf{Imprecise}$$

$$A^{**} = A^* - abcd \cdot \text{Cos}^2\left[\frac{1}{2}(\theta_1 + \theta_2)\right]$$

$$= \sqrt{19800} - (1956)(0.003) \approx 134.8. \quad \textbf{Accurate}$$

Further, if the altitude (n) is equal to one of the sides (a), we are then calculating the area of a trapezium which can be dissected into a right-angled triangle and a rectangle where the area of the this trapezium:

$$A^{***} = \text{Area of triangle} + \text{Area of a rectangle}$$

$$= \frac{1}{2}\left[12 \times \sqrt{(13^2 - 12^2)}\right] + [12 \times 9] = 30 + 108 = \textbf{138}.$$

8.7.4 The diagonals of a (cyclic) quadrilateral

As we saw in the last chapter, a notable discovery of Brahmagupta relates to his work on the properties of a cyclic quadrilateral. Apart from his discovery of the formula for calculating the area of such a figure, he also showed if the sides of a scalene cyclical quadrilateral are known, its diagonals can be calculated. Bhaskaracharya's simpler demonstration of Brahmagupta's discovery is given in [**Verse 198**] and may be restated thus:

> Take two right-angled triangles. Multiply the sides of one of the triangles by the hypotenuse of the other and the sides of the other triangle by the hypotenuse of the first. These are [the two sets of opposite sides of a scalene quadrilateral. Further, sum of the product of the bases (shorter sides) of the triangles and the product of their altitudes is one diagonal (of the quadrilateral). The other diagonal is obtained by adding the product of the one base and the other altitude to the product of one altitude with other base. Thus the diagonals are easily contained (by taking two right-angled triangles).

Given in Figures 8.9(i) and (ii) two right-angled triangles ABC and DEF of (integral sides) a_1, b_1, c_1; a_2, b_2, c_2, and $a_i \leq b_i \leq c_i$.

Form a quadrilateral (Figure 8.9(iii)) with the two right-angled triangles with sides a_2c_1, a_1c_2, b_2c_1 and b_1c_2. The two diagonals are $a_1a_2 + b_1b_2$ are $a_1b_2 + a_2b_1$.

Example 1: Let the two right-angled triangles be

$$(a_1, b_1, c_1) = (3, 4, 5) \quad \text{and} \quad (a_2, b_2, c_2) = (5, 12, 13).$$

The resulting cyclic quadrilateral will have sides equal to:

$$(a_2c_1, a_1c_2, b_2c_1, b_1c_2) = (5 \times 5), (3 \times 13), (12 \times 5), (4 \times 13)$$
$$= (25, 39, 60, 52).$$

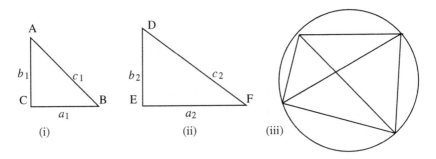

Figure 8.9. Forming a cyclic quadrilateral.

The diagonals: $a_1a_2 + b_1b_2 = (3 \times 12) + (4 \times 5) = \mathbf{56}$ and $a_1b_2 + a_2b_1 = (4 \times 12) + (3 \times 5) = \mathbf{63}$.

8.7.5   Geometry of circles and spheres

In Verse 207, Bhaskaracharya discusses two ways of computing the circumference of a circle for a given diameter. The 'accurate' method involves taking the ratio of the circumference to diameter (i.e. π) as 3927/1250, which is Aryabhata's ratio of 62832/20000 reduced by a factor of 16.[19] The 'practical' method involves taking the Archimedean ratio of 22/7.[20] Rules for finding the area of a circle, the surface area and volume of a sphere follow and there is little that is exceptional in Bhaskaracharya's treatment of these subjects.[21]

It may be remembered that in *Aryabhatiya*, the volume of a sphere was incorrectly calculated as the product of area of the circle and its square root. Bhaskaracharya gives the correct rule for the volume of the sphere as the surface area of the sphere multiplied by the diameter of the circle divided by six.

To illustrate the rules, consider the following problem from *Lilavati*:

Example 1: Mensuration of Circular Objects (**Verse 210**)

What is the area of a disk of diameter 7? What is the area of a net which encompasses a ball of diameter 7? What is its volume?

'Practical' (Approximate) Estimates

Area of the disc $= A = \frac{Cd}{4} = \frac{22}{7} \times \left(\frac{7}{2}\right)^2 = \frac{77}{2} = 38\frac{1}{2}$ square units.

Surface area of the disc $(S) = 4Cd = 4 \times \frac{22}{7} \times \left(\frac{7}{2}\right)^2 = 154$ square units.

Volume of the sphere $= V = \frac{Sd}{6} = \frac{1}{6} \times Sd = \frac{1}{6}(154 \times 7) = 179\frac{2}{3}$ cubic units.

More 'Accurate' estimates of the above can be obtained from taking $\pi = 3927/1250$.

8.7.6   Geometry of chords and arcs

8.7.6.1   *Finding the length of a chord*

Like others who preceded him, Bhaskaracharya did not subscribe to the method of calculating the chord of a circle in terms of the arc of a circle. This was an important difference between Indian and Greek geometry. Following the work of Bhaskara I, discussed in Chapter 6, a new method is introduced in the following verse:

> The circumference decreased and multiplied by the arc is called the 'first'. Now multiply the square of the circumference by 5/4, subtract from it the 'first' and divide the 'first' by this remainder. The result multiplied by four times the diameter is the (length of the) chord.

Expressed in modern notation: Let h be the length of the chord of the arc a. Let d and C be the diameter and circumference of the circle. Then

$$h = \frac{4d(C - a)a}{\left(\frac{5}{4}\right)C^2 - (C - a)a}. \tag{1}$$

This is a modification of an approximation formula found in Bhaskara I's *Mahabhaskariya*, to be discussed in Chapter 12. It is not known how this formula was obtained.

Example 1: [**Verse 220**] Find the length of the chord (h) of a circle whose radius is 120 and arc length (c) one-eighteenth of its circumference. Find the lengths of the chords (h) when the arc length is doubled.

Solution: Here $d = 240$ and $c = \frac{1}{18}C$. The reader is invited to show that the length of the arc is 42.06, if the arc is one-eighteenth of its circumference and 84.12 if the arc is two-eighteenth of the circumference.[22]

8.7.6.2   *Finding the length of an arc given the chord*

$$\text{Length of arc (a)} = a = \frac{C}{2} \pm \sqrt{\frac{C^2}{4} - \frac{\frac{5}{4}C^2 h}{4d + h}}, \tag{2}$$

where (2) is derived after some tedious algebra making a the subject of the formula in (1).

8.7.7 Other 'procedures'

The five traditional procedures dealt with by earlier mathematicians, namely excavations, piles, sawing, heaps and shadows are treated in a summary manner, allocating about five verses to each of the subjects. A few illustrative examples follows in **Verses 224–232**. They include:

1. *The volume of an irregular ditch* involves measuring the breadth at various points and obtaining an average breadth. Similarly, obtain the average length and average depth and calculate the product of the three averages to get the average volume.
2. *The volume of a pyramid and its frustum* is calculating by taking the areas of the top, of the bottom (base) and the rectangle whose adjacent sides are the sum of the lengths and breadths of the top and the base. Dividing the sum of the three areas by 6 gives the mean area of the frustum. The mean area multiplied by the depth (or height) is the volume. This volume divided by 3 is the volume of the pyramid.

   Example 1: A well in the shape of the frustum of a pyramid [where the base is smaller than the top] has a rectangular top of sides 10×12 cubits and a top 5×6 cubits. If the height is 7 cubits, find the volume of the well.

   Solution: Volume of the well $= \frac{7}{6}[120+30+\{(10+5) \times (12+6)\}] =$ **490** cubic cubits.
3. *The volume of a wooden log (the frustum of a cone)*: Number of times that the wooden log is cut $= n$

   In Figure 8.10, ABCD is a trapezium. Suppose AB $= d_1$, DC $= d_2$, EF $= h$.

   Area of ABCD $= \frac{h}{2}(d_1 + d_2)$.

   So the total area of the log is $n \times$ Area of ABCD.

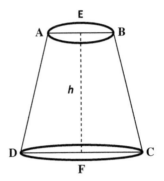

Figure 8.10. Volume of a log.

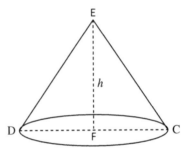

Figure 8.11. A heap of grain.

Example: A log of wood has base 20 *angulas* long, 16 *angulas* top and height 100. If the log is cut into 4 equal pieces sections, find the total surface area of wooden pieces (in cubits). [Note 1 cubit is 576 square *angulas*.]

Solution: Total surface area $= 4 \times \frac{100}{2}(20 + 16)\frac{1}{576} = \mathbf{12.5}$ square cubits.

4. *Volume of a Heap of Grain*

A heap of grain is shaped as a circular cone (Figure 8.11). Its circumference can be measured but not its height without disturbing the heap. [If the grains are big and spherical, their heap has a height which is one-tenth of the circumference. If they are small and spherical, the proportion is one-eleventh and if they are pointed, it is one-ninth. So Bhaskaracharya gives methods to compute the height for three different types of grains].

$$\text{Volume} = \text{height} \times \left(\tfrac{\text{circumference}}{6}\right)^2 = \left(\tfrac{2\pi r}{6}\right)^2 h = \tfrac{\pi^2 r^2 h}{9} \approx r^2 h$$

if π is taken to be 3 as an approximation for practical purpose. A similar problem appears in the premier Chinese text, the *Nine Chapters* appearing around the beginning of the Common Era.

8.8 Shadows

The shadow problems in the *Lilavati* are similar to the ones found in the work of Aryabhata I and his commentator Bhaskara I. However the following problem is worth noting. In **Verse 232** there is a result on the shadow lengths.

8.8.1 Length of a shadow

Divide 576 by the difference of the squares of difference of the shadow lengths and the difference of the hypotenuse. The product of the difference between the hypotenuse and the square root of the above quotients plus 1 is added to and decreased by the difference between the shadow lengths. The resulting quantities divided by 2 give the shadow lengths separately.

[Note: This problem concerns the altitude of the Sun. A pole (of length h) is fixed and a string is tied to its top.[23] Marks are made on the ground where the shadows fall and the corresponding lengths of the string are noted.]

Let the height of the vertical pole be h, its shadows are b, $b + c$ and the corresponding hypotenuses are d, $d + e$. We are given h, c and e and we want to find b when $c > e$.

In modern notation, the shadow length b is obtained from:

$$b = \frac{1}{2}\left[-c + e\sqrt{1 + \frac{4h^2}{c^2 - e^2}}\right].$$

The other shadow is $b + c$.

[*Note*: If h is given as 12 angulas, $4h^2 = 4 \times 144 = 576$. This explains why in the equation above 576 appears in place of $4h^2$.]

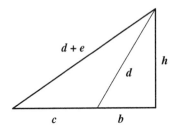

Figure 8.12. Length of a shadow.

8.9   A Digression: The Universality of 'Rule of Three'

This section ends with a digression on the wide applications of the 'Rule of Three' worth quoting in full:

> The [rule of] five or more quantities is completely explained by considering the Rule of Three. Just as this universe is pervaded by Lord Narayana (who removes the suffering of those who worship him and is the sole creator of this universe, in its many forms — worlds and heavens and mountains and rivers and gods and men and demons and so on. In the same way, this whole type of computation is pervaded by the Rule of Three...Knowledgeable persons are fully aware that examples in algebra and arithmetic involving multiplications and divisions are tackled easily by the Rule of Three. For the dull-witted persons like us to master the Rule of Three, it has been explained by the wise with many examples.

It is interesting to speculate whether Bhaskaracharya put himself in the company of the dull-witted.

The next 20 verses or so is concerned with the operation of the pulveriser (or what is now be better known as the 'Euclidean division') for solving linear indeterminate equations. This is a subject discussed in a later section.

The final section consisting of 10 verses provides an introduction of the terminology and content of the subject of permutations and combinations, a subject as we saw earlier, has a long history starting with the Jains. Bhaskaracharya introduces rules for permutations in the following words, modified for sake of clarity:

> To find the number of permutations of given [n] different digits [or objects] by writing 1 in the first place and then $2, 3, 4, \ldots$ to the

number of objects [n] and multiply them. Divide the product of the number of permutations and the sum of the given [n] by the number of the given digits [i.e. by n] by the number of the times [i.e. n times] in a column but leaving one-digit place each time; add them; the result is the sum of the numbers formed [by permuting given n digits].

These rules are illustrated by the following examples.

Example 1: Using (i) 2, 8; (ii) 3, 8, 9; (iii) 2, 3, . . . 9 how many different numbers can be formed. What is the sum of numbers so formed in each case?

Solution:

 (i) Given $n = 2$ so $_2P_2 = 2$ or only two numbers 28 and 82.
 (ii) Given $n = 3$ so $_3P_3 = 3 \times 2 \times 1 = 6$ different numbers 389, 938, 839, ...
(iii) Given $n = 8$ so $_8P_8 = 8 \times 7 \times 6 \times \cdots \times 1 = 40320$.

For $n = 2$, the sum of numbers formed is

$$\frac{(1)(10^2 - 1)(10)}{9} = 110.$$

For $n = 3$, the sum of numbers formed is

$$\frac{(2)(10^2 - 1)(20)}{9} = 4{,}440.$$

For $n = 8$, the sum of numbers formed is

$$\frac{(7)(10^8 - 1)(40320)}{9} = 3{,}135{,}999{,}968{,}640.$$

Example 2:

Statues of Lord Siva (who has 10 arms) are shown (in different temples) holding 10 different weapons (a rope, a trap, a goad, a snake, a drum, a potsherd, a club, a spear, a missile and an bow and arrow) in different hands. Find the total number of statues of Siva showing different combinations. Similarly, how many different statues of Vishnu can be found who has four objects to hold, namely the club, the discus, the lotus and conch-shelf.

Solution: Here, there are $10! = 10 \times 9 \times 8 \times \cdots \times 1 = 3{,}628{,}800$ different Siva statues

$$4! = 4 \times 3 \times 2 = 24 \text{ different Vishnus.}$$

Bhaskaracharya also discusses a rule for permutations of given n digits or objects which allows for repetitions. The rule may expressed in modern terminology as follows:

Let $a_1 = a_2 = a_3 = \cdots = a_i$, $b_1 = b_2 = b_3 = \cdots = b_j$; $c_1 = c_2 = c_3 = \cdots = c_k$ consists of n objects.

Then the number of permutations

$$= \frac{n!}{i!\,j!\,k!\cdots}, \quad \text{where } n = i + j + k + \cdots .$$

Example 3: Leaving aside 0, if the digits 1 to 9 are written six at a time, then find how many different numbers are formed

Solution: Note that $_nP_r = {_9P_6} = 9 \times 8 \times 7 \times 6 \times 5 \times 4 = 60480$.

Given the sum of all digits (n) and the number (r) of blank spaces are given, follow the method given below to get the total number of permutations: $\frac{(n-1)\cdots(n-r+1)}{(r-1)!}$, where the method can be used if $n < r + 9$. This is an example of partitions and the derivation is a standard result in theory of numbers. It is interesting that this was the area in which the Indian mathematician Srinivas Ramanujan (1887–1920) made some notable contributions.

Lilavati remains for centuries one of the most popular mathematics texts in India. Apart from a clear exposition of the common arithmetical operations it includes a number of methods of computations of series using kings and elephants and other objects that ordinary people could relate and understand. There is the final promise that whomsoever masters the text will be happy and prosperous.

In a later insertion an unknown scribe offers the following eulogy:

Bhaskaracharya, the great poet and author of this book, has mastered eight volumes on Grammar, six on Medicine, six on logic, five on Mathematics, four *Vedas*, a triad of three *ratnas* (jewels) and two *Mimamsas* (schools of philosophy). He understands that it is impossible to fathom the ways of the Lord Supreme.

8.10   The Bijaganita

Intended as a more advanced book than the *Lilavati*, it has attracted fewer commentaries and translations into regional languages. Written in a similar style consisting of a number of verses divided into separate topics or sections, the *Bijaganita* ('Seed-computation') starts with an examination of the six operations on positive and negative quantities. There is an interesting discussion of mathematical notation, of the symbols that we came across in the *Bakhshali* Manuscript with the dot over the number replacing the *Bakhshali* cross to signify a negative quantity. One of the notable features in the early section of Bijaganita is its extensive discussion of the six operations with zero. And further to the discussion earlier in *Lilavati*, there is the implication that while in what we perceive as algebra, a division by zero gives infinity while in arithmetic the result is a quantity that is indeterminate.[5] A subject that has attracted the attention of subsequent mathematicians over a 1,000 years, the two alternative answers would figure in subsequent discussions.

It is in the six operations with unknowns that Bhaskaracharya introduces the idea of representing unknown quantities by the names of colours and they are manipulated arithmetically with rules which we can recognise today. Assume that rupa (*ru*) represents 1 and *yavatavat* (or *ya* in short) stand for unknown. Abbreviations of colours — black (*ka*), blue (*ni*), red (*ro*) and yellow (*pi*) — represent different unknowns. Thus for example adding:

$$ya\ 3\ ru\ 4 \text{ and } ya\ 2\ ru\ \dot{3} \text{ gives } ya\ 5\ ru\ 1$$

(remembering that a dot over 3 signifies — 3).

In the next section (Section 4) Bhaskaracharya has an extensive discussion of six operations with surds (*karani*). The next three sections deal with the solution of indeterminate equations which will be covered later in this chapter. Sections 8 and 9 deal with what we would now call algebra. There is little that is new in the treatment of equations in one unknown in Section 8 compared to what we found in *Lilavati*. The next section introduces the possibility of there being more than one solution to a quadratic equation. An example contained in Verse 130 will illustrate:

Example 1: A fifth part of a troop of monkeys, minus three, squared, has gone to a cave; one is seen climbing to the branch of a tree. Tell me how monkeys are there in the troop

Suggested Answer: Bhaskaracharya's equation:

$$\text{yava 1 ya } \dot{5}5 \text{ ru 0 equals yava 0 ya 0 ru } \dot{2}50.$$

(*Note:* yava stands for "the square of the unknown".)

The result is a twofold quantity: 50, 5. In this case the second is not to be taken because of its inapplicability ... [being negative]

An Illustration: Explanation in modern notations:

Let x be the number of monkeys in the troop. Then the equation to be solved for x is:

$$x = \left(\frac{1}{5}x - 3 \right)^2 + 1 \Rightarrow x^2 - 55x + 250 = 0 \Rightarrow x = 50 \text{ or } 5.$$

$x = 5$ is rejected since a fifth part of 5 minus 3 would equal a negative number which does not make sense.

The reader may wish to try out a biquadratic equation problem from *Bijaganita* which yields three solutions of which only two are acceptable.[24]

Example 2: Tell me the numbers which when multiplied by 200 and added to the square of the number and then multiplied by 2 and the resulting number is subtracted from the fourth power of the number is equal to 10,000 minus 1.

The next section deals with equations in more than one unknown. It builds on Brahmgupta's work which we discussed in the last chapter. The solution involves finding one unknown in terms of other unknowns. The method is what we would now call as the method of substitution within a simultaneous equations framework. If there are more unknowns than equations, an armoury of methods to solve indeterminate equations was devised. Some of the methods are discussed in different chapters. However, the next section contains one of the major contributions of Bhaskaracharya.

8.11 Indeterminate Equations: *Chakravala* (Cyclical) Method

Even before Bhaskaracharya's contribution to the subject of solving indeterminate quadratic equations, Brahmagupta had devised his famous *vargaprakriti* procedure for doing so. As pointed out in the last chapter, Brahmagupta's approach needs a trial solution before it could proceed. Bhaskaracharya's *Chakravala* method, a refinement of Brahmagupta, dispenses with such a requirement. It has been attributed to Jayadeva (*c.* 1000 CE). To illustrate the method, consider the following numerical example found in the *Bijaganita* and is known by the misnomer of Pell's Equation:

Solve $Nx^2 + K = y^2$ or in specific form $61x^2 + 1 = y^2$.

Consider the general form of this equation: $61a^2 + K = b^2$.

A solution is: $61 \cdot 1^2 + 3 = 8^2$, where $a = 1$, $K = 3$ and $b = 8$.

We next chose m (an optional number) such that

$$\frac{am + b}{K} = \frac{m + 8}{3} \text{ is a whole number and}$$

$$\frac{m^2 - N}{K} = \frac{m^2 - 61}{3} \text{ is very small numerically.}$$

If $m = 7$, we get

$$a_1 = \frac{am + b}{K} = 5, \quad K_1 = \frac{m^2 - N}{K} = -4$$

and

$$b_1 = \frac{bm + Na}{K} = 39.$$

Therefore

$$61 \times 5^2 - 4 = 39^2.$$

Now apply the *bhavana* principle of Aryabhata, introduced earlier in Chapter 6 to get eventually

$$x = \left(\frac{1}{2}\right)(39.5)(39^2 + 1)(39^2 + 3) = \mathbf{226153980}.$$

$$y = (39^2 + 2)\left[\left(\frac{1}{2}\right)(39^2 + 1)(39^2 + 3) - 1\right] = \mathbf{1766319049}.$$

[Note: $1766319049^2 - (61 \times 226153980^2) = 1$.]

The values calculated for x and y are the smallest integral values satisfying the equation

$$61x^2 + 1 = y^2.$$

It is extraordinary that over 500 years later, the same equation would be proposed by the French mathematician, Fermat, as a challenge problem which was only eventually solved by Lagrange a 100 years later and required the calculation of 21 successive convergents of the continued fractions for the square root of 61. Bhaskaracharya's approach gave the solution in a few steps. As Selenius (1975, p. 180) wrote:

> No European performance in the whole field of algebra at a time much later than Bhaskara's nay nearly up to our times equalled the marvellous complexity and ingenuity of *chakravala*.

8.12   Other Mathematics

8.12.1   The mathematical content of *Siddhantasiromani*

As the title, meaning 'Crest-Jewel of *Siddhantas*', suggests the text was an astronomical treatise although according to some commentators it was the title of a compendium in mathematics and astronomy consisting of four parts — *Lilavita, Bijaganita* and *Ganitadhyaya and Goladhyay* — of which the last two is concerned with astronomy. There is also the view that the title *Siddhantasiromani* stood for the last two parts devoted to astronomy.

The first section of *Siddhantasiromani*, the *Ganitadhyaya*, contains a discussion of computational methods for calculating standard measure such as mean motions, true motions, solar and lunar eclipses which formed the standard fare of astronomical work of that period. The second section, *Goladhyay*, contains subjects relating to *gola* (or the sphere) including the shape and form of the sphere and its relevance for spherical earth. This discussion is interposed with a poetic rendering of the majesty of the seasons and questions to test the knowledge of the readers.

It is in the first section we come across Bhaskaracharya's manipulations of very small quantities akin to our infinitesimal. This makes its appearance in the discussion of a planet's motion which requires a calculation of its mean speed (being the ratio of the integral number of revolutions in a given time period and the number of days in that given time). Bhaskaracharya made the use of the concept of instantaneous motion (*tatkalikagati*) which he computed from the Cosine.[25] It is the similarity between Bhaskara's ideas of motion and the modern concept in differential calculus that has led some to claim that Bhaskara was a pioneer in introducing a key idea in calculus. It may remembered that in *Lilavati* he spoke of the division of zero by itself has being useful in astronomy. However, a word of caution should be sounded here. Does a concept of an infinitesimal increment in certain trigonometric quantities indicate an understanding of the functional concept of the rates of change in general. Irrespective of how this debate will be resolved we must concede that Bhaskaracharya had made a remarkable leap connecting the notion of instantaneous speed and how it can be derived by means of ratios of small increments.

The next notable contribution of Bhaskaracharya in the *Siddhantasiromani* was his use of geometric models in demonstrations in astronomy.[26] He did so extensively in demonstrating results regarding the sphere. Also, in another section of the *Goladhyay*, known as *Jyotpatti*, there are 25 verses stating formulas for determining a variety of Sines, a subject in which the contribution of the Indians was truly revolutionary and deserves a separate chapter. In all these, Bhaskaracharya emphasised the need for demonstrations, as the following passage testify:

> A mathematician knowing only the calculation of the planets...
> Without demonstration of that will not attain greatness in the
> assemblies of the eminent and will himself not be free from doubt...
> Therefore I am undertaking the subject of the sphere as means
> to understanding demonstrations... Like tasty food without ghee,
> like a kingdom deprived of a king, like an assembly without
> a good speaker, so is a mathematician ignorant of the sphere.
> (*Siddhantasiromani, Goladhyay*, 3.54-57).

Bhaskaracharya's works have been influential and have attracted
a number of commentaries. As mentioned earlier, *Lilavati* has been
translated into Persian by Abul Fyzi, a minister at the court of
the Moghul emperor in 1587. His *Bijaganita* was translated into
Persian by Utta ulla Rushdee in 1634. Many commentaries have been
written on the works on the work of Bhaskaracharya. These include
Ganita Kaumudi (1356) and *Bijaganita Vatamsa* (1370) by Narayana
Pandita. *Ganitamrita* (1538) and *Suryaprakasa* (1541) by Suryadasa,
Buddivilasini (1545) of Ganesh Daivajna, and *Vasana Bhasya* of
Ranganatha in the early seventeenth century. However, the best
commentary and one the most influential text of Kerala mathematics
is *Kriyakramakari* (*c.* 1534) of Sankara Variyar and Mahisamangala
Narayana, of whom there will be more in Chapter 10. There were
also commentaries in regional languages, notably in Telugu and
Kannada.

Endnotes

[1] A short list of the contents of the two astronomical texts follows
since we will not be discussing them. *Goladhaya* contains an extensive
treatment of theoretical astronomy, improving the rules suggested by his
predecessors and explaining their rationale There is a clear exposition
of the eccentric and epicyclical theories for the motion of planets. There
was also a discussion of the varieties astronomical instruments. In the
Grahaganita, there is an extensive study of the lunar and solar eclipses,
risings, settings and conjunctions of planets.

[2] Operations with zero (*sunyaganita*), mentioned in an earlier chapter,
has been a characteristic of Indian mathematics texts from the time of
Brahmagupta. While the discussion in the arithmetical texts (*patiganita*)

is limited only to the addition, subtraction and multiplication with zero, the treatment in algebra texts (*bijaganita*) covered such questions as the effect of zero on the positive and negative signs, division with zero and more particularly the relation between zero and infinity (*ananta*). In *Lilavati* of Bhaskaracharya, the eight operations involving zero — addition, subtraction, multiplication and division with zero as well as square, square root, cube and cube root of zero — are listed. A number divided by zero is given, like Brahmagupta, as 'zero-divided' or 'that which has zero as the denominator'. For further details, see Joseph (2015).

3 Bhaskaracharya, while he did not mention the name of Mahavira who preceded him, is clear where the sources of his inspiration came from. He acknowledges that his inspiration lay in the works of Brahmagupta and Sridhara.

4 The invocation, loosely translated reads:

> I offer homage to Lord Ganesa, whose neck is adorned by a playful black snake, bright as a blue and shining lotus (Verse 1).

5 In the discussion of the mathematics of zero, Bhaskaracharya gives an interesting example of a calculation of relevance in astronomy:

> What is the number when multiplied by zero, added to half of itself, multiplied by three and divided by zero amounts to sixty-three?

The solution offered by Bhaskaracharya may be expressed with modern notation as:

$$0 \left[\left(x + \frac{x}{2} \right) \frac{3}{0} \right] = 63$$

$$\left(\frac{3x}{2} \right) 3 = 63 \Rightarrow x = 14.$$

It is clear that Bhaskara had not worked out the correct calculation with the infinitesimal.

6 It is worth quoting the relevant verse (Verse 50) in *Lilavati* on the 'Rule of False Assumption'.

> Any number assumed at pleasure is considered as specified in the particular question, being multiplied and divided, increased or diminished by fractions. Then the given quantity, being multiplied by the assumed and divided by the answer, yields the number sought. This is called the process of supposition.

Expressed in modern notation, the Rule may be expressed thus:

Let x be the number to be operated on by the assumed quantity A to yield $A(x) = z$. Take any x^* and do the same operations to yield z^*, Then

$$x = z \times \frac{x^*}{z^*}.$$

[7] In *Liber Abaci* appears the following instruction:

> There is indeed another method which we use, namely that you put for the unknown thing some arbitrary known number which is divided evenly by the fractions which are put in the problem itself. And according to the posing of that problem, with that put number you strive to discover the proportion occurring in the solution of that problem.

The solution is similar to the one just discussed in *Lilavati*.

[8] The problems involving quadratic equations fall into the following categories:

(i) Given $x^2 \pm bx$, we are required to find x^2. Add $b^2/4$ and find the square root of the result. Next add or subtract (as the case may be) $b/2$ and square the result. This is x^2.

(ii) Given $x^2 \pm (1/n)x^2 \pm bx$, then multiply by $1/(1 \pm 1/n)$ and proceed as in (i).

[9] The negative root will lead to $(x + 1)^2 = -10^2$.

[10] We came across a similar example in Chapter 5 in *Bakhshali* Manuscript where the problem is one of calculating the impurity of gold and we use the 'Rule of Three' to obtain the solution. There is essentially hardly any difference in the approach to solving the problem in both. It is interesting that one could trace this problem from different sources over centuries.

[11] In *Siddhantasekhara*, Śripati (*c.* 1000 CE): states succinctly:

> For addition or subtraction, the surds should be multiplied (by an *optional number* intelligently (selected), so that they become squares. The square of the sum, or difference of their roots, should then be divided by that optional multiplier. Those surds which do not become squares on multiplication (by an optional number), should be put together (side by side).

In this case the optional number (m) is chosen such that c is divisible by $m^2 + 1$. Hence, $m = 2$ is a suitable optional number in Example 1. The reader is invited to try out other combinations of altitude (a) or base (b) and hypotenuse (c).

[12] Note the similarity between this example of a 'Jumping Monkey' and that of Prthudakaswamin's 'Holy Men' discussed in the previous chapter.

[13] No formal proof is provided by Bhaskaracharya. However, it may be remembered that this is a well-known result in Euclidean geometry with a *reductio ad absurdum* proof.

[14] The flavour of the original text is captured by the following more literal translation of Verse 28, which reads:

> One should multiply the sum of the products of the arms adjacent to the diagonals, after it has been mutually divided on either side, by the products of the arms and the counter-arms. For an unequal [cyclical] quadrilateral the two square roots are the two diagonals.

[15] An astrologer/astronomer whose commentaries on Varahmahira's *Brihatsamhita* and Brahmagupta's *Khaṇḍakhadyaka* have been reported by later writers.

[16] Notable extensions in this area are contained in Narayana Pandita's *Ganita Kaumudi* in the fourteenth century and Paramesvara's *Lilavati Bhashya*, a detailed fifteenth-century commentary on Bhaskaracharya's *Lilavati*. The cyclic quadrilateral was an important device used by the Kerala School for deriving a number of important trigonometric results, including

$$\sin^2 A - \sin^2 B = \sin(A + B) \cdot \sin(A - B)$$

and

$$\sin A \cdot \sin B = \sin^2 \frac{1}{2}(A + B) - \sin^2 \frac{1}{2}(A - B).$$

[17] See Sarasvati (1979, pp. 88–106) who gave details of the proofs.

[18] The correct formula for the area a non-cyclical quadrilateral is:

$$\text{Area} = \sqrt{(s - a)(s - b)(s - c)(s - d)} - abcd \cdot \cos^2\left(\frac{\theta_1 + \theta_2}{2}\right).$$

[19] Ganesh Daivajna, a commentator on *Lilavati*, explains the derivation of Bhaskaracharya's value in the following terms. Starting with the sides of an inscribed polygon of 12, successively doubling to 24, 48, 96, 192 to reach 384, and taking the diameter of the circle as 100 units, the ratio of the perimeter of the 384 sided polygon to the diameter will give the approximate value π as 3927/1250 (correct to 4dp).

[20] The value of 22/7 was first mentioned by Aryabhata II in the tenth century CE, although there was some evidence of it being known earlier. For further details, see Hayashi *et al.* (1989, pp. 8–11).

21 If d is the diameter, C is the circumference, A is the area of a circle, the surface area of a sphere is S and the volume V, then the following results were already known before Bhaskaracharya's time.

$$A = \frac{Cd}{4}; \quad S = 4Cd; \quad V = \frac{Sd}{6}.$$

22 Note that that the exact value of arc length $(c) = 1/18\times$ Circumference corresponds to today's $d\sin 10° = 240 \times 0.1736 = 41.664$. So Bhaskaracharya's method of estimating the chord length is fairly accurate.

23 Note the other shadow is obtained from $b + c$ once b is calculated. The height of the pole corresponds to a gnomon of 12 *angulas* (1 *angula* is about 20 mm).

24 In terms of modern notation: the biquadratic equation is $x^4 - 2x^2 - 400x = 9999$. Add $40x^2 + 400x + 1$ to both sides and simplifying gives $x^2 + 1 = 2x + 100$. Solving this quadratic equation we get $x = 11$. Note that the other solution $x^2 + 1 = -(2x + 100)$ requires knowledge of complex numbers and the third one is not acceptable since it is negative (i.e. $x = -9$). This is one of earliest attempt at solving a biquadratic equation.

25 An approximate formula for velocity of a planet in terms of Rsine-differences was first discussed in Bhaskara I's *Laghubhaskariya* and then refined by Manjula (*c.* 932 CE) in his *Laghumanasa* and in Aryabhata II's *Mahasiddhanta* in 950 CE.

26 A good illustration of how Bhaskaracharya's demonstration can lead to a critical scrutiny of other work is the manner in which he shows the erroneous reasoning of Lalla with respect to the his measurement of the size of the earth.

Chapter 9

Navigating the Ocean of Mathematics: Narayana Pandita and Successors

9.1 Introduction

Following Bhaskaracharya, it is generally accepted that the next figure of significance in the horizon (if we exclude Kerala mathematicians) was Narayana Pandita. The title 'Pandita' meant 'learned' and was used to refer to teachers and scholars. All we know about his family was he was the son of a Narasimha, and there is a suggestion, based on the way in which the manuscripts of his works are distributed, that he was from North India. His major works consist of an arithmetical treatise called *Ganitakaumudi* ('Moonlight of Mathematics') and a partially extant algebraic treatise called *Bijaganitavatamsa*, the former notable for its discussion of the subject of magic squares. He also wrote a commentary on Bhaskaracharya's *Lilavati* entitled *Karmapradipika*, anticipating a major text of Kerala mathematics at least by a similarity of name.[1] The author is aware of his limited knowledge of mathematics for he ends *Ganitakaumudi* by describing himself as a humble navigator in the ocean of mathematics. The book was completed on a date corresponding to 10 November 1356. In this chapter we will concentrate mainly on subjects in which he made significant contributions.

9.1.1   The contents of *Ganitakaumudi*

9.1.1.1   *Introduction*

The content and style of this text is similar to *Lilavati*. However, the treatment is more extensive and exhaustive, being about four times the length of Bhaskaracharya's text. Divided into 14 chapters with 475 *sutras* (rules) and 395 illustrative examples, the list of topics include many that we have across before.[2] After a standard treatment of metrology and mathematical operation on numbers, Narayana proceeds to a section on 'Miscellaneous Methods' and the eight procedures from mixtures to shadows that we have come across in earlier texts and his treatment of indeterminate equations using the 'pulveriser', essentially the same as given in Bhaskaracharya's *Bijaganita*. An unusual feature is his examination of seven methods of squaring numbers in *Karmapradipika* not found in any other text. The other novelties are the factorisation of integers, treatment of unit fractions, permutations and combinations and the last but not the least the 'auspicious calculation' or the magic squares. There are other points of departure from Bhaskaracharya and other writers on topics such as cyclic quadrilaterals and mathematical series which will briefly be touched on.

9.1.1.2   *Rules for decomposition into unit fractions*

We first came across unit fractions in the *Sulbasutras* discussed in Chapter 3 where an approximation to $\sqrt{2}$ is obtained as:

$$1 + \frac{1}{3} + \frac{1}{3 \times 4} - \frac{1}{3 \times 4 \times 34}.$$

The rules for expressing a fraction as the sum of unit fractions were discussed by Mahavira (Chapter 6) in the *Ganitasarasangraha*. Narayana continued the work and gave a few more rules in the section on *bhagajati* (reduction of fractions to a common denominator) in the twelfth chapter of the text.[3] The stated rules and illustrative examples (solutions left to the reader) are given here in modern notation:

Rule 1. Express 1 as a sum of n unit fractions:

$$1 + \frac{1}{1 \cdot 2} + \frac{1}{1 \cdot 3} + \frac{1}{3 \cdot 4} + \cdots + \frac{1}{(n-1) \cdot n} + \frac{1}{n}.$$

Example 1: The numerators of 6 terms are each 1. Their sum is approximately 1. Find their denominators

Rule 2: Express 1 as a sum of n unit fractions:

$$1 = \frac{1}{2} + \frac{1}{3} + \frac{1}{3^2} + \cdots + \frac{1}{3^{n-2}} + \frac{1}{2 \cdot 3^{n-2}}.$$

Rule 3: Express fraction p/q as a sum of unit fractions. Pick an arbitrary number i such that $(q+i)/p$ is an integer r, and write

$$\frac{p}{q} = \frac{1}{r} + \frac{i}{qr}.$$

Successive denominators are obtained by decomposing each of the new fractions until all the terms are unit fractions. A method already known to the Egyptians (and now known as the 'greedy' algorithm) involved choosing i as the smallest such integer. Narayana was aware that the solution is not unique and that there are many ways possible, according to one's preferences. The reader may wish to try out the following problem.

Example 2 The sum is 5/6. What are their denominators if their numerators are 1 each and their number of terms is 4?

9.1.1.3 *Finding a factor or a divisor of a number (Bhagadana)*

Narayana introduces this subject and expects people to find pleasure in solving it which could turn even an ordinary person into a mathematician. Although the subject was not explicitly treated in texts before *Ganitakaumudi* possibly because it was considered too elementary, there is a strong likelihood that factorisation of numbers by successive divisions by $2, 3, 5, \ldots$ was well known.[4] Narayana was the first mathematician who devoted a whole chapter to the subject. On this subject the rule stated is clear and well known today:

> If the number is even, divide it by 2 repeatedly until it becomes odd; if 5 is in the unit's place, divide by 5. If the number is neither even nor 5 in the unit's place, the prime numbers $3, 5, 7, 11, \ldots$ should be tried as divisors. However, if the number is a perfect square, its square root is a twice repeated factor.

In what follows, let N be the number whose divisors are to be calculated and $N + k^2 = b^2$ Then $(b + k)$ and $(b - k)$ are the divisors of N. This follows from $(b + k)(b - k) = N$.

The rule states:

> In the case of a non-square number if twice the square-root of its nearest increased by 1 and then lessened by the difference of the non-square number and if the nearest square-root be a square, it is to be taken as the number to be added to the non-square to make it a perfect square.

Let

$$N = a^2 + r \quad \text{and} \quad 2a + 1 - r = b^2.$$

According to this rule

$$N = (a + b + 1)(a - b + 1) \quad \text{or} \quad N + b^2 = (a + 1)^2.$$

Generalising the above method, keep adding the arithmetic sequence of numbers $2a + 1, 2a + 3 \ldots$, with common difference $= 2$ such that their sum minus r becomes a perfect square b^2.

That is

$$(2a + 1) + (2a + 3) + \cdots + (2a + 2m + 1) - r = b^2.$$

It would follow that

$$N = (a + m + 1 + b)(a + m + 1 - b).$$

Such a method, useful if the factors of N are large, was rediscovered by Fermat (1601–1665) 300 years later.

Example 1: From Chapter 11 of *Ganitakaumudi*:

Oh you talented in mathematics, find all possible divisors by which 1161 are exactly divisible.

Solution

Here

$$N = a^2 + r \Rightarrow 1161 = 34^2 + 5 \quad \text{and} \quad 2a + 1 - r = b^2 \Rightarrow 68 + 1 - 5 = 8^2.$$

Now

$$N + b^2 = (a+1)^2 \Rightarrow 1161 + 8^2 = 35^2 \quad \text{or}$$
$$1161(= 35^2 - 8^2) = 43 \times 27 = 43 \times 3 \times 3 \times 3.$$

So all the possible divisors of 1161 are 3, 9, 27, 43, 129, 387, 1161.

Example 2: Factorise $N = 1001$,
Here

$$1001 = 31^2 + 40.$$

We first note

$$(2 \times 31 + 1 - 40 = 23) \text{ is not a square.}$$

Hence we calculate

$$63 + 65 + 67 + \cdots + 89 - 40 = 32^2.$$

Therefore,

$$1001 = (31 + 13 + 1 + 32)(31 + 13 + 1 - 32) = 77 \times 13 = 7 \times 11 \times 13.$$

The reader may wish to list all possible divisors of 1001.

9.1.1.4 *Testing multiplication*

The rule states:

> Divide each of the multiplier and multiplicand by an optional number. Multiply the resulting remainders and then divide the product by the optional number. If the remainder so obtained is equal to the remainder obtained by dividing the product of (the multiplier and multiplier, then) it is correct.

Example 1: Oh Friend! 29 multiplied by 17 equal to 493. Is this correct?

Solution Let multiplier be $m = 29$ and $m^* = 17$ be the multiplicand. Let the optional number be $n = 3$, then the remainders are $r = 2$ and $r^* = 2$. Multiplying $r = 2$ and $r^* = 2$ to get 4 and dividing it by $n = 3$ leaves a remainder $R = 1$. Now dividing 493 by $n = 3$, leaves the remainder $R^* = 1$ which is the same as the result $R = 1$. Therefore, the multiplication is correct. (The reader may wish to check that you get the same result if you choose the optional number $n = 5$).

This method of checking the correctness of multiplication, also known as 'checking by nines', originated in India as testified by Islamic scientist and philosopher Ibn Sina (980–1037) and the Greek monk Planudes (*c.* 1260–1305). Narayana was probably the first Indian mathematician to have explained this method of casting out the desired number, which then spread to other parts of the world.

9.1.1.5 *Meeting of two travellers*

Two travellers set out on journeys from two places A_1 and A_2 distant d at the same time in opposite directions with speeds v_1 and v_2. At what time do they meet?

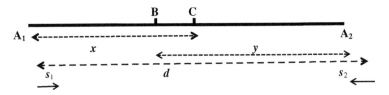

Rule 39 states: The distance divided by the sum of the two speeds will be the time of their first meeting. Twice the quotient obtained by the division of distance by distance is the time of the meeting again after the first meeting.

The travellers start from A_1 and A_2 at speeds s_1 and s_2 (where $A_1 A_2 = d$). They meet first at C where $A_1 C = x$ at time t_1. Traveller 1 continues the journey towards A_2 and traveller 2 continues the journey to A_1. Then they reverse their directions and meet at B where $A_2 B = y$ at time t_2. At C traveller 1 would have travelled a distance of x with speed s_1 and traveller 2 would have travelled $d - x$ with speed s_2.

Therefore, it would follow

$$\frac{x}{s_1} = \frac{d - x}{s_2}.$$

Solving for x we get

$$x = \frac{d s_1}{s_1 + s_2}.$$

Therefore, the time taken for the first meeting

$$t_1 = \frac{x}{s_1} = \frac{d}{s_1 + s_2}.$$

Similarly, it can be shown that if the second meeting is at B where $A_2 B = y$, and the total distance travelled by traveller $I = d + y = D$ and total distance travelled by traveller 2 is $d + d - y = 3d - D$, given the speeds of travellers 1 and travellers 2 are s_1 and s_2 respectively, then

$$\frac{D}{s_1} = \frac{3d - D}{s_2}.$$

Solving for D we get $D = \frac{3 d s_1}{s_1 + s_2}$ so that the time of the second meeting (t_2)

$$t_2 = \frac{D}{s_1} = \frac{3d}{s_1 + s_2}.$$

And the time elapsed between the first and second meetings

$$(t_2 - t_1) = \frac{2d}{s_1 + s_2}.$$

Example: The distance between two towns is 300 *yojanas*. Two messengers started at the same time from their respective towns, one at a speed of 11 *yojanas* per day and the other at 9 *yojanas* per day. O learned one! if you know the answer, tell me quickly the times of their two meetings, the first at the early part of their journeys and the second on their return journeys.

Solution: Here $d = 300$, $v_1 = 11$ and $v_2 = 9$.

$$\text{Time of the first meeting } (t_1) = \frac{d}{s_1 + s_2} = \frac{300}{11 + 9} = \mathbf{15} \text{ days.}$$

$$\text{Time of the second meeting } (t_2) = \frac{3d}{s_1 + s_2} = \frac{900}{11 + 9} = \mathbf{45} \text{ days.}$$

Therefore time between the two meetings

$$= 45 - 15 = \mathbf{30} \text{ days.}$$

A further extension considered is one of the two travellers travelling along a circle. Rule 40 states: "The circumference divided by the difference in the speeds gives the time of the meeting." This is applied in astronomy in investigating when the sun and moon are in conjunction.

9.2 Sequences and Progressions: The Cow Problem

In the third chapter of *Ganitakaumudi* there is both a restatement of certain results on Arithmetic Progression (AP) and certain problems that extend the discussion of the subject. Standard results relating to summation such as

$$\sum r, \quad \sum r^2, \quad \sum r^3, \quad \sum \sum r = \sum \frac{r(r+1)}{2},$$

where summations are from $r = 1$ to n have already been examined in previous chapters.

Narayana considers the following AP where each term (given in square brackets) is a sum of separate APs:

$$[1 + 2 + \cdots + a], [1 + 2 + \cdots + a + (a + 1) + \cdots (a + d)] \ldots$$
$$[1 + 2 + \cdots a + (n - 1)d)].$$

So that the rth term $= 1 + |2| + \cdots + |\{a + (r - 1)d\}|$ is the sum of an AP

It follows that the sum of this AP (excluding the first AP)

$$= \sum_{r=1}^{n} \frac{\{a + (r - 1)d\{a + (r - 1)d\}\{a + (r - 1)d + 1}{2}.$$

Now this is a key result found in *Yuktibhasa* (c. 1550) based on the following results preceding Narayana which were referred to earlier:

$$\sum\sum r = \sum \frac{r(r + 1)}{2},$$

$$\sum\sum r = \sum \frac{r(r + 1)}{2} = \frac{n(n + 1)(n + 2)}{1 \cdot 2 \cdot 3}$$

which is the 'sum of sums' or the 2nd sum.

Narayana generalise this expression to the kth sum as

$$\sum \cdots \sum r = \frac{n(n + 1)(n + 2) \cdots (n + k)}{1 \cdot 2 \cdot 3 \cdots (k + 1)}.$$

This remarkable result played an important role in the derivation of the infinite series for the sine and cosine functions in the Kerala mathematics. This result is elaborated in the *Yuktibhasa* although Narayana's contribution is not referred to in any subsequent work.

The Cow Problem

In *Ganitakaumudi* occurs the following problem:

A cow gives birth to a (she) calf every year. Beginning in its fourth year, each calf produces one calf at the beginning of each year. How many cows and calves are there altogether after 20 years?

The solution is offered in Rule 22:

Subtract the number of years (in which a calf begins giving birth) from the (total) number of years (successively and separately), till the remainder becomes smaller (than what is being subtracted). Those are the number for repeated summations once, (twice) etc., in order. Sum of the summations along with 1 added to the number of years is the numbers of progeny (the original cow is also included).

We need to compute the number of progeny in 20 years of the initial cow. The initial cow could give birth to calf every year for 20 years, which constitute the 'first generation' numbering 20. The calf born in the first year would produce its first offspring in the fourth year, this and the one born in the second year would together produces two offspring in the fifth year, and so on. So the total number of the second generation calves would be

$$V_{17}^{(1)} = 1 + 2 + 3 + 4 + \cdots + 17.$$

Similarly, the total number of third, fourth, fifth, sixth and seventh generation calves would be $V_{14}^{(2)}, V_{11}^{(3)}, V_8^{(4)}, V_2^{(6)}$. There are no more generation within 20 years as the eighth generations would take us to the 22nd year. Therefore the total progeny in 20 years is[5]:

$$V_{20}^{(0)} + V_{17}^{(1)} + V_{14}^{(2)} + V_{11}^{(3)} + V_8^{(4)} + V_5^{(5)} + V_2^{(6)}$$

$$= 20 + \frac{17 \cdot 18}{1 \cdot 2} + \frac{14 \cdot 15 \cdot 16}{1 \cdot 2 \cdot 3} + \frac{11 \cdot 12 \cdot 13 \cdot 14}{1 \cdot 2 \cdot 3 \cdot 4}$$

$$+ \frac{8 \cdot 9 \cdot 10 \cdot 11 \cdot 12}{1 \cdot 2 \cdot 3 \cdot 4 \cdot 5} + \frac{5 \cdot 6 \cdot 7 \cdot 8 \cdot 9 \cdot 10}{1 \cdot 2 \cdot 3 \cdot 4 \cdot 5 \cdot 6} + \frac{2 \cdot 3 \cdot 4 \cdot 5 \cdot 6 \cdot 7 \cdot 8}{1 \cdot 2 \cdot 3 \cdot 4 \cdot 5 \cdot 6 \cdot 7}$$

$$= 2744 + 1 = \mathbf{2745} \text{ cows.}$$

(We should include the original cow that we started with.)

9.3 Geometry

Chapter 4 begins with an examination of certain basis figures such as triangles and quadrangles, the relationship between sides and diagonals of a square and rectangle and finding the areas of certain figures such as triangles, quadrangles and circles. All the results found in Mahavira's *Ganitasarasangraha* and Bhaskaracharya's *Lilavati* are discussed in *Ganitakaumudi*. However, Narayana has an interesting approach and additional material on the geometry of plane figures.

9.3.1 Gross estimation of plane figures

Rule 8: In a triangle or a quadrilateral the product of half the sums of the opposite sides is the gross area. In a triangle, the face (i.e. the side opposite to the base) is considered to be a cipher.

i.e. If a, b, c and d are the sides of a quadrilateral, the rule states that its gross area (A) is approximately equal to:

$$\left(\frac{a+c}{2}\right) \times \left(\frac{b+d}{2}\right).$$

This formula, known as the 'Surveyor's rule', has a history going back to ancient Egypt and has been used by many cultures as a rule for practical measurements in construction and related trades.

Example

If you are skilled in the rules of mathematics, tell the gross area of a equi-bilateral (quadrilateral) whose flank sides are 5 (each), face is 2 and base, 8 and (that of) a equi-trilateral) whose base are 5 (each) and face is 11 (and also that of an 'inequilateral' quadrilateral whose) flank sides are 5 and 7, face is 4 and base, 10.

9.3.2 Gross area of a regular polygon

Rule 15 states:

Subtract the number of sides from the square of the number of sides. Multiply (the difference) by the square of the side. (The product) divided by 12 is the (gross) area of a triangle

In other words, the area of a regular polygon of n sides, each of length s is:

$$A = (n^2 - n)\frac{s^2}{12}.$$

Example

If you are an expert in mathematics, tell me the areas of (regular) figures (whose) sides are 6 each and the number of sides 3. (*Answer:* Area $= 18$).

9.3.3   Gross circumference or area of a circle

Rule 9 states: In a circle, thrice the diameter is the circumference (and the circumference) multiplied by one-fourth of the diameter is the area. Thrice the square of the diameter divide by 4 or (thrice the square of) the circumference divided by 36 is (also) the area.

Let C be the circumference, d the diameter and A, the area of a circle. The constant ratio C/d is universally denoted by π. According to this rule, the 'gross' calculations (taking $\pi = 3$),

$$C \approx 3d; \quad A \approx \frac{Cd}{4}; \quad A \approx \frac{3Cd^2}{4} = \frac{3C^2}{36}.$$

Example 1: Oh learned mathematician, tell me the gross circumference and the area of the circle whose diameter is 10 for day to day use. (*Answer* $C = 30$, $A = 75$).

Rule 16 explains how the Gross area of a curvilinear figure[6] is obtained.

> Multiply the square of a half of the sum of diameters by the number of (equal) circles less one. The product is divided by 9 times the number of circles is the (gross) area (of the curvilinear figure formed by circular arcs) inside (polygons).
>
> i.e. Area $= (nd/2)^2 \times (n-1)/9n$ where d is the diameter of the circles and n is the number of circles.

Example 2: What is the (curvilinear) area inside the regular figures of sides 3 within the connected circles where the diameter of (each of the circles) is 12? [*Answer*: $A = 36$].

Rule 17 states: Area of the circle is added to its one-third. The square root of the sum is the diameter.

$$\text{According to the rule: } d = \sqrt{A + \frac{A}{3}}.$$

Illustrative Example: O friend, if you know tell quickly the diameter of the circle whose area is 75. *Answer*: $d = 10$.

This section concludes with a discussion of gross estimates of areas of an elephant's tooth, a crescent moon, an annular figure and the wheel of a chariot. For example, the gross area of the last item is estimated thus:

Let D and d be the diameters of the outer and the inner circles of a wheel respectively and w be its width.

Rule 14 states that the gross area of the wheel $= 3(dw + w^2)$ by taking $\pi = 3$ and $D = d + 2w$.

9.3.4 Area of a (cyclic) triangle

Area of a triangle equals the product of the sides divided by four times the circumradius (R).

Let ABC be a triangle. Let AD be the altitude from A on BC. Let R be the circumradius of the triangle ABC. Then it can be shown that the area of triangle $ABC = \frac{abc}{4R}$.

The result follows applying Brahmagupta's Theorem to the formula for obtaining the area of the triangle ABC:

$$\text{Area of triangle ABC} = \frac{1}{2} \text{ base BC} \times \text{altitude AD} = \frac{1}{2}a \cdot \text{AD}.$$

Also, in a triangle, the product of the sides (about a perpendicular) divided by the perpendicular gives the circumdiameter of the circle in which the triangle is inscribed

or

$$\text{AD} = \frac{\text{Product of the sides AB and AC}}{2R} = \frac{bc}{2R}.$$

So the area of the triangle

$$\text{ABC} = \frac{1}{2}a \cdot \frac{bc}{2R} = \frac{abc}{4R}.$$

There is also a new formula for the circumradius and area of a triangle. Expressed in modern notation and referring to Figure 9.1

$$R = \frac{1}{2}\sqrt{\text{BC}^2 + \left(\frac{\text{AD}^2 - \text{BD} \cdot \text{DC}}{\text{AD}}\right)^2}.$$

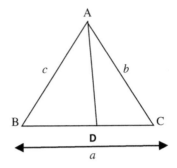

Figure 9.1. Area of a cyclic triangle.

The result follows from applying the Brahmagupta Theorem that gives:

$$R^2 = \frac{\mathrm{AB}^2 \cdot \mathrm{AC}^2}{4\mathrm{AD}^2}$$

and further application of the Pythagorean theorem to Figure 9.1, where $\mathrm{AB}^2 = \mathrm{AD}^2 + \mathrm{BD}^2$ and $\mathrm{AC}^2 = \mathrm{AD}^2 + \mathrm{DC}^2$ results (after some tedious algebra) in

$$R = \frac{1}{2}\sqrt{\mathrm{BC}^2 + \left(\frac{\mathrm{AD}^2 - \mathrm{BD} \cdot \mathrm{DC}}{\mathrm{AD}}\right)^2}.$$

9.3.5   Construction of rational triangles whose sides differ by unity

Rule 118 states the following procedure:

> Divide twice an optional number by the square of the optional number less 3. Add 1 to thrice the square (of the quotient). Twice the square root of the sum is the base. 1 added to and subtracted (from the base) are the flank sides.

In Figure 9.2, we assume that the perpendicular CE and the segments of x to be rational. To find the values of x and y we need to solve the indeterminate equation.[7]

$$\frac{3}{4}x^2 - 3 = y^2.$$

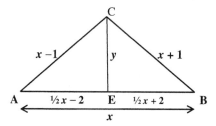

Figure 9.2. Rational triangle whose sides differ by unity.

The smallest non-trivial integral solution is $x_1 = 4$ and $y_1 = 3$. There are an infinite number of solutions (and triangles) consisting of $(x_2, y_2), (x_3, y_3), \ldots$

9.3.6 Geometry of cyclic quadrilaterals

A *cyclic quadrilateral* is a quadrilateral that is inscribed in a circle whose vertices all lie on a single circle. Two important theorems are relevant for exploring the properties of cyclical quadrilateral. The Ptolemy Theorem gives a relationship between the side lengths and the diagonals of a cyclic quadrilateral. The theorem often appears as intermediate steps in problems involving inscribed figures. The Brahmagupta's Formula, as we saw in an earlier chapter, is a formula for determining the area of a cyclic quadrilateral given only the four side lengths.

Narayana extended the work on the geometry of cyclic quadrilateral initiated by Brahmagupta. He begins by restating Brahmagupta's expression for the diagonals of a cyclic quadrilateral, but introduces the concept of a third diagonal which is important in deriving many results including the area of a cyclic quadrilateral. In Rule 48 Narayana states:

> When the topside and the flank side of any four-sided figure are interchanged, we obtain a third diagonal called ('*para*' or an auxiliary diagonal). This interchange of sides without changing the area is possible only in the case of a cyclic quadrilateral.

In *Rules 47–52* he lays out a method of finding the area of a cyclic quadrilateral in terms of the three diagonals.

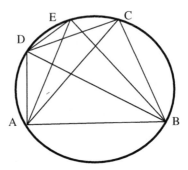

Figure 9.3. Three diagonals.

Divide the sum of the products of the sides about both the diagonals by each other. Multiply the quotients by the sum of the products of opposite sides. Square roots of the products are the diagonals in a quadrilateral. In all (cyclic) quadrilaterals, the new (third) diagonal is obtained by the interchange of its face and flank side. The area of a cyclic quadrilateral is equal to the product of the three diagonals divided by twice the circumdiameter.

In Figure 9.3, let ABCD be a cyclic quadrilateral and E be the vertex obtained by interchanging the sides DC and BC. Then the area of ABCD $= \frac{AC \cdot BD \cdot AE}{2d}$, where d is the circumdiameter.

It is worth reiterating that while AC and BD are the two diagonals, AE is the 'third' diagonal. The new cyclical quadrilateral ABED is generated by interchanging the sides BC and BE with AE being the third diagonal and is equal to the area of quadrilateral ABCD.

The proof follows directly from a previous result relating to the area of a triangle:

$$\text{Area of triangle ABC} = \frac{AB \cdot BC \cdot AC}{2d}$$

and

$$\text{Area of triangle ACD} = \frac{AC \cdot CD \cdot AD}{2d}.$$

Therefore, the area of cyclical quadrilateral

$$ABCD = \frac{AB \cdot BC \cdot AC + AC \cdot CD \cdot AD}{2d} = \frac{AC}{2d}[AB \cdot BC + CD \cdot AD].$$

Applying Ptolemy's theorem that the sum of the product of the opposite sides of the cyclic quadrilateral is equal to the product of its diagonals will give the result that in the cyclic quadrilateral ADEB

$$AB \cdot DE + AD \cdot BE = AE \cdot BD.$$

Hence, the area of the cyclical quadrilateral

$$ABCD = \frac{AC \cdot BD \cdot AE}{2d}.$$

It follows that the product of the three diagonals divided by two times the area is the circumdiameter. Narayana also stated the expressions for the main diagonals of cyclic quadrilaterals which are merely a restatement found in the Brahmagupta's text and discussed earlier. However, the third diagonal can be shown to be

$$AE = \sqrt{\frac{[(AB \cdot BC) + (CD \cdot DA)][(AB \cdot DA) + (BC \cdot CD)]}{[(AB \cdot CD) + (BC \cdot DA)]}}.$$

9.4 Combinatorics: Magic Squares (*Bhadraganita*)

Magic squares have a rich history dating back to the second millennium BCE. A Chinese myth claimed that while the Chinese Emperor Yu was walking along the Yellow River, he noticed a tortoise with a unique diagram on its shell.[8] The Emperor decided to call the unusual numerical pattern *lo shu*. This is one of the earliest recorded magic squares. Magic squares make an appearance in astrology, in divination, in philosophical discourse, in study of natural phenomena and human behaviour, in decoration of porcelain and in art. So in India as in other parts of the world, magic squares served multiple purposes other than the dissemination of mathematical knowledge.[9] For example, Varahamihira used a fourth-order magic square to specify recipes for making perfumes in his book, *Brihatsamhita* (ca. 550 CE). A third-order magic square appeared in the *Siddhayoga* (*c.* 900 CE) which was supposed to ease the pain of childbirth.

The last chapter of *Ganitakaumudi* contains an extensive and exhaustive (75 verses) treatment of magic squares and magic figures.[10] Beginning with a discussion of their origins where he attributes to Lord Siva who taught Manibhadra, the king of Yaksas (a group of semi-deities in the Hindu pantheon around the time of Ramayana), Narayana points to a more mundane reason for mastering *Bhadraganita* (magic squares) for knowing it would help to destroy the arrogance of bad mathematicians who cannot fathom the good mathematics embedded in the subject.[11]

Thakkura Pheru in his *Ganitasarakaumudi* and Narayana Pandita in his *Ganitakaumudi* begin by distinguishing between three categories of $n \times n$ magic squares. A double-even magic square is one where dividing the number of cells in a row by 4 leaves a zero remainder. If the remainder is 2, then the magic square is single-even and if the remainder is 3 or 1, the magic square is an odd. For example, Varahamihira's magic square, shown in Figure 9.4(a), is a double-even square whose constant is 18. It is also a pan-diagonal magic square.[12] This appears in a divination text entitled *Brihatsamhita* and discusses how this magic square can be used to prepare perfumes from 18 original substances.[13] Nagarjuna's magic square shown in Figure 9.4(b) whose square constant is 100 is also a pan-diagonal square. Finally, on a 12th century Jain inscription in Khajuraho (Figure 9.4(c)) has yet another pan-diagonal square whose square constant is 34.

9.4.1 Properties of pan-diagonal magic square

1. Consider a (4×4) magic square with entries $1, 2, \ldots, 16$ which is mapped on to a torus (i.e. a surface formed by revolving a circle about a line) by identifying the opposite edges of the square. Then

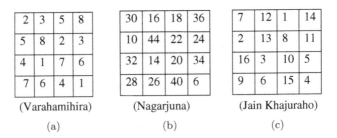

(Varahamihira)   (Nagarjuna)   (Jain Khajuraho)

(a)   (b)   (c)

Figure 9.4. Magic squares from three sources.

the entries of any 2×2 sub-square formed by consecutive rows and columns on the torus add up to 34.

1	12	13	8
15	6	3	10
4	9	16	5
14	7	2	11

For example: $1 + 12 + 15 + 6 = 1 + 12 + 14 + 7 = 34$.

Predating slightly Narayana's work on magic squares is the *Ganitasarakaumudi* of Thakur Pheru who only presents non pan-diagonal magic squares. Moreover, they are 'normal' magic squares of order $n = 3, 4, 5, 6, \ldots$ whose magic sum is $S = 15, 34, 65, 111, \ldots$ and are squares with the sequence of natural numbers $1, 2, \ldots, n^2$. Note that the magic sum is given by

$$S = \frac{n(n^2 + 1)}{2}.$$

However, in a pan-diagonal magic square described by Narayana, there is no sequence of natural numbers and the magic sum doesn't follow the rule for S given above. This is amply illustrated by the pan-diagonal magic squares given in Figures 9.4(a)–9.4(c).

A notable feature of Narayana's work on magic squares is his methodical presentation bringing into play a range of topics from mathematics. Also, fairly early in in his presentation he draws a connection between magic squares and arithmetical progression (AP). He sees a correspondence between the number of cells of a magic square and the number of terms of AP. And the beginning of the chapter, Narayana gives the following rule for finding the magic sum (S). If N is the number of terms, then

$$S = \frac{N^2 + N}{2\sqrt{N}}. \quad \text{If } N = 16, \text{ then } S = \frac{16^2 + 16}{2\sqrt{16}} = 34.$$

If we know the magic sum and the order of the magic square (n), the first thing we need to construct the magic square is the first term (a) and the common difference (d). To find these two quantities,

Narayana makes use of indeterminate analysis to solve the following equation:

$$S = n^2 \left[\frac{1}{2}\{a + [a + (n^2 - 1)d]\} \right] = na + \left(\frac{n}{2}\right)(n^2 - 1)d.$$

There exist an infinite number of integral solutions for (a, d) if S is divisible by the Greatest Common Denominator (GCD) of $\{n, \left(\frac{n}{2}\right)(n^2 - 1)\}$ or that S should be divisible by n when n is odd and by $\left(\frac{n}{2}\right)$ when n is even. An example from Narayana would illustrate this point. Given a 4×4 magic square with $S = 40$, the solution of the equation $40 = 4a + 30d$ will give integral pairs of $(a, d) = (-5, 2), (10, 0), (25, -2)$.

Figures 9.4(a)–(c) and Figure 9.5 are magic squares from earlier times. They are all pan-diagonal magic squares. As mentioned earlier, a pan-diagonal magic square (also known as diabolical magic square) is a magic square with the additional property that the broken diagonals, i.e. the diagonals that wrap round at the edges of the square, also add up to the magic constant. A pan-diagonal magic square remains pan-diagonally magic not only under rotation or reflection, but also if a row or column is moved from one side of the square to the opposite side. As such, an $n \times n$ pan-diagonal magic square can be regarded as having $8n^2$ orientations.

9.4.2   Constructing 4×4 pan-diagonal magic squares

The methods of constructing magic squares is a vast subject and it is not our intention to examine the procedures in any great detail.

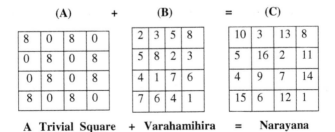

A Trivial Square + Varahamihira = Narayana

Figure 9.5. From Varahamihara to Narayana.

1	8	13	12
14	11	2	7
4	5	16	9
15	10	3	6

Figure 9.6. Knight moves.

Narayana distinguishes between four distinct ways of filling the magic with numbers. Three of these methods are similar to the one by Narayana to explain the formation of a fourth order magic square using a knight's move (in Indian chess known as turaga-*gati* or horse move). To illustrate, consider the following example from *Ganita Kaumudi* (Figure 9.6) of the smallest pan-diagonal magic squares of order 4. Its magic constant of 34 can be seen in a number of patterns in addition to the rows, columns and diagonals. Also any of the 16 2×2 sub-squares, including those that wrap around the edges of the whole square (e.g., $14 + 11 + 4 + 5$ or $1 + 12 + 6 + 15$) add up to 34. And so does the corners of any 3×3 square (e.g., $8 + 12 + 5 + 9$) as well the corners of any pair of horizontally or vertically adjacent numbers, together with the corresponding pair displaced by a $(2, 2)$ vector, e.g., $1 + 8 + 16 + 9$. Thus of the 86 possible sums adding to $34, 52$ of them form regular patterns, compared with 10 for an ordinary 4×4 magic square.

In chess, a knight moves, two cell horizontally (left or right) and then vertically (upwards or downwards). For example, if we start from the top left cell which contains **1** and then go horizontally right two cells (**13**) and vertically downwards by one cell, this would be the position of **2**. Then a movement from 2 downward along the adjacent cell would give **3** and then a knight move horizontally two cells to the right and one cell vertically would give **4**. And this process is continued to fill all the cells in the magic square. There are other methods such as the folding method which will not be discussed here.[14]

9.5   Local and Family Lineages: In Maharashtra and Varanasi

A family lineage of mathematicians and astronomers have been found in Maharashtra/Gujarat, notably Ganesha Daivajna (b. 1507) from Nandigrama (a coastal village about 40 miles south of present-day Mumbai) and Jnanaraja (beginning of 16th century) from Parthapura (about 70 miles from Aurangabad). Later in Varanasi (Benares), some of the descendants of the astronomers from Parthapura and Nandigrama settled and it soon became a magnet for a number of the scholars, some of whom were supported by rulers of the outlying areas.[15]

Ganesha's father was a well-known astronomer named Kesava. He came from a distinguished line of astronomers who composed texts on astronomy and astrology. His *Grahalakaghava* (modelled on his father's work *Grahakatuka*) became an immensely popular text whose fame spread to other parts of India. Of his mathematical works, *Buddhivilasini*, a commentary on Bhaskaracharya's *Lilavati* and *Tithi-Cintamani*, a commentary on the astronomical text, *Siddhantasiromani* are well known. Commentaries by Gangadhara, Visvanatha and Mallari spread the popularity to these texts even further.

To illustrate the contributions of Ganesha, let us just consider first his work on right-angled triangles. It may be remembered that in our discussion of the work of both Brahmagupta and Bhaskaracharya on rational right-angled triangles, the following is one of the solutions in the determination of the other two sides given one side a.

$$a, \frac{1}{2}\left(\frac{a^2}{n} - n\right), \frac{1}{2}\left(\frac{a^2}{n} + n\right),$$

where n is an arbitrarily chosen rational number.

Ganesha provides a rationale for this solution. Let $a, b,$ and c be the sides of a right-angled triangle where a is given and c is the hypotenuse.

From the Pythagorean theorem:

$$a^2 + b^2 = c^2 \quad \text{or} \quad a^2 = c^2 - b^2 = (c+b)(c-b).$$

Let

$$c - b = n. \tag{1}$$

Then

$$c + b = \frac{c^2 - b^2}{c - b} = \frac{a^2}{n}. \tag{2}$$

Subtracting (1) from (2) and simplifying to get

$$(a,\ b,\ c) = \left[a, \frac{1}{2} \left(\frac{a^2}{n} - n \right), \frac{1}{2} \left(\frac{a^2}{n} + n \right) \right].$$

In *Buddhivilasini*, Ganesha's commentary on *Lilavati*, there appears a more complete proof of the Pythagorean theorem improving on the rationale provided by Bhaskaracharya. In Figure 9.7, ABC is a right-angled triangle at C. Let CD be a perpendicular to AB. Now triangles ADC, BDC and ABC are similar.

From similar triangles ABC and ACD,

$$\frac{AC}{AB} = \frac{AD}{AC} \quad \text{or} \quad AD = \frac{AC^2}{AB}. \tag{1}$$

Also from similar triangles ABC and CBD, it can be shown that

$$DB = \frac{CB^2}{AB}. \tag{2}$$

Add (1) and (2) to get

$$AD + DB = AB = \frac{AC^2}{AB} + \frac{CB^2}{AB}$$

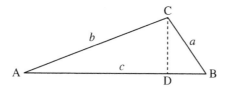

Figure 9.7. An improved proof of Pythagorean theorem.

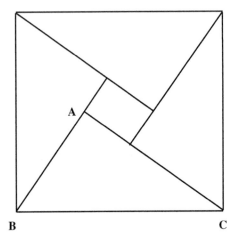

Figure 9.8. Behold!

or

$$AB^2 = AC^2 + CB^2.$$

It is interesting to note that an early historian of Indian mathematics, Colebrooke pointed out that the English mathematician John Wallis gave the same proof 100 of years later in his treatise on angular sections.

Ganesha also gave a visual demonstration of the Pythagorean result found earlier in Bhaskaracharya. In Figure 9.8, the three triangles congruent to a given right angled triangle ABC are placed in such a way to form a square with side BC with a square 'hole' at its centre — the square 'hole' has side equal to $(AC - AB)$. The area of each of the four congruent right-angled triangles $= \frac{1}{2}AB \cdot AC$. The area of the bigger square = Area of the smaller square + the area of the four triangles.

So

$$BC^2 = (AC - AB)^2 + 2AB \cdot AC$$

or

$$BC^2 = AC^2 + AB^2.$$

This proof relies on the reader to be persuaded about the square configuration of the perimeter and in the centre. But by the same token, Ganesh supplies in the earlier proof a deductive proof of the same theorem, suggesting that Indian mathematicians used both formal proof methods interchangeably with informal ones.

Endnotes

[1] *Karanapaddhati* forms part of the corpus of Kerala mathematics and was composed by Puthumana Somayaji in 1733. The flowering of mathematics and astronomy in that region of India forms the subject matter of the next chapter.

[2] The list of topics in order of their appearance consisted of: weights and measures; partnership, sales and interest; sequences and series; geometry of plane figures; excavations; mounds of grain; shadow problems; linear indeterminate equations; quadratic indeterminate equations; factorisation; unit fractions; combinatorics; magic squares. Narayana also wrote his own commentary which provides solutions to all the examples.

[3] It may be remembered that the Egyptian Ahmose Papyrus (1650 BCE) contained a table in which the fraction 2/n has been expressed as a sum of unit fraction for each value of n from 5 to 101. Such a table was used for multiplication and division. For further details, see Joseph (2011, pp. 91–100).

[4] Of course there are references to this method in Mahavira's *Ganitasarasangraha* without stating the rules. Sripati (*c.* 1039 CE) in his *Siddhantasekhara* provides the first known formal treatment of the subject which was acknowledged by Narayana.

[5] Note that if we denote the kth sum of n by $V_n^{(k)}$ which is equal to $^{n+k}C_{k+1}$, we can show that

$$V_n^{(k)} = V_1^{(k-1)} + V_2^{(k-1)} + \cdots + V_n^{(k-1)} = \sum_{r=1}^{n} V_r^{k-1}.$$

[6] The curvilinear figure referred to is obtained when the line joining the centres of the circles form a regular polygon.

[7] This follows from the application of the Pythagorean result and simplifying:

$$(x-1)^2 - \left(\frac{1}{2}x - 2\right)^2 = (x+1)^2 - \left(\frac{1}{2}x + 2\right)^2 = y^2.$$

[8] It is interesting that the title of Nagarjuna's book *Kacchaputa* in which magic squares are discussed also refers to what appears on the shell of a tortoise.

[9] It is interesting in the context of the wide and lasting popularity off the subject of magic squares that the first chapter of Srinivasa Ramanujan's *Notebooks* is on magic squares.

[10] A magic square is generally understood to be a square array of numbers arranged in such a way that the numbers along any row, column or principal diagonal add up to the same total. Depending on the number of ways that the desired results are achieved, the squares may be classified into: semi-magic (rows and columns add up to the desired sum), magic (includes diagonal as well) and pan diagonal (includes broken diagonals as well). In most magic squares of n rows and n columns the n^2 'cells' are occupied by the natural numbers from 1 to n^2. For example, a magic square of four rows and four columns (i.e. of order 4) could contain all the integers from 1 to 16. Magic squares have some interesting mathematical properties. If s is the constant sum of the numbers in each row, column or principal diagonal, and S is the grand total of the numbers in the n^2 cells, then $S = \frac{1}{2}n^2(n^2 + 1)$ and $s = S/n$.

If n is odd, the number of the central cell is given by S/n^2, which is also the mean of the series $1 + 2 + \cdots + n^2$. (For magic squares of even order, for which there is no single central cell, S/n^2 is not a whole number.) This is a key number for all odd-order magic squares, since from this number and the value of n it is possible to work out the partial sums s and total sum S. For example, if $n = 3$ and the middle number is 5, $S = 9 \times 5 = 45$ and $s = 45/3 = 15$.

[11] Although Narayana was one of the earliest writers to introduce the subject of magic squares into Indian mathematics, there are references to magic squares from the ancient Puranic times to discoveries by the legendary figure of Garga, the author of *Gargasamhita* (a book on astrology). These included wearing good-luck amulets inscribed with magic squares prescribed by Garga, to ward off the evil effects produced by bad positioning of Mars. It is therefore no surprise that magic squares appear in a work *Kacchaputa* composed by the alchemist Nagarjuna (*c.* 2nd century CE). Later, Varahamihira (*fl.* 550) in an astrological work *Sarvatobhadra* gave a magic square of the fourth order with a square constant of 18.

Magic squares are only of marginal interest today, forming part of a peripheral area known as 'recreational mathematics'. Yet until 400 years ago, in almost all mathematical traditions, magic squares engaged the interests of notable mathematicians as a challenging object of study.

Their attractions were heightened not only by their aesthetic appeal, but also by their association with divination and the occult; they were engraved on ornaments worn as talismans

[12] A pan-diagonal magic square has the additional property that it has 'broken diagonals' that wrap around the edges of the squares adding up to the square constant of 18. Note that Varahamihira's magic square has the constant of 18 in the case of the following combinations:

   (i) Each of the columns, rows, diagonals, broken diagonals;
   (ii) Each of the four corners $[2, 3, 5, 8]$, $[5, 8, 3, 2]$, $[4, 1, 6, 7]$, $[7, 6, 1, 4]$;
   (iii) The small central square $[8, 2, 7, 1]$;
   (iv) The four corner of the magic square $[2, 8, 1, 7]$;
   (v) The sum of the two central cells of the first row $[3, 5]$ and the central cells of the last row $[6, 4]$;
   (vi) The sum of the two central cells of the first column $[5, 4]$ and the central cells of the last column $[3, 6]$.

[13] The perfume produced contained two parts of *agaru*, three parts of *patra*, five parts of *turuska* and eight parts of *saileya* giving 18 parts altogether. The other rows indicate different combinations of the same ingredients.

[14] For further information about the work on magic squares in India, see Kusuba (1993) and Paramand Singh (1998–2002) and Sridharan and Srinivas (2011).

[15] A recorded example is that of Divakara, a student of Ganesha, who moved to Varanasi and began a family lineage of scholars consisting of his five sons, two grandsons and three great grandsons.

Chapter 10

A Passage to Infinity: The Kerala School and Its Impact

10.1 Introduction

Two powerful tools contributed to the creation of modern mathematics in the seventeenth century: the discovery of the general algorithms of calculus and the development and application of infinite series techniques. These two streams of discovery reinforced each other in their simultaneous development since each served to extend the range and application of the other.

Existing literature would lead us to believe that the methods of the calculus were invented independently by Newton and Leibniz, building on the works of their European predecessors such as Fermat, Taylor, Gregory, Pascal and Bernoulli in the preceding half century. What appears to be less well known is that certain fundamental elements of the calculus, including numerical integration methods and infinite series derivations for π and certain trigonometric functions such as $\sin x$, $\cos x$ and $\tan^{-1} x$ (or π when $x = 1$), were already known in Kerala over 250 years earlier.[1]

For much of the twentieth century, it was believed that mathematics and astronomy in India ceased to make progress after the twelfth century and that Bhaskaracharya, the author of *Lilavati* and *Bijaganita*, was the last great Indian mathematician and astronomer. All works, which appeared after him, were thought to be mere commentaries on earlier works, having no original contributions to make. This was at times associated with a parallel belief that Muslim rulers turned a blind eye to the growth of science and technology, and hardly ever patronised the production of new knowledge. These arguments were a continuation of the view that progress in these disciplines (astronomy and mathematics) was spasmodic with gaps or fissures appearing between the *Sulbasutras* (c. 800–400 BCE) and Aryabhata I (b. 476 CE), and then after 500 years of creativity long silence, only to be broken by the introduction of Western mathematics around the middle of the 19th century.[2]

The situation is somewhat different today. The belief that no significant mathematical treatise appeared in India after the twelfth century has been called into question. In the second half of the 20th century there has been some acknowledgement (for example Baron, 1969; Calinger, 1999; Jushkevich, 1964; Katz, 1992) of the Indian origin of the analysis and derivations of certain infinite series, although none would admit to the beginnings of calculus there. Even such acknowledgement is not universal. For example, Fiegenbaum (1986), biographer of Brook Taylor, mentions a group of European mathematicians who anticipated the famous 'Taylor's theorem' but makes no reference to the Kerala work. Fiegenbaum does, however, issue this disclaimer: "It reflects only our current awareness of the published and unpublished papers of Taylor's predecessors." (p. 72). But, prior to his paper, several publications referring to Kerala mathematics had already appeared in the West of which Whish (1835)[3] was an early example and Rajagopal and his collaborators (1944, 1949, 1952, 1978), and Jushkevich (1964) are more recent examples. There were also histories of Indian mathematics that discussed the Kerala contribution, such as Bag (1979), Saraswati Amma (1967) and Srinivasiengar (1967).

We come across as many as 11 leading mathematicians in Kerala between the fourteenth and the nineteenth century. The last of them was Sankaravarman (nineteenth century) who came from a princely family in northern Malabar. His immediate known predecessor was Pudumana Somayaji (eighteenth century), who came from Trissur. The rest of them were part of a scholarly pedigree, which commenced with Madhava of Sangamagrama whose student was Vadasseri Paramesvara of Alathur who taught his son, Damodara. The teacher–student lineage continued with Nilakantha Somayaji, Jyesthadeva, Chitrabhanu, Acyuta Pisharati, Sankara Variyar and Mahishamangalam Narayanan.[4]

There are several questions worth exploring about the activities of this group of mathematicians/astronomers (hereafter called the Kerala School[5]), apart from technical ones relating to the mathematical content of their work. In this chapter we begin with a survey of the social and historical background of the Kerala School before examining the motivation and the mathematical content of their work. We return finally to the vexed issue of possible transmission of the work of the Kerala School to Europe between sixteen and seventeen centuries.

Figure 10.1 contains a schematic presentation of the principal 'actors' of the Kerala School of Mathematics and Astronomy. The stage for their presentation was highly localised, confined to an area of about a 1,000 square kilometres north of the present-day metropolis of Kochi (or Cochin). Their wider impact may not have gone much beyond this area. The history and genesis of this School and its demise provides a useful backdrop to its mathematical achievement.

10.2   The Origins of the Kerala School

In 683 CE, a meeting was called at a conference at Tirunavaya (in Northern Kerala) to discuss the shortcomings of the astronomical methods of Aryabhata I and his School and to reform the system. It is assumed by some that the legendary figure of Varurici,[6] popularly believed to be the founder of astronomical studies in Kerala, was responsible for setting up this meeting as well as introducing two important changes to prevailing astronomical practices. As the author of 248 *chandravakyas* (or "moon sentences"),[7] he popularised their use to describe, through a series of "nonsense" mnemonic words or phrases, the positions of the moon at regular intervals each day that enabled worshippers carry out their daily observances and

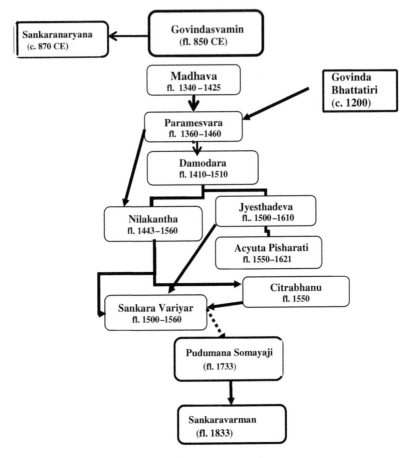

Figure 10.1. The principal actors.

rituals at the right time. He was believed to have introduced a new system of numerical notation, known as the *Katapayadi* into Kerala.[8] Most numerical information (in Kerala) was subsequently recorded through this system of notation.

A more likely participant at the meeting of astronomers was Haridatta (*fl.* 650–700) who introduced the *parahita* system.[9] A modified planetary system based on *Aryabhatiya* of Aryabhata I and its various commentaries, it used the *katapayadi* system of notation for the first time. This system remained popular for fixing auspicious times for rituals and ceremonies even after the introduction of the more accurate Paramesvara's *drgganita* system (i.e. a system based on a revised set of astronomical parameters) in the fifteenth century.

By the beginning of the ninth century, Kerala had undergone significant political changes. The Kulasekharas in the person of Kulasekhara Alvar had reestablished the Cera power in Kerala.[10] The history of this dynasty, often called the Second Cera Empire, has only come to light recently from a study of the inscriptions of that age. It was a dynasty that kept Kerala relatively peaceful and united for 200 years. It was also a period of uninterrupted growth of scientific endeavours, including mathematics and astronomy.

The burgeoning interest in astronomy was given a further boost during the reign of the second in the Kulasekhra dynasty, Sthanu Ravi Kulasekhara (844–870). He established an observatory, containing a giant armillary sphere, in Mahodayapuram (the present-day Kodungallur), under the charge of Sankarnarayana.[11] Sankarnarayana, the court astronomer, was a student of Govindasvamin and like his teacher applied the Aryabhatan planetary model in his calculation, basing it on Bhaskara I's *Laghubhaskariya* (the shorter commentary on *Aryabhatiya*).

Govindasvamin, who lived during the first half of the ninth century, prepared a *bhasya* (or commentary) on the *Mahabhaskariya*, which in turn was a detailed commentary by Bhaskara I on *Aryabhatiya*. This text contains some interesting mathematics, including rules for second-order interpolation to estimate intermediate sine values for different intervals that turns out to be a special case of the general Newton–Gauss interpolation formula.[12]

The age of the Kulasekharas was also notable for progress in the field of education and learning. Schools and colleges attached to temples sprang up in different parts of Kerala. These institutions, known as *salas*, were maintained by the court with the help of private endowments. 100's of youth, including Buddhists and Jains, were given free boarding and lodging to study mainly religious scriptures. In some institutions there were also courses on philosophy (*mimamsa*), grammar (*vyakarana*), law (*dharmasastra*) and astronomy (*jyotisastra*). These *salas* gradually became an integral part of the temple, financed from revenue received by the temple from neighbouring villages.

In 999, the period of peace and stability ended with the start of a prolonged war between the two major powers of the region that was believed to have lasted for over a 100 years. The *Cera–Chola* conflict led to extensive changes and the way of life that emerged in Kerala was very different from what it was before.

The war called for a total mobilisation of the resources of the state. The Nambuthiri Brahmins, who had gradually established themselves at the top of the social hierarchy during centuries of Cera rule, provided valuable assistance to the rulers. A few of them even gave up their traditional pursuits to take up arms. A number of *salas* were turned into military training camps.

Between the period of the end of the Second Cera dynasty and the coming of the Portuguese (1102–1498), four major kingdoms emerged in Kerala, namely Venad, Calicut, Cochin and Kolathunad, with the first two becoming powerful military states. And in the absence of a strong central power, the combined effects of a rigid caste structure, the growing power of the temple and the considerable autonomy enjoyed by local feudal chieftains, led to weak governments in Cochin and Kolathunad. This is particularly relevant for our study since the location of the Kerala School was mainly in the kingdom of Cochin. To simplify exposition, we refer to Kerala of this period as medieval Kerala.

For about 300 years after Govindaswamin no major figure appeared in Kerala. Indeed, hardly any records exist for sketching developments in mathematics and astronomy from the eleventh to thirteenth. However, it was a period of considerable activity in North India where mathematician-astronomers, such as Sridhara (*fl.* 900), Aryabhata II (*fl.* 950), Sripati (*fl.* 1040) and of course the famous Bhaskaracharya (b. 1114), were in the forefront. Some of their work must have slowly percolated into Kerala for their names were known to those who came later.

The next major figure was Govinda Bhattatiri[13] who lived around 1175. Born in the village of Alathur (near present-day Tirur), one of the 32 Brahmin settlements in Kerala established during the seventh and eighth centuries, his major work was *Dasadhyaya*, a commentary on the first ten chapters of the astrological text, *Brhajjataka*, the work of the astronomer Varahamihira (*c.* 505–587 CE). This is generally recognised as one of the most important of the 70 other known commentaries on this text. A lesser known work, *Mahurtaratna*, was referred to by Paramesvara (a major figure of the Kerala School) as an influential astrological treatise. Indeed, a notable family of astrologers in Kerala, the *Kaniyans* of Pazhur, trace their astrological prowess to their supposed descent from Govinda.[14]

10.3 The Kerala School: The Actors

The next century saw the emergence of the founder of the Kerala School, Madhava of Sangamagramma (ca. 1380–1420). Sangama-gramma, a village near Cochin dedicated to a deity of the same name, was one of the 32 Brahmin settlements and located in the present-day Irinjalakkudu. His only surviving works are in astronomy and mainly concerned with refining the *vakya* system of Varuruci. In both *Venvaroha* and *Sphutacandrapti*, Madhava carried out a revision of the *chandravakyas*, calculating the exact positions of the moon, correct to the second, for given intervals of the day. Before Madhava's work, Varurici's "moon-sentences" only gave values correct to the minute. Using the cyclic nature of the lunar *vakyas* in which nine anomalistic (irregular) months equal 248 days, Madhava estimated the lunar longitude at nine equally distant times of a day and computed the longitudes of the planets and of the ascendant.

However, it is from the commentaries of those who came after him, such as Paramesvara, Nilakantha and Jyesthadeva, that we know of Madhava's remarkable contribution to the development of Kerala mathematics. His fame rests on his discovery of the infinite series for circular and trigonometric functions, notably the Gregory series for arctangent, the Leibniz series for π, and the Newton power series for sine and cosine. There are also some remarkable approximations, based on the incorporation of "correction" terms to these slowly converging series that have also been attributed to Madhava.[15]

Madhava's notable student was Paramesvara. He was born in 1380 into a Nambuthiri Brahmin family of Vedic scholars in the village of Alathur, the birthplace of Govinda Bhattatiri who had taught his father. According to Nilakantha, Paramesvara learnt mathematics from Madhava. He wrote a number of commentaries including ones on Aryabhata's *Aryabhatiya*, on Bhaskara I's *Mahabhaskariya* and *Laghubhaskariya*, on *Suryasiddhanta*, on Govindasvamin's commentary on *Mahabhaskariya*, on Manjula's *Laghumanasa* and on Bhaskaracharya's *Lilavati*. His independent works included *Goladipika, Drigganita, Grahanamandana* and *Vakyakaran*, all of which dealt with planetary motions and their calculation. Paramesvara's role in scrutinising and then disseminating the contents of these major texts of Indian astronomy and mathematics should not be underestimated.

Paramesvara's main importance in the development of planetary astronomy in South India is his *drgganita* system, expounded in his text of that name written in 1431. This is an astronomical work in two parts. The first deals with the derivations of the mean positions and equations of the centre of the planets, the corrections made and the method for calculating the orbital arc from the sine tables. The second part is a summary of the first part using the *katapayadi* notation. He also wrote a work called *Goladipika*, dealing with various aspects of spherical astronomy. The two versions together contain a detailed discussion of the great gnomon, shadow and parallax; the construction of an armillary sphere, the apparent and true motions of planets, methods of measuring the circumference of the earth and other topics. Between 1393 and 1432, Paramesvara made a series of observations of eclipses of the sun and the moon and recorded them in three short works. He also wrote on astrology, including a commentary on Govinda's *Muhurtratna*.

The foundation laid by Paramesvara heralded the emergence of the major figure of Nilakantha Somayaji. He was born in 1443 into a Nambuthiri family of *Somatris* (those who performed the *soma* sacrifice) in Trikkantiyur in the Ponnani district. He stayed and studied in the house of Damodara (*fl.* 1450), the son of Paramesvara, and probably taught by Paramesvara himself whom Nilakantha referred to as his *paramaguru* ("principal guru"). His major works include *Aryabhatiyabhasya*, a commentary on three chapters of *Aryabhatiya* and a comprehensive survey of all literature available on this text. His fame, however, rests on his work, *Tantrasangraha*.

This work is in eight chapters, containing 432 verses connected with astronomical calculations and follows the *drigganita* system introduced by Paramesvara. It deals with a variety of subjects including the fixing of the gnomon, calculations of the meridian, of latitude, of the declensions etc., and the prediction of eclipses. To illustrate its range and originality, consider briefly two innovations of Nilakantha that have implications for history of astronomy.[16]

In *Tantrasangraha*, Nilakantha carried out a major revision to the Aryabhatan model for the interior planets, Mercury and Venus. In doing so, Nilakantha arrived at a more accurate specification of the equation of the centre for these planets than any other that existed in Arab or European astronomy before Kepler, who was born about 130 years after Nilakantha. In *Aryabatiyabhasya*, Nilakantha developed a computational scheme for planetary motion which is superior to that of Tycho Brahe (1546–1601) in that it correctly takes account of the equation of centre and latitudinal motion of the interior planets. This computational scheme implies a heliocentric model of planetary motion where the five planets (Mercury, Venus, Mars, Jupiter and Saturn) move in eccentric orbits around the mean Sun which in turn goes round the earth. This model is similar to the one suggested by Brahe when he revised Copernicus's heliocentric model. It is significant that all the astronomers of the Kerala School who followed Nilakantha, including Jysthadeva, Acyuta Pissaroti and Pudumana Somayaji, accepted Nilakantha's planetary model.

The other works which he wrote late in life were either commentaries on his earlier works, such as those on his *Chandravakyayaganita* and *Siddhantadarpana* where the latter consists of a short work in 32 verses dealing with certain important astronomical constants, the theory of epicycles and other matters of topical interest; or works such as *Golasara*, a textbook in three chapters on spherical astronomy; or *Sundararaja Prasnottara*, a work which has not been found but mentioned both by Nilakantha and others, giving answers to questions raised by a Tamil astronomer called Sundararaja. The last work is important because it provides rare evidence of interest in Kerala mathematics and astronomy in other areas of South India.

It is likely that Nilakantha saw his 100th birthday and during his long life he taught many students, some of them who were to become important figures in their own right. His *Tantrasangraha* was a compendium of all the results known up to his time and it generated among those who came after him a number of commentaries, both in Malayalam and Sanskrit, which form an important basis of knowledge of Kerala mathematics and astronomy.

One of Nilakantha's students was Chitrabhanu (*fl.* 1530). He was a Nambuthiri from the *Gautamagotra*[17] family and came from the Brahmin village of Sivapuram (in present-day Trissur). His work, *Karanamrita*, contained four chapters dealing with advanced astronomical calculations within the framework of the *drgganita* system and which constitutes the basis of the traditional calendar (panchagam) even today. He was also the author of *Ekamvimsati Prasnottara* (21 Questions and Answers) solving each of a set of 21 (i.e.$^7C_2$) pairs of simultaneous equations in two unknowns. The 21 pairs arise by taking, at a time, any two of the following equations:

$$x + y = a; \quad x - y = b; \quad xy = c; \quad x^2 + y^2 = d; \quad x^2 - y^2 = e;$$
$$x^3 + y^3 = f; \quad x^3 - y^3 = g.$$

The solutions to 15 of the 21 pairs are fairly straightforward while the remaining six are not.[18]

A student of Chitrabhanu, Mahishamangalam Narayana (*fl.* 1560), completed one of the major texts of the Kerala school, *Kriyakramakari*, in 1556. A commentary on Bhaskaracharya's *Lilavati*, it was started by Sankara Variyar (*c.* 1500–1560),[19] a student of both Nilakantha and Chitrabhanu. It is important in the history of Kerala mathematics and astronomy for its detailed discussion of the works of earlier writers, some of which are not extant, and for providing a rationale and proof of a number of earlier results. Sankara also wrote a commentary on the *Tantrasangraha* of Nilakantha called *Laghuvivrti* (or *Yuktidipika*). This text is said to have been based on *Yuktibhasa* which was originally written in Malayalam by another student of Nilakantha.

Jyesthadeva (*fl.* 1500–1575), the author of the highly influential text of the Kerala School, *Yuktibhasa*, was a Nambuthiri Brahmin from the Alathur village. There are at least three versions of this text of which the Malayalam version became an important source of dissemination throughout Kerala.[20] Based on Nilakantha's *Tantrasangraha*, it is unique in that it gives detailed rationale, proofs or derivations of many theorems and formulae in use among the astronomers/mathematicians of that time.

A student of Jyesthadeva came from the Pisaroti community. They were not Brahmins but performed traditional functions as cleaners and suppliers of flowers and plants for the temple. They were also employed by some Nambuthiri families to give instructions on the calculation of the astrological calendar (*panchagam*) and time reckoning.[21] Acyuta Pisaroti (*c.* 1550–1621) was a considerable scholar who made a mark not only in astronomy but in literature and medicine. He attracted the attention of the ruler Raja Ravi Varma of Venad and earned high praise for his astronomical skills such that a contemporary, Vasudeva, described him as greater than even Lord Siva! His major contribution is found in his work, *Sputanarnaya*, where he introduces for the first time in Indian astronomy, a correction called "Reduction to the ecliptic," around the same time as Tycho Brahe did in Western astronomy.[22] In other works (*Rasigolasphuti, Uparagakrikrama Horarsarochaya and Jatakabharanapphati*), he wrote on the computation of the position of the planets, on the constellations and the celestial and terrestrial spheres, on eclipses and on the moon's shadow. He also wrote a Malayalam commentary on Madhava's *Venvaroha*.

In Charles Whish's 1832 paper, which drew the attention of the world for the first time to the existence of Kerala mathematics and astronomy, appears the following passage: "The author of the *Karanapaddhati* whose grandson is now alive in his 70th year was Pudumana Somayaji, a Nambuthiri Brahmana of Trisivapur (Trissur) in Malabar." A major work in the dissemination of

Kerala mathematics and astronomy not only in Kerala but also in the neighbouring areas of present-day Tamilnadu and Andhra Pradesh, *Karanapaddhati* was recorded in 1733, about 200 years after Jysthadeva's *Yuktibhasa*. All we know is that he was a Nambuthiri who carried out *soma* sacrificial rites and belonged to the Pudumana family of Sivapuram in present-day Trissur.

Karanapaddhati is a comprehensive treatise covering Kerala mathematics and astronomy. It has one unusual feature. It follows generally the older *parahita* system and only advocates the *drgganita* system in the prediction of eclipses. In 10 chapters it explains problems that appear in earlier texts in Indian mathematics, like the *kuttakara* approach to solving indeterminate equations or the derivation of implicit values for π and for sines and cosines of angles. While it covers more or less the same ground as the *Yuktibhasa*, its non-technical clarity in explaining from the first principles, methods of deriving various formulae and construction of tables of astronomical constants meant that it became an important source for commentaries, with two in Malayalam, two in Tamil and one in Sanskrit having been discovered so far. Pudumana Somayaji also wrote, in the fashion of the day, an elementary manual, *Nyayaratna*, for explaining "astronomical rationale to the dull-witted" and practical texts such as *Venvarohastaka* for determining the positions of the moon at regular intervals.

After Acyuta Pisharoti, little in the way of original work was done, although the tradition of providing corrections and contributing to the preparation of the astronomical ephemeris for the daily needs of the faithful observers of *muhurtha* (auspicious times) continued. The compilation of the *panchanga* (five limbs) was periodically subjected to *sphuta* (or refinement). About 100 years after *Karanapaddhati* appeared the last of the known texts of the Kerala School. The author of this book, Sankarvarman of Katattanad in North Kerala belonged to the royal family of that area and was a contemporary of Charles Whish. His book, *Sadratnamala*, written in 1823, contains many of the results of the Kerala School, given without the rationale or derivations found in the earlier texts. Whish met him and described him as "a very intelligent man and acute mathematician." He died six years after Whish's article on Kerala mathematics and astronomy appeared in 1832.

Astronomy provided the main motive for the study of infinite-series expansions and rational approximations of circular and trigonometric functions. For astronomical work, it was necessary to have both an accurate estimate of π and highly detailed trigonometric tables. In this and other areas the members of the Kerala School made significant discoveries.

10.4 The Social Landscape of Medieval Kerala

As stated earlier, the mathematicians/astronomers of the Kerala School were predominantly Nambuthiri brahmins with a few who came from sub-castes, such as the *Variyar* and the *Pisaroti*, traditionally associated with specific functions in the temple. The personnel of the temple consisted broadly of three groups: scholars (*bhatta*), priests (*santi*) and functionaries (*panimakkal*), The scholars and priests were invariably brahmins. Among the non-brahmin functionaries were *Variyars* who looked after the routine tasks of the temple and often kept the accounts of the temple. The *Pisarotis* were a subcaste of non-brahmin priests who officiated during the performance of rituals in their own temples. They sometimes acted as Sanskrit instructors. This group seems to have evolved and acquired a high social status over a long period — a period that saw the establishment of the hegemony of brahmins over the Kerala society through their substantial land holding sanctified by custom (*devasvam*). The brahmins had the resources and leisure to pursue higher learning, including the study and reinterpretation of certain mathematical and astronomical works of the earlier period from the North.

The Nambuthiri Brahmins were a patriarchal people following a strict primogeniture system of inheritance. In addition to the structures of political importance enhancing their social and economic powers, there was a customary device called *sambandham*, a form of sexual alliance with the non-Brahmin castes, particularly women from the *Nair* ruling aristocracy. The eldest son of a Nambuthiri family alone entered into a normal marriage alliance (*veli*) with a Nambuthiri female while his younger brothers, if there were any, could only form a *sambandham* relationship. This arrangement was in a way an interlocking institution of the patriachal Nambuthiri males and matrilienal Nair females. It had the effect of divorcing any family responsibility from the younger sons among the Nambuthiris while at the same time stabilising the system of matrilineal inheritance among the Nairs.

The system of primogeniture kept the eldest son of the Nambuthiri family busy looking after the property and community affairs while his younger brothers were totally unencumbered and had plenty of leisure. Lacking in social status on a par with the eldest brother, the younger ones had the need to attain social respectability through other means. Scholarship, both secular and religious, was one way available to them to make a mark. It is therefore little surprise that a number of well-known mathematicians/astronomers emerged from among this unencumbered section of the Nambuthiris.

There exists numerous records called *granthavaris* recounting the day-to-day accounts, from the late 15th century onwards, of prominent families (*swarupams*), Nambuthiri caste corporations (*yogams* and *sanketams*) and prominent Nair houses (*taravads*). These are in the form of palm leaf manuscripts now being studied by historians for reconstructing the socioeconomic history of pre-colonial Kerala. They contain details about economic transactions, social relationships and cultural practices. A noticeable feature is the importance given to the practice of meticulously documenting the events and accounts of the economic transactions of the day.

The *granthavari* culture of the time represented a document-minded society which replaced its oral-based predecessors. A significant aspect of the *granthavaris* was their metrical precision within a numerate culture. In recording the accounts of receipts and expenditure, they showed rare precision reaching out to the minute fractions of numbers. It was a period of the proliferation of garden lands and multiplication of economic resources. The landlords had to be numerate to keep account of their economic and financial transactions. Such a set up of economic transactions necessitating extensive reckoning practices may have stimulated the growth of higher mathematics during the period.

In the history of Kerala's agrarian expansion, the dissemination of calendrical knowledge had a very vital role to play. The knowledge of solar calendar and the skill of agrarian management of seasons was a crucial source of Brahmin economic domination. Astronomy and mathematics were the two instruments of contemporary seasonal forecasts. The heliocentric calendar was crucial to all socio-cultural practices of the period. This again points to the socio-economic relevance of astronomical/higher mathematical learning during the period.

10.5   The Highlights of the Kerala School

A list of the principal achievements of the Kerala School in the wider context of the general history of mathematics and astronomy would include:

(i) *The first correct formulation of the equation of the centre* for the interior planets, Mercury and Venus, by Nilakantha (1444–1545) in the *Tantrasangraha* about one hundred years before the German astronomer, Johannes Kepler (1571–1630).[23]

(ii) *The discovery of the formula for the 'reduction to the ecliptic'*, as mentioned earlier, first given in the West by Tycho Brahe (1546–1501), but in Kerala by his contemporary, Acyuta Pisharati (1550–1621), in his book *Sphutanirnaya*.

(iii) Nilakantha's commentary on *Aryabhatiya* contains a number of interesting geometrical demonstrations, some of which were known earlier and discussed in earlier chapters. These include showing that:

(1) *The area of a circle is equal to the product of half its circumference and half the diameter*

This was done by cutting up a circle into a large number of equal tapering laminas (Figure 10.2(a)) and on the basis that the base of each lamina (being a small arc segment) will approximate a straight line. Juxtaposition of each pair of two thin laminas

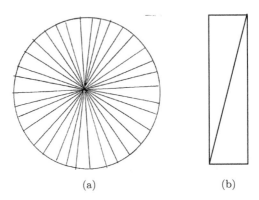

Figure 10.2. Area of a circle.

(Figure 10.2(b)) forms a series of rectangles with the longer side being equal to the radius and the smaller being the arc segment. These rectangles are then re-arranged into a rectangular sheet. From this one can deduce that the area of the circle is approximately equal to the area of the rectangular sheet. Note that the area of the rectangular sheet is the product of the adjacent sides consisting of the product of half the circumference and of the radius (or half the diameter).

(2) *The chord of one-sixth of the circumference of a circle is half the diameter*

The demonstration of this result follows directly from the diagram below. Figure 10.3 contains three equilateral triangles where the side is a chord of length radius. Six of these triangles are inscribed in the circle.

(3) *The sum of an infinite convergent geometric series*

This result was stated by Nilakantha in his commentary on the *Aryabhatiya*.[24] It was used in deriving an approximation for an arc of a circle in terms of its chord. Sarasvati Amma (1963, pp. 325–326) gives it as follows:

> The sum of an infinite series, whose terms (from the second onwards) obtained by diminishing the preceding ones by the same divisor (i.e. by the denominator of the first term) is always equal

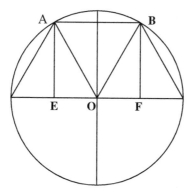

Figure 10.3. Chord and circumference of a circle.

to the numerator of the first term divided by one less the common divisor.

Expressed in modern notation, this is equivalent to the modern rule for the summation of a convergent infinite geometric series. If the first term is a and r is the common ratio (or divisor), the geometric series generated is of the form

$$a + ar + ar^2 + ar^3 + \cdots .$$

The sum of the series assuming, that it is convergent (i.e. $-1 < r < 1$) is

$$S = \frac{a}{1 - r}.$$

[So if $a = 1$ and $r = \frac{1}{2}$, then $S = 2$].

To illustrate the application of the Nilakantha Rule, consider the following infinite convergent geometric series:

$$\left(\frac{1}{4}\right) + \left(\frac{1}{4}\right)^2 + \left(\frac{1}{4}\right)^3 + \cdots .$$

The problem expressed by Nilakantha is as follows:

The entire series of powers of 1/4 adds up to just 1/3. How does one know that the sum increases only up to that [limiting value] and that it actually does increase up to that [limiting value]?

The crucial step in Nilakantha's argument for the derivation of this infinite convergent geometric series is:

As we sum more terms, the difference between 1/3 and sum of the powers of 1/4 becomes extremely small but never diminishes to zero. Only when we take all the terms together do we obtain the equality to 1/3.

Expressed in modern terms:

$$\left(\frac{1}{4}\right) + \left(\frac{1}{4}\right)^2 + \left(\frac{1}{4}\right)^3 + \cdots$$

$$= \frac{1}{3} - \left[\left(\frac{1}{3} - \frac{1}{4}\right) - \left(\frac{1}{12} - \frac{1}{16}\right) - \left(\frac{1}{48} - \frac{1}{64}\right) - \cdots\right]$$

$$= \frac{1}{3} - \left[\frac{1}{(4)(3)} + \frac{1}{(4)(4)(3)} + \frac{1}{(4)(4)(4)(3)} + \cdots\right]$$

$$= \frac{1}{3} - \left[\sum_{n=1}^{\infty} \frac{1}{(3)(4)^n}\right].$$

If n is very large, the terms in the square bracket will become negligibly small and can be ignored. Therefore, the sum of this infinite series equals 1/3 when $n \to \infty$. As we shall see in the Appendix 1 to this chapter, this rule found an application in the derivation of the arctan (and π) series, a major milestone in Kerala mathematics.[25]

(4) *Nilakantha's approach to the summation of an arithmetical series*

Each term of the arithmetical series is represented by a rectangular strip (see Figure 10.4(a)) whose length is equal to the number itself and whose width is one unit. Each of the strips are arranged in a manner given in Figure 10.4(a) such that the "piling up of the rectangles" represents the series and the area of the figure is the sum of the series. Now assume that, as shown in Figure 10.4(b), the piled up rectangular strips are fitted together, with one inverted to allow for such a fit. Now the adjacent sides of the whole rectangle in Figure 10.4b are given by the number of rectangular strips and the

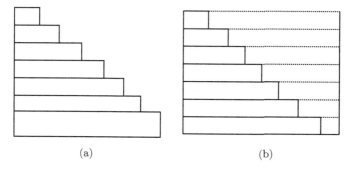

Figure 10.4. Piling up rectangles.

sum of the first term (a) and the last term (l) of the series. So the area of the whole rectangle is $n(a + l)$ and therefore the sum of the arithmetic series is $\frac{1}{2}n(a + l)$.

Nilakantha uses a similar form of geometrical reasoning in his *Aryabhatiyabhasya* to obtain the sum of a series of triangular numbers,[26] the sum of sums,[27] etc. This technique was then extended by Sankara Variyar in his *Kriyakramakari*.[28]

(5) *The discovery of Newton-Gauss interpolation formula* for a particular case (to second order) by Govindasvamin (*c.* 800–850), much earlier than either of the names associated with the formula. This formula may be expressed in modern notation as:

$$f(x + h) = f(x) + n\Delta h f(x) + \frac{n(n + 1)}{2}[\Delta f(x) - \Delta f(x - h)],$$

where Δ is the finite difference operator and h is the step-length.[29]

A consistent strand running through the work of Indian mathematicians, at least from the time of Aryabhata I (b. 476 CE), was the search for more accurate methods of interpolation, particularly in respect of sine values for intermediate angles. In fact, Brahmagupta (b. 598 CE) used a second difference interpolation formula which was rediscovered nearly a 1,000 years later and labelled as Newton–Stirling formula.

(6) In Paramesvara's commentary on Bhaskara II's *Lilavati* occurs a number of formulae relating to a cyclic quadrilateral and in particular in *obtaining the circumradius of a cyclic quadrilateral whose sides are known*. The cyclic quadrilateral was an important device used by the Kerala School for deriving trigonometric results, including:

$$\sin^2 A - \sin^2 B = \sin(A + B) \cdot \sin(A - B),$$

$$\sin A \cdot \sin B = \sin^2 \frac{1}{2}(A + B) - \sin^2 \frac{1}{2}(A - B).$$

As mentioned in an earlier chapter, Paramesvara states a formula for the diagonals of a cyclic quadrilateral (first discussed in the *Brahmasphutasiddhanta* of Brahmagupta) thus:

> The three sums of the products of the sides, taken two at a time, are to be multiplied together and divided by the product of the sums of the sides taken three at a time and diminished by the fourth. If a circle is drawn with the square root of this quantity as radius the whole quadrilateral will be situated in it.

A rationale for this rule is provided both in the *Kriyakramakari* and in the *Yuktibhasa*, These result made their first appearance in Europe in 1782 in the work of l'Huilier.

(7) *The power series for the sine and cosine*

They are stated in Nilakantha's *Aryabhatiyabhasa*, derived in Jyesthadeva's *Yuktibhasa* and attributed to Madhava in both texts. In modern notation, and expressed in degrees, the two power series are:

$$\sin \theta = \theta - \frac{\theta^3}{3!} + \frac{\theta^5}{5!} - \frac{\theta^7}{7!} + \cdots,$$

$$\cos \theta = 1 - \frac{\theta^2}{2!} + \frac{\theta^4}{4!} - \frac{\theta^6}{6!} + \cdots.$$

It has been suggested that Madhava may have used the first of the above power series to construct a sine table of 24 values (i.e. the values of cumulative series obtained by dividing the quadrant of a circle into 24 equal parts) which corresponds to the present-day angular measurement of 3 degrees and 45 minutes. A more plausible explanation is that Madhava developed the '*Suryasiddhanta*-Aryabhata' method, to be discussed in Chapter 11, to arrive at such an accurate table. The values are correct in almost all cases to the 8th or 9th decimal place. These power series make their first appearance in Europe in 1676 in a letter written by Newton to the Secretary of the Royal Society.[30]

(8) Nilakantha's *Aryabhatiyabhasya* and Jysthadeva's *Yuktibhasa* contains the following approximations with the Malayalam "*iti Madhava*" ("thus said Madhava") appearing at the end of the latter text and the Sanskrit "*tatraha Madhavah*" in the former text. The following passage is from the Nilankantha:

"Place the (sine and cosine) chords nearest to the arc whose sine and cosine chords are required and obtain the arc difference to be subtracted or added. For making the correction, 13751 should be divided by twice the arc difference in minutes and the quotient is to be placed as the divisor. Divide the one (say sine) by this (divisor) and add to or subtract from the other (i.e. cosine), according to whether the arc difference is to be added or subtracted. Double (this result) and do as before (i.e. divide by the divisor). Add or subtract the result (so obtained) to or from the first sine or cosine to get the desired sine or cosine chords."

Expressed in modern symbolic notation, the above rule may be expressed as:

$$\sin(\theta + h) \approx \sin\theta + \frac{h}{r}\cos\theta - \frac{h^2}{2r^2}\sin\theta,$$

$$\cos(\theta + h) \approx \cos\theta - \frac{h}{r}\sin\theta - \frac{h^2}{2r^2}\cos\theta.$$

These results are but special cases of one of the most familiar expansions in mathematics, the *Taylor series* (named after Brook

Taylor, 1685–1731), up to the second power of the small quantity u:

$$f(\theta + u) = f(\theta) + u \cdot f'(\theta) + \left(\frac{u^2}{2}\right) f''(\theta) + \cdots,$$

where u (in radian measure) $= h/r$

(9) In the *Yuktibhasa* and other Kerala texts, a key result relates to the *power series of circular and trigonometric function.*[31] The demonstration in the *Yuktibhasa* follows along the lines explained in the Appendix 1 to this chapter, the main difference being that the explanation uses modern notation.

10.6   Motivation and Method

The direct inspiration for Kerala mathematics was the works of Aryabhata I and his commentators, notably Bhaskara I. As stated earlier, at the age of 23, Aryabhata composed his seminal text *Aryabhatiya*. The influence of the astronomical and mathematical ideas in the text, both inside and outside India cannot be overestimated. This was particularly so regarding the work of the Kerala School, founded by Madhava (ca. 1340–1425).

A primary mathematical motivation for the Kerala work on infinite series arose from a recognition of the impossibility of arriving at an exact value for the circumference of a circle given the diameter. Nilakantha explained in his *Aryabhatiyabhasya* (a commentary on Arybhata's *Aryabhatiya*) why only an approximate value of the circumference could be obtained:

> ...If the diameter can be measured without a remainder, the circumference measured by the same unit (of measurement) will leave a remainder. Similarly, the unit which measures the circumference without a remainder will leave a remainder when used for measuring the diameter. Hence, the two measured by the same unit will never be without a remainder. Though we try very hard we can reduce the remainder to a small quantity but never achieve the state of 'remainderlessness'. This is the problem.

This explanation was prompted by a passage in *Aryabhatiya*. Verse 10 of the section on *Ganita*, which states:

Add 4 to 100, multiply by 8, and add 62,000. The result is *approximately* the circumference of a circle whose diameter is 20,000.

It was the word "approximately"[32] that gave food for thought. And the strategy recommended in the *Kriyakramakari*:

Thus even by computing the results progressively, it is impossible theoretically to come to a final value. So, one has to stop computation at that stage of accuracy that one wants and take the final result arrived at ignoring the previous results.

To understand the Kerala method it is necessary to remember that there have historically been two main approaches to calculating the circumference. The first approach, going back to the time of Eudoxus (*c.* 375 BCE) and Archimedes (*c.* 250 BCE) if not earlier,[33] is to inscribe and/or circumscribe the circle in regular polygons. The end result, in terms of modern mathematics, is a recursion relation involving square roots which if repeated, gives increasingly accurate approximations for the circumference.[34]

The Kerala School adopted a second approach based on infinite series and integrals. Here, the circumference is obtained from devising a limiting procedure that allows the series to converge (however slowly) to the circumference as the number of terms grows.[35]

In the course of this demonstration, the following series of the sum of integral powers of natural numbers were already well-known in Indian mathematics (and stated in earlier chapters)[36]:

$$1 + 2 + 3 + \cdots + N = \frac{N(N+1)}{2}, \tag{I}$$

$$1^2 + 2^2 + 3^2 + \cdots + N^2 = \frac{N(N+1)(2N+1)}{6}, \tag{II}$$

$$1^3 + 2^3 + 3^3 + \cdots + N^3 = \frac{N^2(N+1)^2}{4}. \tag{III}$$

It was then realised that for large N (i.e. smaller steps in the rectification of the circle) the following result applied:

$$1^k + 2^k + 3^k + \cdots + N^k \approx \frac{N^{k+1}}{k+1},$$

so that in the limit [for large N] we can replace

$$\sum_{n=1}^{N} \left(\frac{n}{N}\right)^k \approx \frac{N}{k+1}.$$

However, the Kerala mathematicians did not have the modern calculus techniques for proceeding any further. Instead, they adopted a geometrical approach which involved finding the length of an arc by approximating it to a straight line. Known as the method of direct rectification, it involves summation of very small arc segments and reducing the resulting sum to an integral. In both the *Yuktibhasa* and the *Kriyakramakari* a set of verses elaborating the above are stated as originating with Madhava.

The Kerala derivation is based on an interesting geometric technique. The tangent is divided up into *equal* segments while at the same time forcing a sub-division of the arc into *unequal* parts. This is required since the method involves the summation of a large number of very small arc segments, traditionally achieved by the 'method of exhaustion,'[37] where there was a sub-division of an arc into *equal* parts. The adoption of this "infinite series" technique rather than the "method of exhaustion" for implicitly calculating π was not through ignorance of the latter in Kerala mathematics. But, as Jysthadeva points out in the *Yuktibhasa*, the former avoids tedious and time-consuming root-extractions at a time when mechanical aids to calculations were not available.[5]

The Kerala derivation deployed an ingenious iterative re-substitution procedure[39] to obtain the binomial expansion for the expression $1/(1 + x)$ and then proceeding through a number of repeated summations (*varamsamkalithas*) of series, arrived at what must be the most remarkable part of the derivation, an intuitive leap as it were leading to the asymptotic formula, expressed in modern notation as[40]:

$$\lim_{n\to\infty} \frac{1}{n^{k+1}} \times \sum_{i=1}^{n} i^k = \frac{1}{k+1}, \quad k = 1, 2, 3, \dots. \tag{1}$$

It was soon realised that the series (derived in Appendix 1)

$$\frac{\pi}{4} = 1 - \frac{1}{3} + \frac{1}{5} - \cdots \tag{2}$$

was not useful for making accurate estimates of the circumference for given a specified diameter (i.e. estimating π) because of the slowness of the convergence of the series. This gave impetus to developments in two directions: (i) rational approximations by applying corrections to partial sums of the series, and (ii) obtaining more rapidly converging series by transforming the original series. There was considerable work in both directions, as shown in *Yuktibhasa* and *Kriyakramakari* and examined in Appendix 2 of this chapter. What this work exhibits is a measure of understanding of the concept of convergence, of the notion of rapidity of convergence and a clear awareness that convergence can be speeded up by transformations.

As an illustration of the remarkable efficiency of some of the corrections suggested, consider the following example from *Yuktibhasa*. What is required is to evaluate the circumference of a circle with a diameter of 10^{11}. *Without* the correction and with the number of terms on the right-hand side of (2) as many as 50, the implicit value for π is even not correct to two decimal places. However, incorporating one of the end corrections[41] gives the circumference as $3.1415926529 \times 10^{11}$ which is correct to 8 places. And this search for greater accuracy continued for a long time, so that as late as the nineteenth century, Sankaravarman, the author of *Sadratnamala* estimated the circumference of a circle of diameter 10^{18} as:

$$314{,}159{,}265{,}358{,}979{,}324 \qquad \text{correct to 17 places.}$$

It is clear that in deriving the arctan series, the Kerala School showed both an awareness of the principle of integration and an intuitive perception of small quantities and operations with such quantities. However, having come near to the formulation of the crucial concept of the 'limit' of a function, they seem to have shied away from developing the methods and algorithms of calculus, being perfectly content with the geometrical approach which their European counterparts would eventually replace with calculus.[42]

10.7　Kerala Mathematics to Europe: Possible Transmissions*

The basis for establishing transmission of scientific or mathematical ideas is often *direct* evidence of translations of the relevant manuscripts. The transmission of Indian mathematics and astronomy since the early centuries CE via Islamic scholars to Europe has been established by direct evidence. The transmission of Indian computational techniques was in place by at least the early 7th century for by 662 CE it had reached the Euphrates region (Berggren, 1986, p. 30). Indian Astronomy was transmitted westwards to Iraq, by a translation into Arabic of the *Siddhantas* around 760 CE and then into Spain (Subbarayappa and Sarma 1985, p. xxxviii).

The impetus to search for other types of evidence to 'prove' the conjecture of transmission of the Kerala 'calculus' was initially inspired partly by the emergence of a cluster of results between the 14th and 16th centuries which were more or less replicated in Europe two centuries later. The other stimulus arose from a test initially proposed by Neugebauer (1962, pp. 166–167), based on the multiple criteria of chronological priority of discovery, availability of accessible communication routes, and the existence of significant methodological similarities, To that we could add two other criteria: opportunity and motivation.

The first criterion regarding *priority* of Kerala calculus is proved beyond any doubt. The second is validated by the fact that a communication route between the South India and the Arabian Gulf (via the port of Basra) had been in existence for centuries

---

*This section is based on a joint paper by Almeida and Joseph (2009) entitled "*Kerala Mathematics and its Possible Transmission to Europe*".

(Potache, 1989). The arrival of the Portuguese Vasco da Gama to the Malabar coast in 1499 heralded a direct route between Kerala and Europe via Lisbon. Thus, after 1499, despite its geographical location, which prevented easy communication routes with the rest of India, Kerala was linked with the rest of the world and, in particular, directly to Europe.

As far as methodological similarities are concerned, consider the following key result discussed earlier which appears in the *Yuktibhasa* (1530).

$$\lim_{n \to \infty} \frac{1}{n^{k+1}} \sum_{i=1}^{n} i^k = \frac{1}{k+1}, \quad k = 1, 2, 3, \ldots \tag{1}$$

The result assumes a practical knowledge of integration for it permits the evaluation of the area under the parabola $y = x^k$, or, equivalently, the calculation $\int x^k dx$. And this is precisely what Fermat, Pascal, Wallis and other European mathematicians did in the 17th century to evaluate the area under the parabola. Indeed, the 'distance' between formal differentiation and the practical methods deployed in the Kerala calculus is insignificant. Furthermore, our conjecture of the transmission of Kerala mathematics is given credence by the fact that John Wallis used a numerical induction technique uncannily similar to the one used in the *Yuktibhasa*.[43]

There is circumstantial evidence to indicate that Jesuits could well have been the possible conduit of transmission of the Kerala mathematics to Europe. The well-known Matteo Ricci was in the first batch of Jesuits trained in the new mathematics curriculum of the Collegio Romano by Clavius.[44] He also studied cosmography and nautical science in Lisbon. Ricci was then sent to India in 1578 and spent a year at the Cochin Jesuit College in Kerala. Following him there appeared other Jesuit scientists trained in Rome by notable mathematicians such as Clavius and Grienberger who then spent time in Kerala. These included Johann Schreck and Antonio Rubino.[45] The former had studied with the French mathematician Viete, well known for his work in algebra and geometry. Other Jesuits just as qualified as the ones mentioned above also went to India.[46]

And at some point in their stay in India a number were posted to the Malabar region which included the region of Cochin, the epicentre of developments in the Kerala mathematics. Further, in order to not only aid conversions but to gather local knowledge, the Jesuits became proficient in languages like Malayalam, Telegu and Tamil.[47]

Ricci and others sought Indian calendrical knowledge.[48] They could not but have noticed the discrepancies between their calendar and the local calendar. For example, the Jesuits were accustomed to the idea that festivals like Christmas fell on a fixed day of their calendar so they could not have failed to notice that the major Indian festivals like Dussehara, Diwali, Holi, Sankranti, etc., did not fall on the same days of the Julian calendar each year. However, the Jesuit missionaries prior to Matteo Ricci probably were not sufficiently versed in astronomy to have understood the difference even between the sidereal year (the basis of the Indian calendar) and the tropical year (the basis of the Julian/Gregorian calendar); and so could not have been expected to understand the intricacies of the Indian calendar. They could easily have acquired written knowledge and sent it back to either Maffei in Portugal or to Clavius in Rome for analysis. It should be emphasised that the Jesuits had much more than a casual interest in the calendar. For at just about that time, Matteo Ricci's teacher, Christopher Clavius was busy sitting on the commission that ultimately introduced the Gregorian calendar in 1582, an event that had been preceded by centuries of controversy. The 1545 Council of Trent had already acknowledged the error in the Julian calendar, and had authorised the Pope to correct it. So Matteo Ricci's interest in the Indian calendar need not have been a casual one, but was an effort preceded by years of preparation and study, and came at a time when the Jesuit interest in both India and the calendar was at a peak. And it is not beyond plausibility that the Patriarch of Antioch, a member of the calendar reform committee who was knowledgeable about the Indian calendar gave information to Clavius about the Indian calendar.

Indeed, it appears that the Jesuits tried to formalise this policy by including local sciences such as astrology (or *jyotisa*) in the curriculum of the Jesuit colleges in the Malabar Coast (Wicki, 1948-vol III, p. 307). Further, the early Jesuits were also active in the transmission of local knowledge back to Europe. Evidence of this knowledge acquisition is contained in the collections *Goa* **38**, **46** and **58** found in the Jesuit historical library in Rome (ARSI). The last collection contains the work of Father Diogo Gonsalves on the judicial system, the sciences and the mechanical arts of the Malabar region. This work started from the very outset of the Portuguese presence in India.

If the early and late Jesuits were involved in learning the local sciences then, given the academic credentials of the Jesuits such as Ricci, Schreck and Rubino of the middle period, it is a plausible conjecture that this work was carried out by others. For example there is de Menses who, writing from Kollam in 1580, reports that, on the basis of local knowledge, the European maps contain inaccuracies.[49] Antonio Rubino wrote, in 1610, similarly about inaccuracies in European mathematical tables for determining time.[50] Then there is the letter from Schreck, in 1618, of astronomical observations intended for the benefit of Kepler. The latter had requested the eminent Jesuit mathematician Paul Guldin to help him to acquire these observations from India to support his theories. (Baumgardt, 1951, p. 153; Iannaccone, 1998, p. 58).

Whilst this does not establish the fact that these Jesuits knew about the existence or even obtained manuscripts containing the Kerala mathematics, it does indicate that their scientific investigations about the local astronomy and calendrical sciences could have lead them to an awareness of this knowledge. There are reports (D'Elia, 1960, p. 15) that the Brahmins were secretive and unwilling to share their knowledge. However this was not an experience shared by others. As mentioned earlier, in the mid-seventeenth century, Fr. Diogo Gonsalves, who knew the local language Malayalam well, wrote a book about the administration of justice, sciences and mechanical arts of the Malabar. This book is to be found in the MS *Goa* **58** in the Jesuit Historical library (ARSI), Rome. Around the same time a Brahmin spent eight years translating Sanskrit works for Fr. Frois (Ferroli, op cit, p. 402).

How might the Jesuits have obtained key manuscripts of Indian astronomy and mathematics such as the *Tantrasangraha* and the *Yuktibhasa*? Clearly such manuscripts would require the Jesuits being in close contact with scholarly Brahmins or Kshatriyas who, in the case of the latter in Kerala were mainly members of the royal family. And the Jesuits working on the Malabar Coast had close relations with the royal family of Cochin[51] despite their aggressive evangelical work. Furthermore, around 1670, King Rama Varma granted them special privileges that resulted in their greater influence in the royal court (Wicki, op cit, vol XV, p. 224). There is considerable evidence that the royal family of Cochin contained some notable students of astronomy and mathematics. For example, Whish (1835, p. 521) reports that his contemporary and the author of *Sadratnamala*, Sankaravarman, was the younger brother of the Raja of Cadattanada near Tellicherry. Srinivasiengar (1967, p. 145) refers to a book, *History of Sanskrit Literature in Kerala*, written in Malayalam which identifies the King of Cochin, Raja Varma, as being aware of the chronology and content of *Karanapaddhati*, a key treatise of Kerala mathematics and astronomy mentioned earlier. In 1948, Rama Varma Thampuran, who jointly published an exposition in Malayalam of the *Yuktibhasa*, was one of the royal princes of Cochin. Sarma (1972) identifies the valuable contribution to the analysis of Kerala astronomy by Rama Varma Maru Thampuran. Mukunda Marar, the son of the last king of Cochin, worked with the eminent mathematician from Madras, Rajagopal, to produce an important expository paper on Kerala mathematics. (Rajagopal and Marar, 1944, pp. 65–82). Years later, Marar stated in a personal communication that the kings of Cochin would have, at the least, been aware both of the astronomical methods for astrological prediction and of the manuscripts that contained these methods. Finally, various authors, from Whish onwards in the 19th century to Rajagopal and Rangachari (1978, 1986) in the 20th century have fulsomely acknowledged the help of the members of the royal household in supplying manuscripts on Kerala mathematics in their possession. This cumulative evidence of the continuing tradition of scientific scholarship among some of the royal personages together with the undoubted Jesuit influence in the royal courts of Kerala may indicate how the Jesuits could have acquired the knowledge of Kerala mathematics and astronomy.

10.8   Conclusion

The work of Kerala School of Mathematics and Astronomy has now entered popular discussion recently mainly because of the controversies regarding the origins of calculus and the possibilities of its transmission. While the latter question remains open without definite conclusions in the absence of direct evidence, a clearer assessment is possible in the case of the former. It is worth pointing out that developments in Indian mathematics from the time of Aryabhata I had anticipated the emergence of modern mathematics and in particular calculus. In the course of this book, we have outlined discoveries which included: (i) notions of zero and infinity (beginning with the Jains and the Buddhists and culminating with Brahmagupta and Bhaskaracharya); (ii) irrationals such as square root of 2 and π and the associated iterative approximations (beginning with the *Sulbasutras* and culminating with Narayana Pandita); (iii) second-order differences and interpolations in computing sines (traced back to the Aryabhata I and Brahmagupta); (iv) the notion of instantaneous velocity (notably contributions of Bhaskaracharya) and culminating with its application in astronomy; (v) summations of infinite geometric series (beginning with Pingala through Mahavira and Virasena and culminating with Nilakantha); (vi) estimating surface area and volume of a sphere (starting with an incorrect solution offered by Aryabhata I and corrected by Bhaskaracharya) and (vii) notions of summations and repeated summations (beginning from Aryabhata I to Narayana Pandita). Drawing on these contributions, the Kerala mathematicians were able to derive the infinite series of circular and trigonometry functions.

Finally, whether the origins of the calculus occurred in India or elsewhere is ultimately a definitional question. The Indian contribution is summed up admirably by Rajagopal and Marar seventy years ago:

"There are two points which emerge from a consideration of the [mathematics of the Kerala school]... In the first place, it employs relations which would appear not to have been noticed in Europe before modern forerunners and followers of the calculus started investigations... Our second point is not unconnected with the first. The Hindu mathematicians achieved, without the aid of calculus, results which, for us, are treated best by means of the calculus... This is not to gainsay the fact that (i) the Hindus' proof of [infinite] series shows their awareness of the principle of integration as we ordinarily use it nowadays (ii) their intuitive perception of small quantities like $O(1/n^p)$, $n \to \infty$, in a certain role is as good as a practical knowledge of differentiation." (Marar and Rajagopal, 1944).

An intriguing question remains: How would one explain this remarkably creative episode in the history of Indian mathematics. The explanation may well be found from a whole multitude of factors which are difficult to unravel. Some of these factors are summed up succinctly by the historian, M. G. S. Narayanan in a recent paper entitled 'Social Background of Science in Medieval Kerala':

The existence of efficient and scholarly Brahmin leadership in state and society produced by a theocratic oligarchy; the absence of serious challenge from the Jains and Buddhists from the other side of the Western Ghats; the protection from medieval Muslim invasion provided by high mountains of the Western Ghats; the unusual commercial prosperity generated by the coastal sea trade in the pockets of Jews, Syrian Christians, Arab Muslims and the Chinese which compensated for the feudal anarchy and unrest — all these guaranteed the continuity of culture in Kerala and promoted scientific researches while the rest of India was suffering from wars and invasions which uprooted royal dynasties and created a serious break in culture (Calicut, 2014, p. 20).

Endnotes

[1] It is a source of some surprise that while innovations in mathematics allegedly declined in other parts of India, the lamp continued to burn brightly in Kerala, where work of an original kind outpaced similar work in Europe by two or three centuries. The quality of the mathematics

available from the texts that have been studied is of such a high level compared with the earlier period that it may seem difficult to bridge the gap between the two periods. No one can invoke a 'convenient' external agency, like Greece or Europe to explain the Kerala phenomenon. We will return to this interesting question in the concluding paragraph of this chapter.

[2] Even within the 500 years of renaissance, following Brahmagupta (b. 598 CE) there were two centuries of relative silence before we meet the next prominent mathematician, Mahavira (ca. 860 CE), and another two centuries before Bhaskaracharya (b. 1114 CE) and a similar period before the appearance of Narayana Pandita (c. 1350 CE). The chronology is therefore assumed to be pitted with 'holes'.

[3] The story of the discovery of Kerala mathematics sheds some fascinating light on the character of the historical scholarship of the period (Joseph, 1995). In 1832, Charles Whish read a paper to a joint meeting of the Madras Literary Society and the Royal Asiatic Society in which he referred to five works of the period, 1450–1850: The *Tantrasangraha* ("A Digest of Scientific Knowledge") of Nilakantha (1444–1545), the *Yuktibhasa* ["An Exposition of the Rationale"] of Jyesthadeva (*fl.* 1500–1610), *Kriyakramakari* ("Operational Techniques") of Sankara Variyar (*c.* 1500–1560) and Narayana (*c.* 1500–1575), *Karanapaddati* ("A Manual of Performances in the Right Sequence") of Pudumana Somayaji (*c.* 1660–1740) and *Sadratnamala* ("A Garland of Bright Gems") of Sankaravarman (1800–1838).

[4] It should be noted Acyuta Pisarati and Sankara Variyar were not Brahmins but belonged to families traditionally associated with temple services (the *ambalavasi*). Neither did Sankaravarman, belonging to a non-Brahmin royal lineage, who may be seen as bringing to end this productive episode in the history of mathematics. It is interesting to note that the founder Madhava may not have been a Nambuthri Brahmin but belonged to another sub-caste of Brahmins known as *Emprans*. The other seven mentioned above were all Nambuthris.

[5] Madhava began a school that had the following direct teacher-student lineage:

Madhava (*fl.* 1380–1420) ==> *Paramesvara* (*fl.* 1380–1460) ==> Damodara (*fl.* 1450) ==> *Nilakantha* (b. 1444) =====> Citrabhanu (*fl.* 1530) ======> Narayana (*fl.* 1529) and Sankara Variyar (*fl.* 1556) Also Damodara ==> *Jyesthadeva* (*fl.* 1500–1575) ==> *Acyuta Pisaroti* (d. 1621)

The names in italics are generally recognised as the major figures of the Kerala School.

⁶ The author of one of the *Sulbasutras*, Katyayana, was also known as Varurici which would place him around 500 BCE! An important personage in the court of Vikramaditya was called Varurici which would mean that he lived about 200 CE. There is a popular legend in Kerala that he married a *Parayan* (low caste) woman and had twelve children by her, each of the children being brought up by a different caste. The eldest is supposed to have lived between 343–378 CE!

⁷ *Chandravakyas* are a list of numbers relating to the orbit of the moon around the earth. These numbers are represented using the *katapayadi* system of notation (explained in the next Endnote) and appear as words or phrases.

⁸ As stated in Chapter 2, *Katapayadi* was a refinement of an earlier system of numerical notation, namely Aryabhata's alphabet-numeral system. In *Katapayadi*, the Sanskrit consonants क (*k*) to झ (*jh*) indicate one to nine, and so does त (*t*) to ध (*dh*); प (*p*) to म (*m*) stand for one to five and य (*y*) to ह (*h*) for one to eight A vowel not preceded by a consonant stands for zero. In case of conjunct consonants, only the last consonant has a numerical value. The number-words were read from right to left so that the letter denoting the units was given first and so on. If such a system was applied to English, the letter $b, c, d, f, g, h, j, k, l, m$ would represent the numbers 0 to 9. So would $n, p, q, r, s, t, v, w, x, y$. Let the remaining letter z represent 0. The vowels, a, e, i, o, u serve the function of helping to form meaningful words. Thus the word, "Madras," would be represented by 9234 and "love" by 86 read in the usual manner. It was a system devised to facilitate memorisation, since for any particular number, different memorable words could be made up with different chronograms. For details on the origin and use of the *Katapayadi* system, see Madhavan (1991).

⁹ The terms '*parahita*' means 'for the benefit of the common man'. The intention was to simplify astronomical computations so that everyone could do it. The simplification was both in computations and in representation of numerical notation which substituted the Arybhata's numerals with the *Katapayadi* system.

¹⁰ The Ceras were one of the three empires that flourished in South India around the beginning of the Common Era before the emergence of the Indo-Aryan influence from the North. The other two were the Chola and Pandyan empires.

¹¹ Ravi Varman was a keen student of astronomy whose insightful questions were answered by Sankaranaryana in his book named *Vivarana*, an influential commentary on *Laghubhaskariya* of Bhaskara I, which was 'for the enlightenment of *mandabuddhis* (or people who are backward in intelligence)'.

[12] Govindasvamin's procedure is an extension of Brahmagupta's interpolation procedure found in his book *Khanda Khadyaka* (665 CE) and discussed in the Chapter 11 in this book. For further details, see Gupta (1969).

[13] The Bhatattatires, a sub-group of Nambuthri Brahmins, became the custodians of Vedic knowledge. They performed all the duties associated with the sacred fires (*agnihotris*), except sacrifices. After an intensive study of the *Vedas*, they were required to learn logic (*tarka*), religious philosophy (*vedanta*), grammar (*vyakarna*) and rituals (*mimasa*) and then teach these disciplines at the *salas* attached to the temples.

[14] This connection is supposed to date back to a night which he spent with a woman who belonged to the Kaniyan family at Pazhur. The progeny of their union is the ancestor of this family of astrologers.

[15] The attribution to Madhava usually took the form of either a statement such as "*... ata eva Madhavopyaha...*" ("*...* hence Madhava said...") preceding the quotation of a result or a statement such as "*Iti jyacapayah karya grahanam Madhavotitam*" ("*...* computation of the arc from the sine and cosine is given by Madhava"). It is possible that Madhava wrote a text on mathematics and astronomy which was quoted by later writers. Unfortunately, this text is no longer extant.

[16] The discussion that follows on Nilakantha's contribution is based on Ramasubramanian *et al.* (1994) to which reference should be made for details.

[17] The family tree of every Hindu is supposedly traceable to an ancestral sage (or *rishi*). An early text (*Baudhayana Shrautasutra*) mentions only eight original *rishis* as founders of of *gotras*: Bharadwaja, Jamadagni, Gautama, Atri, Viswamitra, Vashistha, Kashyapa and Agastya. Later, many more names were added to the list of *gotras*. Marriage was not allowed under the Hindu religion among people from the same *gotras*.

[18] For further details, see Hayashi and Kusuba (1998) and Mallayya (2011).

[19] The *Variyars* were a group of non-Brahmin temple officials who assisted the Brahmin priests in their religious rituals. A number of them were skilled in astrology and many were learned in Sanskrit. There is one story that they were descendants of a Brahmin and a Sudra woman.

[20] There are close similarities between this text and Sankara Variyar's *Kriyakramakari* and *Yuktidipika*, where the former is a commentary on the *Lilavati* and the latter on the *Tantrasangraha*. However, Sankara acknowledges the source of some material in the latter text is the work of Jysthadeva ("the Brahmin of Parakroda").

[21] There were two common methods of telling time. At a point early in a child's education, the two methods were taught in the form of

verses to be memorised. The *ativakyam* showed how to tell the time of the day by measuring the length of the shadow before and after noon. The *nakhstvakyam* showed how to reckon time at night by the position of stars and particularly by the time at which certain stars rose. This required considerable knowledge of astronomy and hence the method was only sketched out with further elaboration at an older age. At a later age, an *acharya* (teacher), usually a Pisaroti, gave them further instruction on the use of water clocks where the basic unit of time was *narika* (24 minutes) or the time that a typical vessel took to sink.

[22] In astronomical calculations, the longitude of a planet is measured along the ecliptic, while, in fact, its motion takes place along its own orbit, which deviates slightly from the ecliptic. For an accurate computation of the planet's position, this deviation has to be corrected. Acyuta Pisaroti, for the first time in Indian astronomy, gave a formula for the reduction to the ecliptic in the case of the moon in his work, *Sputanirnaya* and a simpler version of that formula in another work, *Uparagakriyakrama* (Procedures for computing eclipses). In Europe, a similar formula for the reduction to the ecliptic was formulated by the Danish astronomer, Tycho Brahe in his *Astronomiae instaurtae Proggymnasmata* which was published a year after his death in 1602.

[23] For further details, see Ramasubramanian *et al.* (1994).

[24] Virasena (*fl.* 816 CE), a Jain mathematician, had first introduced the idea of 'fractional doubling' or in modern terminology, for 2^x the fractional values of x yields the number which, when halved x times, results in 1. He also gave the sum of the infinite geometric series $(1/4) + (1/4)^2 + (1/4)^3 + \cdots$ without proof. Nilakantha provided the proof and showed that this expression could be used in expressing a small arc in terms of its corresponding chord.

[25] Nilakantha also uses this series to prove the following relation between the arc of a circle (*capa*), the sine (*jya*) and the versine (*sara*):

$$Capa = \left[\left(\frac{4}{3}\right) sara^2 + jya^2\right]^2.$$

[26] The term 'triangular' originates with the Greeks for whom there were certain numbers which could be represented with a triangular array of dots:

Triangular numbers can also be seen as the sums of consecutive natural numbers beginning with

$$1 = 1; \, 3 = 1+2; \, 6 = 1+2+3; \, 10 = 1+2+3+4; \, 15 = 1+2+3+4+5;$$
... and T_n

where

$$T_n = 1 + 2 + 3 + \cdots + (n-1) + n.$$

Adding gives

$$2T_n = (n+1) + (n+1) + (n+1) + \cdots + (n+1) + (n+1)$$

There are n groups of $(n+1)$, so we see that $2T_n = n(n+1)$, or

$$T_n = \frac{n(n+1)}{2}.$$

[27] Expressed mathematically, the formula for the sum of sums, given the notation of the previous endnote, is:

$$T_1 + T_2 + T_3 + \cdots + T_n = n \left[\frac{(n+1)}{2} \frac{(n+2)}{3} \right],$$

where $T_r = \frac{r(r+1)}{2}$, where $r = 1, 2, \ldots$

[28] For further details, see Mallayya (2002).

[29] For further details, see Gupta (1969. pp. 86–98).

[30] An intriguing question is why the sine series was needed in Europe when all work there related to trigonometric tables which were expressed in degrees rather than radians.

[31] In attempting to increase the accuracy of estimating circumference for a given diameter, the binomial series expansion is required. The derivation of this series in found in both the *Yuktibhasa* and the *Kriyakramakari*. Given three positive numbers a, b and c where $2x$ for fractional values of x to yield the number which, when halved x times, results in 1, and if $(b-c)/c = \boldsymbol{x}$, it can be shown through a process of iteration that

$$\frac{a}{1-x} = a - ax + ax^2 - ax^3 + \cdots + (-1)^m ax^m + \cdots.$$

[32] "Approximately" would seem to be an imprecise translation of the Sanskrit word '*asana*' which also contains the notion of "unattainability".

[33] In Problem 49 of the Ahmose Papyrus of ancient Egypt (*c.* 1650 BC) appears a labelled diagram showing a square with four isosceles triangles removed leaving an octagon which is interpreted to have an area equal

to that of a circle inscribed in the square. For further details, see Joseph (2011, 110–112).

[34] In modern notation this relation can be expressed as:

$$x_0 = 1, \quad x_{n+1} = \frac{\sqrt{1 + x_n^2} - 1}{x_n}, \quad \pi =_{n \to \infty} 4 \lim 2^n x_n.$$

[35] In terms of modern notation,

$$\pi = 4 \lim \frac{1}{N} \sum_{n=1}^{N} \left[\frac{1}{1 + \left(\frac{n}{N}\right)^2} \right],$$

where the sum tends towards

$$\int_0^1 \frac{dx}{1 + x^2}.$$

Taking the limit of the above gives the infinite series

$$C = 4D \left(1 - \frac{1}{3} + \frac{1}{5} - \frac{1}{7} + \cdots \right)$$

for the circumference of a circle C of diameter D.

[36] According to the standard histories of mathematics, the Pythagorean Greeks knew (I) in the sixth century BCE. In the third century BC, Archimedes in his book *On Spirals* gave a "formula" for the sum of squares whose equivalent modern formulation is given in (II). Archimedes applied this to deduce the area inside what we now call an Archimedean spiral by the classical Greek method of exhaustion. The ability to sum yet higher powers was key to finding areas and volumes of other geometric objects. We find a number of works in which there was an understanding of the method of finding sums of cubes, including the works of Nicomachus of Gerasa (first century BCE), Aryabhata in India (499 CE) and al-Karajı in the Arab world (*c.* 900–1000 CE). The first evidence of a general relationship between various exponents is in the work of ibn al-Haytham (965–1039), who needed a formula for a sum of fourth powers in order to find the volume of a general paraboloid of revolution. Although not stated in full generality, his discovery was essentially the recursive relationship:

$$(n + 1) \sum_{i=1}^{n} i^k = \sum_{i=1}^{n} i^{k+1} + \sum_{p=1}^{n} \left(\sum_{i=1}^{p} i^k \right).$$

For further details, see Berggren (2007, 587–592).

[37] The basic approach is to inscribe or circumscribe a regular polygon. The problem then is to find the side of the polygon as a multiple of the diameter. An interesting extension of the method, first suggested by Eudoxus of Cnidus (*fl.* 375 BC) and then extensively used by Archimedes, was a rigorous alternative to "taking the limit" which the Greeks avoided given their well-known "horror of the infinite." It is based on the simple observation that if a circle is enclosed between two polygons of n sides, then, as n increases, the gap between the circumference of the circle and the perimeters of the inscribed and circumscribed polygons diminishes so that eventually the perimeters of the polygons and the circle would become identical. Or in other words, as n increases, the difference in the area between the polygons and the circle would be gradually exhausted. For further details, including the Chinese approach to this problem, see Joseph (2011, 261–271).

[38] In the sixth chapter of Jyesthadeva's *Yuktibhasa*, there is a section entitled 'Circle from a Square' which is a detailed description of this approach.

[39] A familiar result from elementary algebra is that the series $1 - x + x^2 - x^3 + \cdots +$ represents for $x < 1$ an infinite decreasing geometric progression with common ratio $-x$, and sum equal to:

$$\frac{1}{1 - (-x)} = \frac{1}{1 + x} = (1 + x)^{-1}.$$

[40] This asymptotic relation made its first appearance in Europe in the works of Roberval (1634) and Fermat (1636).

[41] The correction used is to incorporate the following as the last term:

$$F_c(n) = \frac{(n^2 + 1)}{(4n^3 + 5n)},$$

where n is the number of terms.

These and other corrections, $F_a(n)$ and $F_b(n)$, are discussed in Appendix 2 of this chapter.

[42] It is interesting in this context to note about five hundred years after Madhava, Ramchandra wrote a book in 1850, entitled *A Treatise on the Problems of Maxima and Minima* in which he claimed that he had developed a new method, consistent with the Indian tradition of mathematics, to solve all problems of maxima and minima by algebra and not calculus. This book was republished in England with the help of the British mathematician, Augustus De Morgan. For further details, see Chapter 13 in this book.

[43] For a discussion of Wallis' 'induction' technique, see Scott (1981, p. 30) and Jesseph (1999, p. 42).

[44] Cronin (1984, p. 22) states that "... (I)n 1575 Ricci entered a new phase of his studies; philosophy and mathematics, Aristotle and Euclid. The advanced course was taught by a young German, Christopher Clavius, the most brilliant mathematician of his day... He showed special aptitude for this course winning notice as a mathematician of promise."

[45] For further details, see Baldini (1992) and Iannaccone (1998)'.

[46] A translation of a passage of Baldini (1992, p. 70) reads: "It can be recalled that many of the best Jesuit students of Clavius and Greienberger (beginning with Ricci and continuing with Spinola, Aleni, Rubino, Ursis, Schreck and Rho) became missionaries in Oriental Indies. *This made them protagonists of an interchange between the European tradition and those Indian and Chinese, particularly in mathematics and astronomy, which was a phenomenon of great historical meaning*" (My emphasis).

[47] For evidence, see J. Correia-Afonso, 1997, p. 47, 64; J. Correia-Afonso, 1969, p. 58, 92; Wicki, 1948–, vol XV p. 34).

[48] In a letter to Giovanni Pietro Maffei (a chronicler of activities of Jesuits in the Orient) he writes that he requires the assistance of an "intelligent Brahmin or an honest Moor" to help him understand the local ways of recording and measuring time or *jyotisa*. See Venturi (1913, p. 24).

[49] "I have sent Valignano a description of the whole world by many selected astrologers and pilots, and others in India, which had no errors in the latitudes, for the benefit of the astrologers and pilots that every day come to these lands, because the maps of theirs are all wrong in the indicated latitudes, as I clearly saw" (Wicki, 1948–, volume XI, p. 185).

[50] "... (C)omparing the real local times with those inferable from the ephemeridis [tables] of Magini, he [Rubino] found great inaccuracies and, therefore, requested other ephemeridi" (Baldini, 1992, p. 214).

[51] See for example, Wicki, 1948–, vol X, p. 239 and p. 838; vol XV p. 591.

Appendix 1.   The Rationale and the Method: A Study of the Madhava–Gregory Series

In the 14th century, Madhava of *Sangamagrama* discovered a remarkable connection between circumference of a circle and the reciprocals of all whole odd numbers that exists. In modern notation, the result may be expressed as a converging infinite series of the form

$$\frac{\pi}{4} = 1 - \frac{1}{3} + \frac{1}{5} - \frac{1}{7} + \cdots .$$

However, the convergence of this series is very slow,[1] and it would require a large number of terms on the right-hand side of the above series to approach a reasonably accurate value of π. The desire for speeding up the convergence motivated some of the most innovative work on the part of the Kerala mathematicians. This series was also discovered in Europe by Gregory (1671) and Leibniz (1773).

In the derivation of this infinite series, the Kerala mathematicians were dependent on a 'toolkit', containing the four main results listed below. Only the first two will be examined here. The last two, already referred to in earlier chapters, are sufficiently known to require further elaboration:

(i) Summation of a geometric series;
(ii) Establishing the behaviour of a certain quotient taken to its limit;
(iii) Properties of similar triangles;
(iv) The Pythagorean theorem.

A.1.1 Summation of a geometric series:

The demonstration of this summation, as outlined in the *Yuktibhasa*, may be restated as follows.

Let r be a number such that $0 < r < 1$. We need to show that

$$1 + r + r^2 + r^3 + \cdots = \frac{1}{1 - r} \quad \text{(Infinite Series)}$$

and

$$1 + r + r^2 + r^3 + \cdots + r^k = \frac{1 - r^{k+1}}{1 - r} \quad \text{(Finite Series)}.$$

Proof (A visual demonstration for $r < 1$ only).
In Fig. A.1.1, either

$$(1 - r) + (r - r^2) + (r^2 - r^3) + \cdots = 1,$$

or

$$(1 - r)(1 + r + r^2 + r^3 + \cdots) = 1.$$

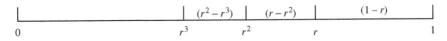

Figure A.1.1. A visual demonstration.

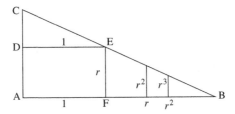

Figure A.1.2. Geometric demonstration.

Therefore

$$1 + r + r^2 + r^3 + \cdots = \frac{1}{1-r}.$$

Implied in the development of this argument in the *Yuktibhasa*, this is a more conventional geometrical demonstration.

In Fig. A.1.2, triangle EFB is similar to triangle ABC by scale factor r, and AC and AF are each of length 1. The other smaller triangles are similar with scale factors r^2, r^3, and so on. Since triangles ABC and DEC are similar, it follows that

$$\frac{AB}{DE} = \frac{AC}{DC}. \tag{A.1.1}$$

But AC = 1, DE = 1 and DC = $1 - r$.

Therefore (A.1.1) reduces to

$$AB = \frac{1}{DC} = \frac{1}{1-r}.$$

But

$$AB = 1 + r + r^2 + r^3 + \cdots .$$

So

$$1 + r + r^2 + r^3 + \cdots = \frac{1}{1-r}. \tag{A.1.2}$$

In the finite case,

$$(1 - r) + (r - r^2) + (r^2 - r^3) + \cdots + (r^k - r^{k+1}) = 1 - r^{k+1}.$$

Factor out $(1 - r)$ from all terms on left-hand side and re-arrange, we get:

$$1 + r + r^2 + r^3 + \cdots + r^k = \frac{1 - r^{k+1}}{1 - r}. \qquad (A.1.3)$$

In Indian mathematics, the problem of investigating the behaviour of a quotient taken to its limit emerged from an interest in series which has a long history probably starting from the Vedic times, with notable work by Jains followed by the synthesis of the result by the Aryabhatan School.[2] It may have also arisen, rather unusually, from an interest peculiarly Indian.

A Digression: *Sunyaganita* (i.e. 'Operations with Zero'): What is zero divided by zero?

The question may be rephrased as:

$$\text{Evaluate} \quad \frac{x^2 - 1}{x - 1} \quad \text{as } x \to 1. \qquad (A.1.4)$$

Note that when $x = 1$, (A.1.4) becomes $0/0$. This is not possible since division is a form of inverse multiplication. So $\frac{0}{0} = 0 \times ? = 0$, where ? can be any number except 0.

So what is the solution to (A.1.4)?

Recall from above:

$$\frac{1 - x^{k+1}}{1 - x} = 1 + x + x^2 + \cdots + x^k.$$

For $k = 1$ and where $x \to 1$,

$$\frac{1 - x^2}{1 - x} = 1 + x = 2.$$

For $k = 2$ and where $x \to 1$,

$$\frac{1 - x^3}{1 - x} = 1 + x + x^2 = 3.$$

For the general case k and where $x \to 1$,

$$\frac{1 - x^k}{1 - x} = 1 + x + \cdots + x^k = k + 1.$$

A.1.2   The Madhava–Gregory series: derivation:

Recall the infinite series given at the beginning of this appendix, which is

$$\frac{\pi}{4} = 1 - \frac{1}{3} + \frac{1}{5} - \frac{1}{7} + \cdots .$$

Today its derivation is a simple affair using calculus and involves five steps. In Fig. A.1.3, Apq is a section of a circle of unit radius OA

Step 1: $\partial\theta = \text{arc}\,pq \approx pm = \frac{\partial(\tan\theta)}{(1+\tan^2\theta)}$ where $OA = Op = 1$.

Step 2: So $\theta = \int \frac{dt}{1+t^2} = \int (1 - t^2 + t^4 - \cdots)dt = t - \frac{t^3}{3} + \frac{t^5}{5} - \cdots$, where $t = \tan\theta$.

Step 3: It is next shown that the last term of the series in Step 2 is given by

$$(-1)^{n+1} \cdot \frac{t^{2n+2}}{(1 + t^2)}dt \to 0 \quad \text{as } n \to \infty \text{ if } |t| \leq 1.$$

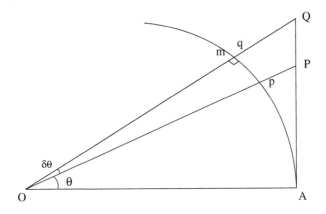

Figure A.1.3. A modern derivation.

Step 4: Or, $\arctan\theta$ can be represented as an infinite series of the form:

$$\arctan\theta = \tan\theta - \frac{\tan^3\theta}{3} + \frac{\tan^5\theta}{5} - \cdots \quad \text{if } |\tan\theta| \leq 1. \quad (*)$$

Step 5: For $\tan\theta = 1$ or $\theta = 45° = \frac{\pi}{4}$ radians, the above series becomes

$$\frac{\pi}{4} = 1 - \frac{1}{3} + \frac{1}{5} - \cdots . \quad (**)$$

The series (*) was first investigated in Europe by the Scottish mathematician, James Gregory, in 1671. Two years later, the German philosopher and mathematician, Gottfried Wilhelm Leibniz, discovered (**) using a different approach to that of Gregory.[3] But nearly three centuries earlier, both series were known and used in Kerala.

It is now proposed to show how the Kerala mathematicians derived the series given in (**).

Figure A.1.4(a) represents a circle of unit radius (i.e. $AC = CD = 1$) with the angle ACB at the centre of the circle equalling 45°. AB is the tangent at point A. It would follow that since $AB = 1$, circumference $(C) = 2\pi$ and arc $AD = \frac{1}{8}C = \frac{\pi}{4}$. Figure A.1.4(b) represents the line segment AB being divided into a number of sections (say n) of *equal* length. Note that the length of each section is $\frac{1}{n}$. Also, note that at the same time arc AD is cut into n *unequal* sections. The objective is to estimate the lengths of each of these small arcs and sum these estimates to obtain in turn an estimate of the length of arc AD which we know equals $\frac{\pi}{4}$.

Without loss of generality, consider the special case of $n = 5$, as shown in Fig. A.1.4(c). We wish to estimate the portion of the arc labelled EF that corresponds to the line segment $\frac{3}{5}$ to $\frac{4}{5}$, labelled GH. In terms of Fig. A.1.4(d), the objective is to estimate FI which will be a good approximation for the arc FE when GH is very small.

Proof. The right-angled triangles GJH and CAH are similar. So

$$\frac{GJ}{AC} = \frac{GH}{CH}. \quad (A.1.5)$$

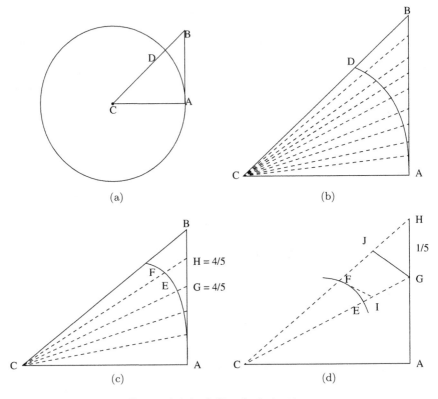

Figure A.1.4. A Kerala derivation.

Since $AC = 1$ and $GH = \frac{1}{5}$, it follows that

$$GJ = \frac{\frac{1}{5}}{CH}. \qquad (A.1.6)$$

Also since right-angled triangles CFI and CJG are similar, it follows that

$$\frac{FI}{GJ} = \frac{CF}{CH} \Rightarrow FI = \frac{GJ \cdot CF}{CJ}.$$

But $CF = 1$. So

$$FI = \frac{GJ}{CJ}. \qquad (A.1.7)$$

When n is large, GH is small and so is JH. Therefore CH is a good approximation to CJ, or

$$\frac{CH}{CJ} \approx 1.$$

Using this fact, rewrite (A.1.7) as

$$FI = \frac{GJ}{CJ} \cdot \frac{CH}{CH}.$$

So

$$FI = \frac{GJ}{CH}. \tag{A.1.8}$$

Combining (A.1.8) with the earlier result given in (A.1.5), we get

$$FI = \frac{\frac{1}{5}}{CH} = \frac{\frac{1}{5}}{CH^2}. \tag{A.1.9}$$

Now, note that CH is the hypotenuse of the right-angled triangle CAH.

So

$$CH^2 = CA^2 + AH^2 = 1 + \left(\frac{4}{5}\right)^2. \tag{A.1.10}$$

Combining (A.1.9) and (A.1.10) we get

$$FI = \frac{\frac{1}{5}}{1 + \left(\frac{4}{5}\right)^2}.$$

So an estimate of arc FE is:

$$\text{arc FE} \approx FI = \frac{\frac{1}{5}}{1 + \left(\frac{4}{5}\right)^2}. \tag{A.1.11}$$

In the general case, the numerator $\left(\frac{1}{5}\right)$ will be replaced by $\left(\frac{1}{n}\right)$ and the $\left(\frac{4}{5}\right)$ in the denominator of (A.1.11) will be replaced by a fraction that describes the length AH. In the present case of $n = 5$, the length of the arc AD in Fig. A.1.4(c) is the sum of the estimates of five short arcs where each estimate will correspond to that calculated for arc FE in (A.1.11). In each case the numerator is $\frac{1}{5}$ and the denominator is different, each going from $1 + \left(\frac{1}{5}\right)^2$ to $1 + \left(\frac{5}{5}\right)^2$.

The total estimate of arc length AD will be:

$$\frac{\frac{1}{5}}{1+\left(\frac{1}{5}\right)^2} + \frac{\frac{1}{5}}{1+\left(\frac{2}{5}\right)^2} + \frac{\frac{1}{5}}{1+\left(\frac{3}{5}\right)^2} + \frac{\frac{1}{5}}{1+\left(\frac{4}{5}\right)^2} + \frac{\frac{1}{5}}{1+\left(\frac{5}{5}\right)^2}. \quad \text{(A.1.12)}$$

What we are really interested is what happens to the summands as n increases from $n = 5$ to an arbitrary large number. Equation (A.1.12) can then be generalised to:

$$\frac{\frac{1}{n}}{1+\left(\frac{1}{n}\right)^2} + \frac{\frac{1}{n}}{1+\left(\frac{2}{n}\right)^2} + \frac{\frac{1}{n}}{1+\left(\frac{3}{n}\right)^2} + \cdots + \frac{\frac{1}{n}}{1+\left(\frac{n}{n-1}\right)^2} + \frac{\frac{1}{n}}{1+\left(\frac{n}{n}\right)^2}.$$

$$\text{(A.1.13)}$$

Factoring out the numerator $\frac{1}{n}$ in (A.1.13) gives:

$$\frac{1}{n}\left[\frac{1}{1+\left(\frac{1}{n}\right)^2} + \frac{1}{1+\left(\frac{2}{n}\right)^2} + \frac{1}{1+\left(\frac{3}{n}\right)^2} + \cdots \right.$$

$$\left. + \frac{1}{1+\left(n-\frac{1}{n}\right)^2} + \frac{1}{1+\left(\frac{n}{n}\right)^2}\right]. \quad \text{(A.1.14)}$$

Recall the earlier result for the summation of a geometric series:

$$\frac{1}{1-r} = 1 + r + r^2 + r^3 + \cdots.$$

Replace r by $-s$ where $s > 0$

$$\frac{1}{1+s} = 1 - s + s^2 - s^3 + \cdots. \quad \text{(A.1.15)}$$

Applying (A.1.15) to each term in the bracket in (A.1.14), we get

$$\frac{1}{1+\left(\frac{1}{n}\right)^2} = 1 - \left(\frac{1}{n}\right)^2 + \left(\frac{1}{n}\right)^4 - \left(\frac{1}{n}\right)^6 + \cdots$$

$$\frac{1}{1+\left(\frac{2}{n}\right)^2} = 1 - \left(\frac{2}{n}\right)^2 + \left(\frac{2}{n}\right)^4 - \left(\frac{2}{n}\right)^6 + \cdots$$

$$\frac{1}{1+\left(\frac{3}{n}\right)^2} = 1 - \left(\frac{3}{n}\right)^2 + \left(\frac{3}{n}\right)^4 - \left(\frac{3}{n}\right)^6 + \cdots$$

$$\vdots$$

$$\frac{1}{1+\left(\frac{n-1}{n}\right)^2} = 1 - \left(\frac{n-1}{n}\right)^2 + \left(\frac{n-1}{n}\right)^4 - \left(\frac{n-1}{n}\right)^6 + \cdots .$$

The last term in the brackets in (A.1.13) reduces to

$$\frac{1}{n}\left[\frac{1}{1+\left(\frac{n}{n}\right)^2}\right] = \frac{1}{2n}$$

which tends to 0 as n increases.

Adding the above set of equations column wise, we get the following:

The first column sum of $(n-1)$ 1's gives $n-1$.

The second column sum, after factoring out the denominator n^2, gives:

$$-\frac{1}{n^2}[1^2 + 2^2 + 3^2 + \cdots + (n-1)^2].$$

Similarly, the subsequent column sums, after factoring out of the denominator n^2, can be written as:

$$\frac{1}{n^4}[1^4 + 2^4 + 3^4 + \cdots + (n-1)^4]$$

$$-\frac{1}{n^6}[1^6 + 2^6 + 3^6 + \cdots + (n-1)^6]$$

$$\vdots$$

Substituting the column sums given above in (A.1.14) gives:

$$\frac{n-1}{n} - \frac{1}{n^3}[1^2 + 2^2 + 3^2 + \cdots + (n-1)^2]$$

$$+ \frac{1}{n^5}[1^4 + 2^4 + 3^4 + \cdots + (n-1)^4]$$

$$- \frac{1}{n^6}[1^6 + 2^6 + 3^6 + \cdots + (n-1)^6] + \cdots . \quad \text{(A.1.16)}$$

From an earlier result regarding the behaviour of a quotient taken to the limit, we established that as n approaches infinity:

$$\frac{1}{n^{k+1}}[1^k + 2^k + 3^k + \cdots + (n-1)^k] = \frac{1}{k+1}. \qquad \text{(A.1.17)}$$

Applying (A.1.17) to (A.1.16) gives

$$-\frac{1}{n^3}[1^2 + 2^2 + 3^2 + \cdots + (n-1)^2] = -\frac{1}{3}$$

$$\frac{1}{n^5}[1^4 + 2^4 + 3^4 + \cdots + (n-1)^4] = \frac{1}{5} \qquad \text{(A.1.18)}$$

$$-\frac{1}{n^7}[1^6 + 2^6 + 3^6 + \cdots + (n-1)^6] = -\frac{1}{7}$$

$$\vdots$$

Therefore, the length of arc AD or $\frac{\pi}{4}$ is given by the infinite series:

$$\frac{\pi}{4} = 1 - \frac{1}{3} + \frac{1}{5} - \frac{1}{7} + \cdots .$$

Note that this is equivalent to

$$C = 4d - \frac{4d}{3} + \frac{4d}{5} - \frac{4d}{7} + \cdots ,$$

which in turn was what *Kriyakramakari* reports Madhava as saying that the way to measure the circumference of a circle is to: "Multiply the diameter by 4. Subtract from it and add to it alternately the quotients obtained by dividing four times the diameter to the odd integers $3, 5$, and so on."

A.1.3   Conclusion

The major breakthrough in Kerala mathematics was the appearance of mathematical analysis in the form of infinite series and their finite approximations relating to circular and trigonometric functions. The primary motivation for this work was a mixture of intellectual curiosity and a requirement for greater accuracy in astronomical computations. Demonstrations of these results may not be completely rigorous by today's standards, but they are nonetheless correct. And these demonstrations may well be chosen for a modern mathematics class room because the approach is more intuitive and therefore more convincing.[4]

Endnotes

[1] As we saw earlier, with 19 terms on the right-hand side, the ratio of circumference over diameter (π) does not achieve even a one decimal place accuracy.

[2] It is an interesting point of comparison that in European mathematics this quotient arose from two sets of problems: (i) finding the area under a curve and (ii) constructing a tangent to a curve (Joseph, 2009, pp. 117–119).

[3] In an attempt to discover an infinite series representation of any given trigonometric function and the relationship between the function and its successive derivatives, Gregory stumbled on the arc tan series. He took, in terms of modern notation,

$$d\theta = \frac{d(\tan \theta)}{1 + \tan^2 \theta}$$

and carried out term by term integration to obtain his result: a procedure not dissimilar to the modern derivation given earlier. Leibniz's discovery arose from his application of fresh thinking to an old problem, namely quadrature of a circle. In applying a transformation formula (similar to the present-day rule for integration by parts) to the quadrature of the circle, he discovered the series for π. It must be pointed out, however, that the ideas of calculus such as integration by parts, change of variables and higher derivatives were not completely understood then. They were often dressed up in geometric language with, for example, Leibniz talking about "characteristic triangles" and "transmutation".

[4] Geometric intuition and logically deductive reasoning formed the basis of "proofs" both in India and later in Europe. By the end of the 19th century, geometry fell out of favour to be replaced by arithmetic and set theory. Thus Bolzano and Dedekind tried to prove that infinite sets exist by arguing that any object of thought can be thought about and thus give rise to a new thought object. Today we reject such proofs and use an axiom of infinity. Starting from a practical orientation and serving practitioners of the astronomical arts, the subject of analysis by its peculiar logic developed eventually into a highly abstract and rarefied entity for the delectation of primarily the professional mathematician.

Appendix 2.   Improving Accuracy Through Corrections

A.2.1   Introduction

In applying an infinite series approach to estimate the circumference, the Kerala mathematicians came across a serious difficulty. The problem is that the series relating to the circumference converges far too slowly. For example, summing the first 20 terms on the right-hand side of the circumference (or π) series gives the inaccurate estimate of π as 3.194. The problem was tackled by the Kerala School in two directions: (a) deriving rational approximations by applying end corrections to partial sums of the series; and (b) obtaining more rapidly converging series by transforming the original series. There was considerable work in both directions as shown in both the *Yuktibhasa* and the *Kriyakramakari*.

A.2.2   Approximations through end corrections

In *Tantrasangraha*, Nilakantha gives the following corrections to the circumference formula $[C(n)]$ which he credits to Madhava:

$$C(n) = 4 \left[d - \frac{d}{3} + \frac{d}{5} - \cdots + (-1)^{n-1} \frac{d}{2n-1} + (-1)^n d \cdot F(n) \right],$$
$$(A.2.1)$$

where d is the diameter of the circle and n the number of terms, and the last term is the correction. Note that without the correction and if $d = 1$, then (A.2.1) becomes the Madhava–Leibniz series which is given earlier:

$$C(n) = 4 \left[1 - \frac{1}{3} + \frac{1}{5} - \cdots \right]. \qquad (A.2.2)$$

Three types of corrections are as follows:

$$F_a(n) = \frac{1}{4n}, \qquad (A.2.3a)$$

$$F_b(n) = \frac{n}{4n^2 + 1}, \qquad (A.2.3b)$$

$$F_c(n) = \frac{4^2 + 1}{4n^3 + 5n}. \qquad (A.2.3c)$$

No explanation is given as to how these corrections were arrived at. However, in **Kriyakramakari**, there appears an attempt at a rationale for the incorporation of corrections (A.2.3a) and (A.2.3b) and with the statement that the methods were known to the **acharya** (teacher) who could be either Madhava or Nilakantha. Start with (A.2.1), which is rewritten as follows:[1]

$$C(n) = 4d \left[1 - \frac{1}{3} + \frac{1}{5} - \cdots + (-1)^{n-1} \frac{1}{2n-1} + (-1)^n F(n) \right].$$

Replace n by $(n+1)$ to form $C(n+1)$ and subtract $C(n)$ from $C(n+1)$. For the correction $F(n)$ to be "ideal", $C(n)$ must be equal to $C(n+1)$ when n is large, i.e.,

$$F(n) + F(n+1) - \left(\frac{1}{m} \right) = 0. \quad \text{where } m = 2n + 1. \quad \text{(A.2.4)}$$

The problem is therefore to find a function $F(n)$ which satisfies (A.2.4). Obviously, the condition is satisfied if $F(n) = F(n+1) = \frac{1}{2m}$. But as the authors of *Kriyakramakari* point out, this is not a realistic assumption. So the problem, in terms of modern exposition, is to minimise the expression:

$$F(n) + F(n+1) - \left(\frac{1}{m} \right) \quad \text{for large } n.$$

By a process of trial and error, the authors then proceed to consider different values for $F(n)$ which would accomplish this purpose. If

$$F_a(n) = \frac{1}{4n} = \frac{1}{(2m-2)},$$

then

$$F_a(n+1) = \frac{1}{(2m+2)}.$$

So that

$$F_a(n) + F_a(n+1) = \frac{1}{(2m-2)} + \frac{1}{(2m+2)} = \frac{m}{(m^2-1)},$$

or

$$\frac{m}{(m^2-1)} - \frac{1}{m} = \frac{1}{(m^3-m)} = [4n(n+1)(2n+1)]^{-1}$$

is the "error" (**sthaulya**) that remains from the use of the correction

$$F_a(n) = \frac{1}{4n} \quad \text{given in (A.2.3a)}.$$

For large n, the incorporation of this correction gives a small numerical error, i.e.,

$$\lim\{[4n(n+1)(2n+1)]^{-1}\} \to 0 \quad \text{as } n \to \infty.$$

The authors of **Kriyakramakari** added the following:

> "[The *acharya*] was not satisfied with the [size] of this error and tried to find out [more accurate] corrections." (p. 388)

A series of adjustments was made to the denominator of $F_a(n)$ with varying degrees of success until "the *acharya* added four that is divided by itself." One can only assume that this means an addition of $\frac{4}{4n}$ to the denominator of $F_a(n)$, that is

$$F_b(n) = \frac{1}{\left(4n + \frac{4}{4n}\right)} = \frac{n}{(4n^2+1)}.$$

The authors of *Kriyakramakari* then proceed to calculate the error (*sthaulya*) in the case of the correction $F_b(n)$. Using a similar approach to the one used in calculating the "error" in the case of $F_a(n)$ and assuming that $m = 2n+1$, the error (E) in $F_b(n)$ can be shown to be:

$$E = \frac{4}{m^5 + 4m} = \frac{4}{[(2n+1)^5 + 4(2n+1)]}.$$

As n increases, $E \to 0$ faster than the error of $F_a(n)$.

No explanation is given in any of the texts of Kerala mathematics as to how the last of the corrections [i.e., $F_c(n)$, as given in (A.2.3c)] is derived. However, Hayashi *et al.* (1990) offer an explanation of how all corrections were derived by the *acharya* (taken to be Madhava) which is both ingenious and convincing.

A.2.3 Rapidly converging series for circumference

Another approach to obtaining accurate estimates of the circumference is to derive more rapidly converging series. The *Yuktibhasa* contains a number of such series either obtained from the transformation of the original series or by extending the theory of the "errors" discussed in the last section.

Grouping the terms of the series

$$C = 4d \left[1 - \frac{1}{3} + \frac{1}{5} - \frac{1}{7} + \frac{1}{9} - \frac{1}{11} + \cdots \right] \tag{A.2.5}$$

two by two gives

$$C = 4d \left[\left(1 - \frac{1}{3} \right) + \left(\frac{1}{5} - \frac{1}{7} \right) + \left(\frac{1}{9} - \frac{1}{11} \right) + \cdots \right].$$

Hence

$$C = 8d \left[\frac{1}{(2^2 - 1)} + \frac{1}{(6^2 - 1)} + \frac{1}{(10^2 - 1)} + \cdots \right]. \tag{A.2.6}$$

Regrouping (A.2.5) two by two but leaving out the first term, we get:

$$C = 4d \left[1 - \left(\frac{1}{3} - \frac{1}{5} \right) - \left(\frac{1}{7} - \frac{1}{9} \right) - \left(\frac{1}{11} - \frac{1}{13} \right) - \cdots \right].$$

Hence

$$C = 4d - \frac{8d}{(4^2 - 1)} - \frac{8d}{(8^2 - 1)} - \frac{8d}{(12^2 - 1)} - \cdots . \tag{A.2.7}$$

Taking half the sum of (A.2.6) and (A.2.7) and incorporating the correction term $F_a(n)$ gives:

$$C(n) = 2d + \frac{4d}{(2^2 - 1)} - \frac{4d}{(4^2 - 1)} + \cdots + (-1)^n \cdot \frac{4d}{2} [(2n + 1)^2 + 2]. \tag{A.2.8}$$

There were other formulas which are developed along similar lines as the derivation of "errors" discussed in the last section. These results include[2]:

$$C = \sqrt{12}d \left[1 - \frac{1}{3 \times 3} + \frac{1}{3^2 \times 5} - \frac{1}{3^3 \times 7} + \cdots \right], \qquad \text{(A.2.9)}$$

$$C = 3d - 4d \left[\frac{1}{(3^3 - 3)} - \frac{1}{(5^3 - 5)} + \frac{1}{(7^3 - 7)} - \cdots \right]. \qquad \text{(A.2.10)}$$

A.2.4   Conclusion

It is clear from the above discussion that interest in the application of rational approximations to infinite series had taken the Kerala mathematicians a long way. They showed some understanding of convergence, of the notion of rapidity of convergence and an awareness that convergence can be speeded up by transformations, several of which were worked out with considerable dexterity. And their interest in increasing the accuracy of their estimates continued for a long time, so that as late as the early decades of the 19th century, the author of *Sadratnamala* estimated the ratio of the circumference to diameter correct to 17 places of decimals.

Endnotes

[1] In the *Yuktibhasa*, $F_a(n)$ appears only as a step in the derivation of $F_b(n)$. Only $F_b(n)$ and $F_c(n)$ are given explicitly. The expressions of $F_b(n)$ and $F_c(n)$ as continued fractions reveal the fact that the incorporation of $F(n)$ permits an interpretation of the three correction terms as the first three convergents of an infinite continued-fraction. We will not examine this interpretation here. For a brief discussion of this point, see Rajagopal and Rangachari (1986, p. 90).

[2] This is based on the following passage from the *Yuktibhasa*:

> Multiply the square of the diameter by 12 and extract the square root of the product. This is the first term of a sequence in which each of the successive terms is divided by the odd numbers $1, 3, 5, \ldots$ and then from the second term onwards a further division by successive powers of 3 produces the final terms of this sequence. The series is now formed by subtracting and adding alternately the successive terms, starting from the second term. See Rajagopal and Rangachari (1986, p. 114).

Appendix 3.   Power Series for Sines and Cosines

A.3.1   Introduction

In a commentary on Nilakantha's *Tantrasangraha*, by an unknown student of Jyesthadeva, is found the following descriptions of the power series for sine and versine $(1 - \text{cosine})$ without any derivations.[1]

(A) The arc is repeatedly multiplied by the square of itself and divided (in order) by the square of each and every even number increased by itself and multiplied by the square of the radius. The arc and the terms obtained from these repeated operations are to be placed one beneath the other in order, and the last term subtracted from the one above, the remainder from the term then next above, and so on, to yield the [*bhuja*] *jya* (or Indian Sine) of the arc.

(B) The radius is repeatedly multiplied by the square of the arc and divided (in order) by the square of each and every even number diminished by itself and multiplied by the square of the radius, with the first term involving only 2. The resulting terms are placed one beneath the other in order, and the last term subtracted from the one above, the remainder from the term next above and so on, to yield *ukramajya* or *sara* (or Indian versine) of the arc.

Expressed symbolically, where r is the radius, a is the length of the given arc and $n! = n(n-1)(n-2)\ldots 1$, the first three even numbers given by (A) is equivalent to:

$$\frac{a \cdot a^2}{1!(2^2 + 2)r^2} = \frac{a^3}{3!r^2},$$

$$\frac{a^3 \cdot a^2}{3!(4^2 + 4)r^4} = \frac{a^5}{5!r^4},$$

$$\frac{a^5 \cdot a^2}{5!(6^2 + 6)r^6} = \frac{a^5}{7!r^6}.$$

Hence,

$$\text{Indian Sine} = r\sin\theta = a - \frac{a^3}{3!r^2} + \frac{a^5}{5!r^4} - \frac{a^7}{7!r^6} + \cdots .$$

Substituting $a/r = \theta$ gives:

$$\sin\theta = \theta - \frac{\theta^3}{3!} + \frac{\theta^5}{5!} - \frac{\theta^7}{7!} + \cdots . \qquad (A.3.1)$$

Using the above notation and denoting Indian versine (*sara*) by $(r - r\cos\theta)$, the first three even numbers given in (B) can be written as:

$$\text{Indian versine } (r - r\cos\theta) = \frac{a^2}{2!r} - \frac{a^4}{4!r^3} + \frac{a^6}{6!r^6} \cdots .$$

Substituting $a/r = \theta$ and simplifying gives:

$$\cos\theta = 1 - \frac{\theta^2}{2!} + \frac{\theta^4}{4} - \frac{\theta^6}{6!} + \cdots . \qquad (A.3.2)$$

The series given in (A.3.1) and (A.3.2) are usually named after Newton. They make their first appearance in European mathematics in a letter from Newton to Oldenburg in 1676 and then provided a firmer algebraic basis by De Moivre (1708–1738) and Euler (1748). The series should be more appropriately named after Madhava to whom it is attributed by the later members of the Kerala School.

A.3.2   The background and alternative formulations

The seventh chapter of *Yuktibhasa* contains a detailed discussion of the sine and cosine of an arc of a circle. Starting with two well-known results from the Aryabhatan era, namely:

(i) $r\sin(\pi/6) = \frac{1}{2}r$
(ii) $r\sin(\pi/2) = r$

and applying the formulae

$$r\sin\theta = \frac{1}{2}\sqrt{(r\sin 2\theta)^2 + r^2(1 - \cos 2\theta)}^{\,2}$$

and

$$r \sin \left(\frac{\pi}{2} - \theta \right) = \sqrt{r^2 - (r \sin \theta)^2},$$

the values of $r \sin \frac{c\pi}{48}$ are generated for $c = 1, 2, \ldots, 24$. Using yet another known result: the cosine of an arc of a circle is equal to its complementary arc (i.e. the portion of the arc required with the given arc to complete a quadrant), the values of $r \cos \frac{c\pi}{48}$ are also generated for $c = 1, 2, \ldots, 24$. A problem arises in calculating the sine and cosine of an arc whose length is not an exact multiple of $\frac{\pi r}{48}$. The length of an arc can lie anywhere between 0 and $2\pi r$. It was known then that in order to obtain the sine and cosine of an arc of any length, it was sufficient to evaluate the sine and cosine of the portion of the arc that lies in a quadrant of the circle.[2] It was the search for finding the length of any arc (which were not exact multiples of $\frac{\pi r}{48}$) lying in the first quadrant that led to the work on infinite sine and cosine series. In the next section of this appendix we will examine how these series were derived in Kerala.

It should be noted that the main motivation behind the evaluation of arc-lengths was primarily astronomical, as can be seen from the following representation of the sine and versed-sine series apart from the ones given in the last Appendix. In Nilakantha's *Aryabhatiyabhasya* (and reproduced in the *Yuktibhasa*) appears the following passages, again attributed to Madhava:

"For a sequence of five numbers $(3, 5, 7, 9, 11)$, the first number is multiplied by the square of the given arc and divided by the square of 5400. The quotient is subtracted from the second number. The result of the subtraction is next multiplied by the square of the given arc and divided by the square of 5400 and the quotient that results is subtracted from the third number. This process of multiplication and division followed by subtraction is repeated till all the five numbers are gone through. The final result is then multiplied by the cube of the given arc and divided by the cube of 5400. If the quotient then resulting is subtracted from the given arc, what remains will be the required [*bhuja*] *jya* (or Indian sine)" (Rajagopal and Rangachari, 1978, 98–99 p. 195).

Expressed in symbolic notation and with the following additional symbols, where all angular measures are given in radians:

Let $c = 90 \times 60 = 5400$ *ilis* ('minutes') be the first quadrant of a circle of circumference 21600 *ilis*; let a_i be the ith odd number where $i = 3, 5, 7, 9, 11$; and let $s = r\theta$ be the sine-chord.

Then

$$jya\theta = r\sin\theta = s - \left(\frac{s}{c}\right)^3 \left\langle a_3 - \left(\frac{s}{c}\right)^2 \left\{ a_5 - \left(\frac{s}{c}\right)^2 \right. \right.$$
$$\left. \left. \times \left[a_7 - \left(\frac{s}{c}\right)^2 \left(a_9 - \left(\frac{s}{c}\right)^2 a_{11} \right) \right] \right\} \right\rangle$$

or

$$jya\theta = r\sin\theta$$
$$= s - a_3 \left(\frac{s}{c}\right)^3 + a_5 \left(\frac{s}{c}\right)^5 - a_7 \left(\frac{s}{c}\right)^7 + a_9 \left(\frac{s}{c}\right)^9 - a_{11} \left(\frac{s}{c}\right)^{11}.$$

Now substitute the individual values of $a_i = 3, 5, 7, 9, 11$; $c = 5400$ and $s = r\theta$ in the above expression and simplify to get the power series for sine θ

$$\sin\theta = \theta - \frac{\theta^3}{3!} + \frac{\theta^5}{5!} - \frac{\theta^7}{7!} + \frac{\theta^9}{9!} - \frac{\theta^{11}}{11!} \cdots$$

which is the same as (A.3.1) in the last section.

The verse that follows in the *Yuktibhasa* gives the cosine series:

"For a sequence of the first six even numbers (i.e. $2, 4, 6, 8, 10, 12$), the first number is multiplied by the square of the given arc and divided by the square of 5400 . The quotient is subtracted from the second number. The result of the subtraction is next multiplied by the square of the given arc and divided by the square of 5400 and the quotient that results is subtracted from the third number. This process of multiplication and division followed by subtraction is repeated till all the six numbers are gone through. The final result will be the required *utkramajya* (Indian versine) of the arc" (*Yuktibhasa*, p. 195).

Or, denoting the first six even numbers as a_j where $j = 2, 4, 6, 8, 10, 12$

$$r - r\cos\theta = \left(\frac{s}{c}\right)^2 \left\langle a_2 - \left(\frac{s}{c}\right)^2 \left\{ a_4 - \left(\frac{s}{c}\right)^2 \right. \right.$$
$$\left. \left. \times \left[a_6 - \left(\frac{s}{c}\right)^2 \left(a_8 - \left(\frac{s}{c}\right)^2 a_{10} \right) \right] \cdots \right\} \right\rangle$$

$$r - r\cos\theta = \left(\frac{s}{c}\right)^2 \left[a_2 - \left(\frac{s}{c}\right)^2 \left\{ a_4 - \left(\frac{s}{c}\right)^2 \left(a_6 - \left(\frac{s}{c}\right)^2 \right. \right. \right.$$
$$\left. \left. \left. \times \left[a_8 - \left(\frac{s}{c}\right)^2 \left[a_{10} - \left(\frac{s}{c}\right)^2 a_{12} \right] \right] \right\} \right]$$

or more generally,

$$r - r\cos\theta = a_2 \left(\frac{s}{c}\right)^2 - a_4 \left(\frac{s}{c}\right)^4 + a_6 \left(\frac{s}{c}\right)^6 - a_8 \left(\frac{s}{c}\right)^8$$
$$+ a_{10} \left(\frac{s}{c}\right)^{10} - a_{12} \left(\frac{s}{c}\right)^{12} + \cdots .$$

Substitute the values of a_i's and $s = r\theta$ in the above and simplify to get:

$$\cos\theta = 1 - \frac{\theta^2}{2!} + \frac{\theta^4}{4!} - \frac{\theta^6}{6!} + \frac{\theta^8}{8!} - \frac{\theta^{10}}{10!} + \frac{\theta^{12}}{12!} - \cdots$$

which is the same as (A.3.2) in the last section

In Indian geometry of the time, the circumference of a circle was divided into $360 \times 60 = 21600$ *ilis* (or 'minutes'). An *ili* of an arc faces an angle of one minute at the centre of a unit circle.[3] In terms of *ilis*, the radius is reckoned as 3437' 44" 43"'.[4]

Now

$$\frac{C}{4} = 5400 = \frac{1}{2}\pi r,$$

Define[5]

$$a_1 = 5400 = r\left(\frac{1}{2}\pi\right),$$

$$a_2 = \frac{(5400)^2}{2!r} = \frac{r}{2!}\cdot\left(\frac{1}{2}\pi\right)^2,$$

$$a_3 = \frac{(5400)^3}{3!r} = \frac{r}{3!}\cdot\left(\frac{1}{2}\pi\right)^3,$$

$$a_4 = \frac{(5400)^4}{4!r} = \frac{r}{4!}\cdot\left(\frac{1}{2}\pi\right)^4,$$

$$a_5 = \frac{(5400)^5}{5!r} = \frac{r}{5!}\cdot\left(\frac{1}{2}\pi\right)^5,$$

$$a_6 = \frac{(5400)^6}{6!r} = \frac{r}{6!}\cdot\left(\frac{1}{2}\pi\right)^6,$$

$$\vdots$$

$$a_{12} = \frac{(5400)^{12}}{12!r} = \frac{r}{12!}\cdot\left(\frac{1}{2}\pi\right)^{12}.$$

It is believed that Madhava used this result to construct a sine table of 24 values for θ (i.e. the values of cumulative series obtained by dividing the quadrant of a circle into 24 equal parts) starting from $\theta = 224$ *ilis*, 50 *vilis* and 22 *tatparas* which corresponds to the present-day angular measurement of 3 degrees and 45 minutes. The values of his sine table are correct in almost all cases to the eighth or ninth decimal place. A similar development led to the construction of an equally accurate cosine table.

A.3.3   Derivations of sine and cosine series

In Figure A.3.1, the arc PX of a circle radius r subtends an angle θ at the centre O. Let the arc PX be length x and let $PP_1 = \delta x$ be a small increase in the arc length with corresponding increase of angle $\delta\theta$. P_m is the mid-point of the arc PP_1. Q_1, Q_m and Q are the perpendiculars from P_1, P_m and P respectively to the radius OX.

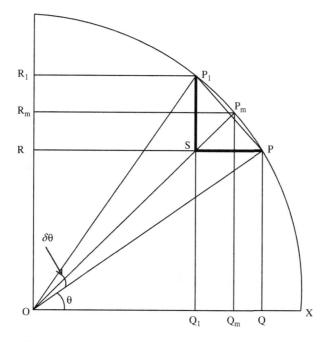

Figure A.3.1. Derivation of sine and cosine series (1).

Applying the 'rule of three' (or properties of similar triangles) to triangles P_1SP and OQ_mP_m we have:

$$P_1S = \text{increase in } r\sin\theta = \frac{OQ_m \times P_1P}{OP_m}. \qquad (A.3.3)$$

$$PS = \text{decrease in } r\cos\theta = \frac{P_mQ_m \times P_1P}{OP_m}. \qquad (A.3.4)$$

In what follows we translate the exposition in the *Yuktibhasa* into modern mathematical language. So if $r = 1$, then $\delta x \approx \delta\theta$ and the above reduce to the familiar

$$\delta(\sin\theta) \approx \cos\theta\,\delta\theta \qquad (A.3.5a)$$

$$\delta(\cos\theta) \approx -\sin\theta\,\delta\theta. \qquad (A.3.5b)$$

The formulae $\sin\theta$ and $\cos\theta$ in the *Yuktibhasa* utilises the derivatives above. However, we have already seen in the mathematical exposition so far that there is a marked difference between Indian and modern (Greek-derived) mathematics. This not only applies to epistemology but also to philosophy and to terminology. In order to present an exposition that a reader versed only in modern mathematics can make sense we have 'translated' the mathematics of the *Yuktibhasa* using modern terminology. However, this will not diminish the fact that the geometric and analytical thinking of the Kerala mathematicians has no parallel in Western (Greek) mathematics — their treatment of infinity and infinitesimal processes is a case in point. All this is even more marked in construction of infinite series expansions. Consequently a full appreciation of this analysis in the *Yuktibhasa* can only be achieved by bearing these differences in mind. This analysis is based on the interpretative works of Sarasvati Amma (1963a) and Bag (1976).

In Figure A.3.2, the radius of the circle is r. Let the arc PX of length x, which subtends an angle θ at the centre, be subdivided into

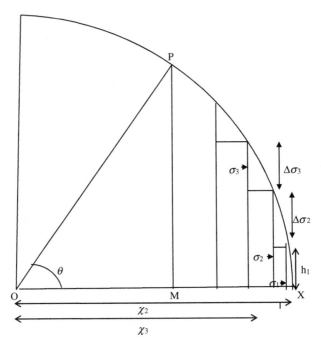

Figure A.3.2. Derivation of sine and cosine series (2).

x equal parts each of length $\delta x =$ one unit of measure. Let σ_i and $\Delta\sigma_i$, $1 \le i \le x$, be the sine chords and sine differences, respectively, at these points. Let χ_i, $1 \le i \le x$, be the cosine chords at the mid points of the arcs.

Now

$$\Delta\sigma_1 = \sigma_1,$$

So the sum of the difference of the sine differences

$$= [(\Delta\sigma_1 - \Delta\sigma_2) + (\Delta\sigma_1 - \Delta\sigma_3) + \cdots + (\Delta\sigma_1 - \Delta\sigma_x)]$$

$$= [(\sigma_1 - \Delta\sigma_2) + (\sigma_1 - \Delta\sigma_3) + \cdots + (\sigma_1 - \Delta\sigma_x)]$$

$$= [n\sigma_1 - (\Delta\sigma_2 + \Delta\sigma_3 + \cdots + \Delta\sigma_x)]$$

$$= [(\Delta\sigma_1 - \Delta\sigma_2) + (\Delta\sigma_1 - \Delta\sigma_3) + \cdots + (\Delta\sigma_1 - \Delta\sigma_x)] \approx r\theta - \sigma_x$$

$$= r\theta - r\sin\theta. \tag{A.3.6}$$

Now the height $H(= \text{PM})$ of the arc PX $=$ sum of the positive decreases in cosines. So using (10b) we have

$$H = \frac{(\sigma_1\delta x + \sigma_2\delta x + \sigma_3\delta x + \cdots + \sigma_x\delta x)}{r}. \tag{A.3.7a}$$

A first approximation is to take $\sigma_i = i\delta x = i$ (as $\delta x = 1$), $1 \le i \le x$. That is, *the sine chords are assumed to be equal to the lengths of the arcs.* Noting that this is an over-estimation, a first approximation for H is

$$H = \frac{(1 + 2 + 3 + \cdots + x)}{r}. \tag{A.3.7b}$$

Now equation (A.3.7a) leads to the following approximation

$$1^k + 2^k + \cdots + x^k \approx \frac{x^{k+1}}{(k+1)}. \tag{A.3.8}$$

Using equation (A.3.17) with $k = 1$ in equation (A.3.16a) gives

$$H = \frac{(1 + 2 + 3 + \cdots + x)}{r} \approx \frac{x^2}{2r}. \tag{A.3.9}$$

On the other hand, the sine differences can be calculated another way. Using (10a) we can write

$$\Delta\sigma_1 \approx \chi_1 \delta x = \chi_1 \quad (\text{as } \delta x = 1).$$

Now the *Yuktibhasa* makes the approximation $\chi_1 \approx (r - h_1)$, where h_1 represents the height at the mid point of the first arc element (shown in Figure A.3.2). So that

$$\Delta\sigma_1 \approx (r - h_1) \tag{A.3.10a}$$

similarly,

$$\Delta\sigma_i \approx \chi_i \delta x \approx (r - h_i), \quad 1 \le i \le x, \tag{A.3.10b}$$

where h_i represents the height at the mid point of the ith arc element. Thus

$$(\Delta\sigma_1 - \Delta\sigma_i) \approx (h_i - h_1), \quad 1 \le i \le x. \tag{A.3.10c}$$

The *Yuktibhasa* states that it is reasonable to assume that h_1 is approximately equal to 0 and h_x is approximately equal to H (the height of the arc PX). Thus

$$(\Delta\sigma_1 - \Delta\sigma_x) = h_x = H \approx \frac{x^2}{2r} \tag{A.3.11}$$

similarly,

$$(\Delta\sigma_1 - \Delta\sigma_i) = h_i = \frac{(x - i)^2}{2r}, \quad 1 \le i \le x, \tag{A.3.12}$$

so that

$$\sum_1^x (\Delta\sigma_1 - \Delta\sigma_i) = \sum_1^x \frac{(x - i)^2}{2r}. \tag{A.3.13}$$

Using (A.3.17) with $k = 2$ and (A.3.15) we have

$$r\theta - r\sin\theta \approx \frac{1}{2} \times \left(\frac{x^3}{3r^2}\right) = \frac{x^3}{3!r^2}. \tag{A.3.14}$$

Initially, the sine chords were assumed to be *equal* to the lengths of the arcs (when, in fact, they are less than the lengths of the arcs).

The excess of the arc length over the sine chord implied by the above is now applied to the left side of equation (16):

$$H = \frac{(1 + 2 + 3 + \cdots + x)}{r} = \frac{(x + (x-1) + (x-2) + \cdots + 1)}{r},$$

so that the second approximation is

$$H = \frac{1}{r} \times \left[\left(x - \frac{x^3}{3!r^2} \right) + \left(x - 1 - \frac{(x-1)^3}{3!r^2} \right) \right.$$
$$\left. + \left(x - 2 - \frac{(x-2)^3}{3!r^2} \right) + \cdots \right)$$
$$= \frac{(x + (x-1) + (x-2) + \cdots)}{r}$$
$$- \frac{\left[\frac{x^3}{3!r^2} + \frac{(x-1)^3}{3!r^2} + \frac{(x-2)^3}{3!r^2} + \cdots \right)}{r}.$$

Thus the correction for the height

$$H = \frac{\left[\frac{x^3}{3!r^2} + \frac{(x-1)^3}{3!r^2} + \frac{(x-2)^3}{3!r^2} + \cdots \right]}{r}$$

or

$$rH = \frac{x^4}{4!r^3}; \quad \text{using (A.3.17) with } k = 3. \tag{A.3.15}$$

With this correction we now have,

$$\sum_{1}^{x} (\Delta\sigma_1 - \Delta\sigma_i) = \sum_{1}^{x} \frac{(x-i)^2}{2r} - \sum_{1}^{x} \frac{(x-i)^4}{4!r^3}.$$

Using (A.3.8) with $k = 2$ and $k = 4$ and (21) we have

$$r\theta - r\sin\theta \approx \frac{x^3}{3!} - \frac{x^5}{5!}. \tag{A.3.16a}$$

In the next step, the correction term is introduced to correct the deficiency of the approximation for the sine chord. So that the next approximation is

$$r\theta - r\sin\theta \approx \frac{x^3}{3!} - \frac{x^5}{5!} - \frac{x^7}{7!}. \tag{A.3.16b}$$

Continuing in this way, ad infinitum, we have

$$r\theta - r\sin\theta = \frac{x^3}{3!} - \frac{x^5}{5!} + \frac{x^7}{7!} - \cdots + \frac{(-1)^n x^{2n-1}}{(2n-1)!} + \cdots. \tag{A.3.17}$$

Substituting $x = r\theta$ gives the familiar series:

$$\sin\theta = \theta - \frac{\theta^3}{3!} + \frac{\theta^5}{5!} - \frac{\theta^7}{7!} + \cdots + \frac{(-1)^n \theta^{2n+1}}{(2n+1)!} + \cdots. \tag{A.3.18}$$

The familiar infinite series for $\cos\theta$ is constructed in likewise manner. It should be remembered these developments were first expounded by Madhava in the 15th century. However, in the vast majority of published books on the history of mathematics the series expansion for $\sin\theta$ and $\cos\theta$ are attributed to Newton in his works in 1669 (*Analysis with infinite series*) and in 1671 (*Method of fluxions and infinite series*).

A.3.4   Rational approximations of sine and cosine functions

In Nilakantha's *Aryabhatiyabhasya* (page 73) and Jysthadeva's *Yuktibhasa* (page 178) appears the following passage with "*iti Madhava*" ("thus said Madhava") appearing at the end of the verse in the latter text:

> Consider the sine (cosine) chord (of the arc which is a multiple of 225 *ilis* (minutes) closest (in length) to the arc whose sine(cosine) is required. Find the arc-difference. For making the correction 13751 should be divided by twice the arc-difference expressed in *ilis* and the quotient is to be considered as the 'divisor'. Divide sine (cosine) by this (divisor) and subtract from or add to the other (sine or cosine) based on the sign of the arc-difference. Double this (result) and do as before (i.e. divide by the 'divisor'). Add or subtract the result (so obtained) to or from sine or cosine to get the desired sine or cosine chords.

Expressed symbolically, let h be the arc-difference. The 'divisor' $(d) = 13751/2h$.

Then

$$\sin x + \left(\cos x - \frac{\sin x}{d}\right)\left(\frac{2}{d}\right) \approx \sin(x+h)$$

and

$$\cos x - \left(\sin x + \frac{\cos x}{d}\right)\left(\frac{2}{d}\right) \approx \cos(x+h).$$

Now

$$d = \frac{2r}{h} \quad \text{since } r = \frac{21600}{2\delta} = 3437.75\delta = \frac{13751}{4}.$$

Substitute $d = 2r/h$ in the expressions above to get:

$$\sin(x+\delta h)\sin x + \left(\frac{h}{r}\right)\cos x - \left(\frac{h^2}{2r^2}\right)\sin x$$

$$\cos(x+\delta h)\cos x - \left(\frac{h}{r}\right)\sin x - \left(\frac{h^2}{2r^2}\right)\cos x.$$

These results are but special cases of familiar expansions in mathematics, the Taylor series, up to the second power of the small quantity u

$$f(x+u) = f(x) + u \cdot f'(x) + \left(\frac{u^2}{2}\right)f''(x) + \cdots$$

where u (in radian measure) $= \frac{h}{r}$.

A.3.5   Construction of sine tables: the Kerala contribution

The procedure for computation of sine and cosine tables of length $l = 3 * 2^m$ at equal arc intervals $h = \frac{1800}{l}$ for $m = 0, 1, 2, 3, 4, 5$, etc. consists of the following steps.

Step-I: Take $R = \dfrac{21600 \times 113}{2 \times 355}$ computed using $\pi = \dfrac{355}{113}$.

Step-II: Start with

$$\theta = 30°, S_0 = R\sin 30° = \frac{R}{2} \quad \text{and} \quad C_0 = R\cos 30° = \sqrt{R^2 - S_0^2}.$$

Step III: Denote

$$S_i = R\sin\left(\frac{\theta}{2^i}\right) \quad \text{and} \quad C_i = R\cos\left(\frac{\theta}{2^i}\right)$$

and calculate

$$S_i = \sqrt{S_{i-1}^2 + (R - C_{i-1})^2} \quad \text{and} \quad C_i = \sqrt{R^2 - S_i^2},$$

$$\text{for } i = 1, 2, 3, \ldots, m,$$

where $m = 3, 4, 5, \ldots$ according as the arc interval h is $225'$, $112.5'$, $56.25', \ldots$

Step IV: Now for initiating the computation of Rsine and Rcosine tables at interval of

$$h = \frac{\theta}{2^m} = \frac{30 \times 60}{2^m},$$

the first tabular Rsine $= J_1$ is the value S_m and the last Rsine is $J_l = R\sin lh = R$, the half diameter, where

$$l = \frac{90 \times 60}{\left(\frac{\theta}{2^m}\right)} = \frac{90 \times 60 \times 2^m}{30 \times 60} = 3 \times 2^m, \quad \text{since } \theta = 30° = 30 \times 60°.$$

Step V: Compute

$$J_{l-1} = \sqrt{R^2 - J_1^2}, \Delta J_{l-1} = J_l - J_{l-1} \quad \text{and} \quad \lambda = 2\left(\frac{\Delta J_{l-1}}{R}\right).$$

Step VI: Compute $\Delta J_{l-k} = \lambda \times J_{l-(k-1)} + \Delta J_{l-(k-1)}.$

Step VII: Now for $k = 2, 3, 4, \ldots l - 2$ compute $J_{l-k} = J_{l-(k-1)} - \Delta J_{l-k}$ and

$$C_{l-k} = \sqrt{R^2 - J_{l-k}^2}.$$

Sine tables of lengths $3, 6, 12, 24, 48, 96, 192, 384, \ldots$ may be computed using this *Golasāra* algorithm. It is quite interesting to note that Nīlakantha has referred to the last and the first Rsine differences by the terms *antya* and *adi khanda* without mentioning that the last Rsine is the 24th. So it may be inferred that Nīlakantha's rule for determination of the Rsines successively gives a general method for constructing sine and cosine tables. A computer program has been developed on the basis of this method and a sample output generated are available from its author, Dr Mallayya. By comparing the values obtained with the corresponding modern values it can be shown that the *Golasāra* method gives by and large fairly accurate values even up to 20 decimal places computed. Cosine tables may also be constructed similarly.

Endnotes

[1] No information is available in the text on the ultimate source of these results, although there is extraneous evidence to indicate that they originated with the founder of the Kerala School, Madhava.

[2] What was known may be summarised in modern notation as:

$$\sin(90° + \theta) = \sin(270° + \theta) = -\cos\theta,$$
$$\sin(180° + \theta) = -\sin\theta.$$

The cosines of quantities outside the first quadrant may be evaluated similarly.

[3] An *ili* is further subdivided into 60 *vilis* (*or vikalas*) and a *vili* is further subdivided into 60 *talparas* and a *talpara* into 60 *pratalaparas*.

[4] Or more precisely, the radius is taken as 3437 *ilis*, 44 *vilis*, 48 *talparas* and 22 *pratalparas* (or in decimal notation 3437.74677) which is approximately equal to the radian measure, 57 degrees, 17 minutes and 44.8 seconds correct to a *talpara* (or 3437.77 in decimal notation).

[5] In the original text of *Yuktibhasa* from which the passage is quoted later, each $a_i (i = 1, 2, \ldots, 12)$ is assigned an individual name according to the *Katapyadi* notation and the associated numerical values. Thus, for example, the last three coefficients, a_{10}, a_{11} and a_{12} were given the names *stripsuna* (5.2 *ilis*), *vidvan* (44 *vilis*) and *stena* (6 *vilis*) respectively.

Chapter 11

Indian Trigonometry: From Ancient Beginnings to Nilakantha*

11.1 A Preliminary Note

The historical context of the emergence of trigonometry (literally, the 'measuring of triangles') is found in the need to explain the non-uniform motion of the planets. It is now generally known that the planets move in elliptical orbits around the Sun and the Moon moves in an elliptical orbit around the Earth and hence the orbits have an eccentricity and it was this that had to be taken into account in studying the exact location of the heavenly bodies.

The origins of trigonometry as a discipline may be traced back to Hipparchus of Nicaea (*fl.* 150 BCE) and embedded in the history of the simple word "sine", a mistranslation of an Arabic transliteration of a Sanskrit mathematical term! The complex etymology of "sine" reveals trigonometry's roots in Babylonian, Greek, Hellenistic, Indian, and Islamic mathematics and astronomy. An aid to the study of astronomy/astrology, the basic problem to be tackled was to estimate, for a given arc of a circle, the length of the chord that connects the endpoints of that arc. It was soon noticed that the chord length depends on both the length of the arc and the radius of the circle. For the ancients (Greeks as well as the others), an angle

---

*This chapter is based on a preprint authored by Madhukar Mallayya (2009) who carried out a comprehensive survey for a AHRB Research project on Medieval Kerala Mathematics.

of 90° was not a right angle (as we understand it today) but the quarter of the circumference of a circle. Or generally, 'degrees' was more a measure of the length of an arc rather than the size of an angle.[1] Thus, for a given circle of circumference of 360°, the radius of the circle was $360/2\pi = 57.2957$, if π was approximated by the Aryabhatan value of 3.1416 as given in Verse 10 of the *Aryabhatiya* and discussed in Chapter 5. Or if the circumference was measured in minutes (60 minutes = 1 degree), then the total circumference is $360 \times 60 = 21,600$ minutes and the corresponding radius is 3437.7387... or approximately 3438 minutes. This was known fairly early in Indian trigonometry with the radius of 3,438 minutes taken as a standard measure in the construction of trigonometric tables.[2]

The earliest uses of trigonometric functions were related to the chords of a circle, and the recognition that the length of the chord subtended by a given angle θ was (in modern terms) $2r \times 2\sin(\theta/2)$ where r is the radius. The Greek astronomer and mathematician Hipparchus produced the first known table of chords in 140 BCE. His work was further developed by astronomers Menelaus (ca. 100 CE) and Ptolemy (ca. CE 100), who relied on Babylonian observations and methodology.[3]

One of the earliest trigonometric tables constructed contained values for the length of the chord for a given arc (normally denoted by *Crd α*). In terms of modern notation and as shown in Figure 11.1, the chord for a given arc is twice the sine of half the angle multiplied, by the radius of the circle taken here to be 3438. On the basis of this relationship, a table for chords can be constructed.[4]

11.2   Basic Indian Trigonometry: Introduction and Terminology

Figure 11.2 below represents a circle of radius R. The following terminology and relationships may be deduced by reference to this diagram.

(i) The arc PQ of the circle was known as *capa*.
(ii) The chord PQ was known as *samastajya*.

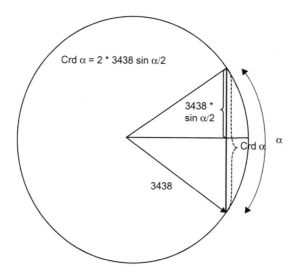

Figure 11.1. The length of a chord (based on Bressoud, 2002, p. 2).

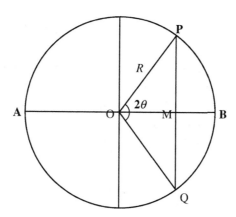

Figure 11.2. Indian trigonometry.

(iii) The half-chord PM (or Indian Sine) was known as *ardhajya* or *jyardha* or *bhujajya* and often shortened to *jya*.

(iv) The relationship between the Indian sine and the modern sine is a simple one:

$$\text{Indian sine } (jya) = R\theta = R\sin\theta$$

(where $R\theta$ = length of the arc of angular measure and sine θ = modern sine).

[*Note*: The Indian sine, often denoted as Sine with a capital 'S', is *not* a ratio but a linear measure]

(v) OM was called *kotijya* (or *kojya*) and equals $R\cos\theta$, where $\cos\theta$ is the modern cosine.

(vi) Now MB = OB − OM = $R − kojya = R − R\cos\theta = R(1−\cos\theta)$. This was known as *sara* or *utkramajya* which is the presently labelled as Rversine θ.

(vii) The radius R of the circle was referred variously as *vyasadala* or *vyasardha or sinjini*.[5]

11.3   The Work of Aryabhata I (b. 476 CE)

The construction of a table of successive Rsines (or *jyas*) and their differences for purposes of interpolation (or for calculating the intermediate values) has been a common feature of Indian astronomical texts from the time of Aryabhata I. The tables were then used for accurate computations of planetary positions. As mentioned earlier, taking the circumference of a circle in angular measure of $360° \times 60 = 21600$ minutes led to an estimated radial value of approximately 3438 minutes, using the Aryabhathan value of 3.1416 as the ratio of circumference to diameter. The circumference was then divided into 24 equal parts so that each part covered an angular measure of $3° 45'$ or $225'$.

Verse 11 in *Aryabhatiya* has the following instruction for the computation of the (Rsine) table:

> Divide a quadrant of the circumference of a circle [into as many parts as desired]. Then from [right-handed] triangles and quadrilaterals, once can find as many Rsines of equal arcs as one likes, for any given radius.

In Figure 11.3(a), a circle is divided into 12 equal parts (*rasis*) with their corresponding arcs and chords. If we take one of these parts and label it as in Figure 11.3(b), then CD is the arc and BD = $R\sin\theta$ is the Indian sine.

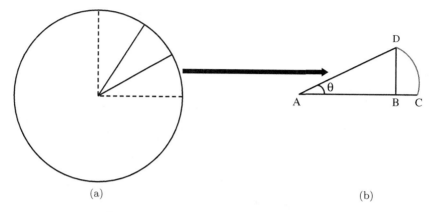

Figure 11.3. Computation of Rsine tables.

In one of the earliest and most notable of the commentators of *Aryabhatiya*, Bhaskara I, gives three examples to illustrate the above verse. The first one concerns the computations of six Rsines at intervals of 15° in a circle of radius R. The next involves the computation of 12 Rsines at intervals of 7° 30 minutes in a circle of radius R. The last example provided the proto-type for later Indian sine tables, the construction of 24 Rsines at the equal intervals of 3 degrees and 45 minutes in the circle of radius R.

Different astronomers used different values for R. In his sixth century text, *Panchasiddhantika*, Varahamihira uses $R = 120$ in the computation of 12 Rsines at intervals of 7 degrees 30 minutes. Brahmagupta computes 6 Rsines at intervals of 15 degrees in a circle of radius $R = 150$. But from the time of Aryabhata I the quadrant was usually divided into 24 arcs of 225 arc- minutes which gives the trigonometric radius of $\frac{60 \times 360}{2\pi} \approx 3438$ minutes, where $\pi \approx 3.1416$ is the Aryabhatan value. The choice of this radius would enable one to use the same unit of length for arcs and their sines since the sines of small arcs are nearly equal to the arcs themselves (i.e. 3°45′ expressed as base 60 = 225′).[6]

One of the remarkable aspects of Indian work on sines is the early appearance of these different methods of tabulating sines. Taking the first entry of the table as sin (225′) = 225, several Siddhantas, including *Paitamahasiddhanta* (ca. sixth century CE) and the later

Suryasiddhanta (?), contain rules which may be termed as first difference method.

Verse 12 in *Aryabhatiya* contains the first explicit statement of the rule:

The first *R*sine divided by itself and then diminished by the quotient gives the second *R*sine-difference. The same first *R*sine diminished by the quotients obtained by dividing each of the preceding *R*sines by the first *R*sine gives the remaining *R*sine-differences.

Expressed in modern notation, let $J_1, J_2, \ldots, J_{24}$ denote the twenty four *R*sines (*Jya*) and $d_1, d_2, d_3, \ldots, d_{24}$ denote the twenty four sine-differences. The above rule (and especially the second sentence) may be interpreted in different ways. A widely favoured interpretation is as follows:[7]

Expressed in modern notation, the first *R*sine-difference (d_1) is given by:

$$d_1 = R\sin\theta = R\theta = h \approx \mathbf{225'} \quad \text{since } \sin\theta \approx \theta \text{ if } \theta \text{ is small}$$

$$(\text{note } \sin 0° = 0).$$

The second *R*sine-difference d_2 is given by:

$$d_2 = R\sin\theta - \frac{R\sin\theta}{R\sin\theta} = R\sin\theta - 1 \approx 225 - 1 = 224'.$$

The third *R*sine-difference d_3 is given by the formula

$$d_3 = R\sin\theta - \left[\frac{R\sin\theta}{R\sin\theta} + \frac{R\sin 2\theta}{R\sin\theta}\right] \approx 225 - 3 = \mathbf{222'}.$$

The fourth *R*sine-difference d_4 is computed using the formula

$$d_4 = R\sin\theta - \left[\frac{R\sin\theta}{R\sin\theta} + \frac{R\sin 2\theta}{R\sin\theta} + \frac{R\sin 3\theta}{R\sin\theta}\right] \approx 225 - 6 = \mathbf{219'} \ldots.$$

Thus it follows general, the $(n+1)$th *R*sine-difference $(\mathbf{d_{n+1}})$ is given by[8]:

$$R\sin\theta - \left[\frac{R\sin\theta + R\sin 2\theta + R\sin 3\theta + \cdots + R\sin(n\theta)}{R\sin\theta}\right].$$

Table 11.1. Aryabhata's Sine Table (R = 3438 minutes).

No (k)	Arc in angular measure ($k\theta'$)	Rsine-diff (d_k)	Rsine(J_k)	Sine values $S_k = J_k/R$	Modern sine value (s_k)	Error $\varepsilon_k = S_k - s_k$
1	225	225	225	0.065445026	0.065403129	0.00004190
2	450	224	449	0.130599186	0.130526192	0.00007299
3	675	222	671	0.195171611	0.195090322	0.00008129
4	900	219	890	0.258871437	0.258819045	0.00005239
21	4725	51	3372	0.980802792	0.98078528	0.00001751
22	4950	37	3409	0.991564863	0.991444861	0.00012000
23	5175	22	3431	0.997963933	0.997858923	0.00010501
24	5400	7	3438	1	1	0

This expression was then used to generate the Rsine-differences successively from the preceding ones and also to find the Rsines successively. As an illustration, Table 11.1 gives the first four and last four entries in such a table.

Table 11.1 also allows a comparison to be made between these sine values (after adjustments obtained by dividing the Rsine values (J_k, an abbreviation for *Jya*) by the value of $R = 3438'$) and corresponding modern sine values. It is the remarkable accuracy of a table computed from methods that are at least 1,500 years old that is very striking.

11.4   The Suryasiddhanta

The *Sūryasiddhanta* (*c.* 400 CE; ii, 15–22), uses a simpler formula to get a list of 24 Rsines, the first being stated as an eighth part of a*rasi*.[9] Since a *rasi* comprises 30° an eighth part of a *rasi* is equal to 3° 45 or 225 minutes. The 24 Rsines are then obtained successively in the following manner.

The eighth part of a *rasi* is the first Rsine. Divide this by itself, subtract the quotient and add to itself. This is the second. By the first, divide each of the Rsines obtained in order, subtract the sum of all the quotients from the first and add the last of the Rsine obtained. The twenty four Rsines are thus obtained successively.

Elaborating on this rule, the Kerala mathematician Paramesvara (1430) gave a formula for computing the Rsine-differences successively from the preceding Rsine-differences according to which the first Rsine divided by itself and diminished by the quotient to give the second Rsine-difference. The other Rsine-differences are obtained successively by subtracting from the preceding difference, the quotient of division of sum of all the preceding differences by the first Rsine.

In other words, if $d_1, d_2, d_3, \ldots$ are the successive Rsine-differences then,

$$d_1 = R\sin\theta \left(= \frac{30°}{8} = 225'\right),$$

$$d_2 = d_1 - \frac{d_1}{R\sin\theta},$$

$$d_3 = d_2 - \frac{d_1 + d_2}{R\sin\theta}$$

$$\cdots$$

or in general,

$$d_{n+1} = d_n - \frac{d_1 + d_2 + d_3 + \cdots + d_n}{R\sin\theta}.$$

Since

$$d_1 = R\sin\theta \quad \text{and} \quad d_k = R\sin k\theta - R\sin(k-1)\theta \quad \text{for } k = 1, 2, 3, \ldots, n,$$

the above implies

$$d_{n+1} = d_n - \frac{\begin{array}{c} R\sin\theta + \{R\sin 2\theta - R\sin\theta\} + \{R\sin 3\theta - R\sin 2\theta\} \\ + \cdots + \{R\sin n\theta - R\sin(n-1)\theta\} \end{array}}{R\sin\theta}$$

or simplifying the expression, we get

$$d_{n+1} = d_n - \frac{R\sin(n\theta)}{R\sin\theta}.$$

This simple formula could then be used to compute successively the *R*sine-differences and also the *R*sines.

From these differences, the *Sūryasiddhanta*, using this method, obtained the twenty four *R*sines at intervals of 225 minutes They are identical to the ones found in the Table 11.1.

11.5   The Work of Varahmahira (*fl.* 500 CE)

A near contemporary of Aryabhata used a different but interesting method of computing sine values. Varahamihira takes $\sqrt{10}$ as the value of circumference to diameter ratio[10] and assumes the radius to be 120 minutes for computing the *R*sines of arcs at interval of $3°45'$ ($= 225'$) of angular measure. For a circumference of angular measure 21600 minutes, the radius will be 3415.259873 instead of 3438, the one used most often in Indian astronomy. However, this variation in the value of R does not affect the *R*sine computation in any way because the *R*sines derived does not depend on the ratio of circumference to diameter.

Varahamihira provides a discussion of *R*sines and construction of trigonometric tables in his *Panchasidhantika*. In it appears:

> The square of half diameter is termed *dhruva*. Its fourth part *mesa* (is of first *rasi*); *dhruvakarani* diminished by that of *mesa* is of two *rasis*. The square root (of a *mesakarani*) is the *R*sine.

Since the diameter is 240 minutes, *dhruva* (or square of radius) $=$ $R^2 = 14400$, and one *rasi* is $30°$, so *R*sine of one *rasi* is $R\sin 30°$. Now,

$$mesakarani- = \frac{R^2}{4} = 3600 \quad \text{or} \quad (R\sin 30°)^2 = \frac{R^2}{4} = 3600.$$

According to the above-stated rule,

> The square of *R*sine of two *rasis*
>
> $= dhruvakarani - mesakarani$
>
> $= R^2 - (R\sin 30°)^2 = 14400 - 3600 = 10800$

or

$$(R\sin 60^0)^2 = R^2 - (R\sin 30^\circ)^2 = 10800.$$

Taking square roots,

$$R = \sqrt{14400} = 120'$$

$$R\sin 30^\circ = \sqrt{\frac{R^2}{4}} = \frac{R}{2} = 60'$$

$$R\sin 60^\circ = \sqrt{R^2 - (R\sin 30^\circ)^2}$$

$$= \sqrt{R^2 - \frac{R^2}{4}} = R\frac{\sqrt{3}}{2}$$

$$= \sqrt{10800}$$

$$= 103.9230485'$$

$$\approx 103'55''.$$

For other tabular sines, the following methods were used.

$$\sin^2 \theta = \frac{1}{4}[\sin^2 2\theta + \{120 - \sin(90 - 2\theta)\}^2],$$

$$\sin^2 \theta = 60[120 - \sin(90 - 2\theta)],$$

where $R = 120'$ and the sines are in minutes.

Using these results, a table of the 24 Rsines can be computed. Corresponding values of modern sines are then computed from them by dividing by R (taken as 120). A comparison of these values with modern values would show that the accuracy is even more remarkable than Aryabhata's Sine Table, sections of which are given in Table 11.1.

11.6   Bhaskara I's sine Approximation Formula

Bhaskara I, in his *Mahabhaskariya*, gave two methods for computation of Rsines. One is without using Rsine differences and the other using Rsine-differences. It is the first of the methods that was truly innovative. Chapter 7 of that work contains three verses (vii, 17–19) which give an approximation to the trigonometric sine function by means of a rational fraction. The formula is amazingly accurate leading to a maximum error of less than one percent. The formula (expressed in modern terms) is:

$$\sin x = \frac{4x(180 - x)}{[40500 - x(180 - x)]},$$
   (*)

where x is measured in degrees and $0 \le x \le 180$
   or measured in radians:

$$\sin x = \frac{16x(\pi - x)}{[5\pi^2 - 4x(\pi - x)]}.$$
   (**)

Bhaskara attributes this formula to Aryabhata although it is not found in *Aryabhatiya*.

Table 11.2 gives the values of the Bhaskara approximation and that of the values from an electronic calculator for x in

Table 11.2. Sine table from Bhaskara I's approximation formula.

Degrees	$\sin x$	$\dfrac{4x(180 - x)}{(40500 - x(180 - x))}$	error
0	0	0	0
5	0.087155743	0.088328076	−0.00117233
10	0.173648178	0.175257732	−0.00160955
15	0.258819045	0.26035503	−0.00153598
20	0.342020143	0.343163539	−0.0011434
25	0.422618262	0.423208191	−0.00058993
30	0.5	0.5	0
35	0.573576436	0.573041637	0.000534799
40	0.64278761	0.641833811	0.000953799
45	0.707106781	0.705882353	0.001224428
50	0.766044443	0.764705882	0.001338561
55	0.819152044	0.817843866	0.001308178
60	0.866025404	0.864864865	0.001160539
65	0.906307787	0.905374716	0.000933071
70	0.939692621	0.93902439	0.000668231
75	0.965925826	0.965517241	0.000408585
80	0.984807753	0.984615385	0.000192368
85	0.996194698	0.99614495	4.97482E-05
90	1	1	0

the range $0 \le x \le 90$. Bhaskara's equation is simple, elegant and, moreover, gives values of the sine function accurate for most practical purposes.[11]

The intriguing question is: How did Bhaskara I come up with this function? The start of the story may lie in the fact that the modern sine curve has roots $x = 0$ and $x = 180$ and its curve 'looks' like a quadratic function $f(x) = x(180 - x)$. However, $\sin 90° = 1$ while $f(90) = 8100$ so there must have been a scaling down of the function. And this scaling factor is not uniform because $\sin 30° = \frac{1}{2}$ while $f(30) = 4500$ which implies a scaling down factor of 9000 while the corresponding values for $x = 90$ imply a scaling down factor of 8100. In addition, $\sin 60° = \frac{1}{2}\sqrt{3}$ and $f(60) = 7200$ imply a scaling factor of $\frac{14400}{\sqrt{3}}$.

Thus the scaling function is not linear as the difference scaling-down factors is -900 between $x = 30$ and $x = 90$, an average 'gradient' of $-900/60 = -15$, while the difference scaling down factors is approximately -686 between $x = 60$ and $x = 90$, an average 'gradient' of $-686/30 = -22.87$.

Thus Bhaskara must have conjectured that the scaling down function was quadratic. Let

$$\sin x = \frac{x(180 - x)}{ax^2 + bx + c}. \qquad (***)$$

Now Indian mathematicians since the time of Aryabhata had known the values of the sines of a large number of angles. In particular they knew that $\sin 90° = 1$, $\sin 30° = \sin 150° = \frac{1}{2}$, $\sin 60° = \frac{1}{2}\sqrt{3}$. Substituting $x = 30°$ and $x = 150°$ in the above expression $(***)$ gives

$$\frac{1}{2} = \frac{4500}{900a + 30b + c} = \frac{4500}{22500a + 150b + c}.$$

Equating the denominators gives $21600a = -120b$ or $b = -180a$. Substituting $b = -180a$ and $x = 30$ in the expression $(***)$ we obtain

$$1 = \frac{8100}{8100a - 16200a + c}.$$

Thus $c = 8100a + 8100$. With this value for c and with $x = 60$ we get[12]

$$\frac{\sqrt{3}}{2} = 0.866 = \frac{7200}{3600a - 10800a + 8100a + 8000}$$
$$= \frac{7200}{8100 - 900a} = \frac{8}{9 - a}.$$

This means that

$$a = \frac{8}{0.866} - 9 = 0.238 \quad \text{correct to 3 decimal places.}$$

So Bhaskara I must have used the approximation $a = \frac{1}{4}$. Using this approximation and simplifying gives:

$$f(x) = \frac{x(180 - x)}{\left(\frac{1}{4}x^2 - \frac{180x}{4}\right) + \frac{8100}{4} + 8100} = \frac{4x(180 - x)}{40500 - x(180 - x)}.$$

This is identical to the Approximation Formula given above as (*).

However, following the work of Aryabhata, Indian mathematicians calculated sine values for any angle in radians. It will now be shown that Bhaskara's Approximation Formula may be expressed in radians as:

$$\sin \frac{\pi}{n} = \frac{16(n - 1)}{5n^2 - 4n + 4}. \qquad (**)$$

To see how the formula may have been derived we write, using the same reasoning as before, $\sin \frac{\pi}{n}$ as the ratio of two quadratics

$$\frac{a_1 + a_2 n + a_3 n^2}{b_1 + b_2 n + b_3 n^2},$$

where a_i and b_j for $i, j = 1, 2, 3$ are constants.

The following properties of $\sin \frac{\pi}{n}$ holds when $n \geq 1$

(1) $\sin \frac{\pi}{n} = \sin \frac{\pi}{m}$, where $m = \frac{n}{n-1}$.

(2) $\sin \frac{\pi}{n} \to 0$ as $n \to \infty$.

(3) $\sin \frac{\pi}{n} = 0, 1, 0.5$, when $n = 1, 2, 6$ respectively.

Using (2) we deduce that $a_3 = 0$

Then using (1) we get

(4) $\frac{(n-1)\{n(a_1+a_2)-a_1\}}{b_1-(b_2+2b_1)n+n^2(b_1+b_2+b_3)} \equiv \frac{a_1+a_2n+a_3n^2}{b_1+b_2n+b_3n^2}.$

Finally, using (3) and the fact that $a_3 = 0$ we derive the following relations

(5) $a_1 + a_2 = 0; b_1 + 2b_2 + 4b_3 = a_1 + 2a_2; b_1 + 6b_2 + 36b_3 = 2a_1 + 12a_2.$

Substituting $a_1 + a_2 = 0$ into (4) yields $b_1 + b_2 = 0$. Whilst substituting the last two equations in (5) and using the derived relations $a_1 = -a_2$ and $b_1 = -b_2$ gives

$$4b_3 - b_1 = a_2,$$

$$36b_3 - 5b_1 = 10a_2,$$

Solving these equations we obtain

$$b_3 = \frac{5}{16}a_2 \quad \text{and} \quad b_3 = \frac{5}{4}b_1.$$

Substituting these relations between a_i and b_j for $i, j = 1, 2, 3$ into (***) gives:

$$\frac{\frac{16}{5}(n-1)}{n^2 - \frac{4}{5}n + \frac{4}{5}} \equiv \frac{16(n-1)}{5n^2 - 4n + 4}$$

which is the approximation formula in terms of radians (**)

An interesting reworking of (**) was given by the sixteenth century Indian mathematician Ganesa Daivajna:

$$\sin \frac{\pi}{n} \approx \frac{n^2 - (n-2)^2}{n^2 + \frac{1}{4}(n-2)^2}.$$

Making n the subject of the formula, would give a procedure for calculating an angle for a given sine for all values where $n \geq 2$.

Different methods based on different assumptions have been suggested. For further details, see Gupta (1986).

11.7 The Work of Brahmagupta (b. 598 CE)

Brahmagupta's rule is as follows:

> Multiply half the difference of the *gata khanda* (tabular difference d_g passed over) and *bhogya khanda* (the difference d_b to be passed over) by the residual arc in minutes) and divide by 900'. The result is added to and subtracted from half the sum (of d_g and d_b) according to whether this half sum is less than or greater than the tabular difference to be crossed. The result obtained is the true functional difference to be crossed.

According to this a better estimate of the functional difference or, true functional difference to be crossed is

$$d = \frac{1}{2}(d_g + d_b) + \frac{1}{2}(d_g - d_b)\frac{\theta}{h} \quad \text{or} \quad d = \frac{1}{2}(d_g + d_b) - \frac{1}{2}(d_g - d_b)\frac{\theta}{h},$$

depending on $\frac{1}{2}(d_g + d_b)$ is less or greater than d_b or d_g is less than or greater than d_b where $h = 900'$ and $d_b = d_{g+1}$.

The interpolated value is $f(x + hu) = f(x) + ud$, where d is as given above and $\theta = hu$.

Hence,

$$f(x + hu) = f(x) + \frac{u}{2}(d_g + d_b) - \frac{u^2}{2}(d_g - d_b),$$

if $d_g > d_b$

This gives

$$f(x + hu) = f(x) + \frac{u}{2}(d_g + d_{g+1}) - \frac{u^2}{2}(d_g - d_{g+1}).$$

From this we get,

$$f(x+hu) = f(x) + \frac{u}{2}\{\Delta f(x-h) + \Delta f(x)\} - \frac{u^2}{2}\{\Delta f(x-h) - \Delta f(x)\}$$

or

$$f(x + hu) = f(x) + \frac{u}{2}\{\Delta f(x - h) + \Delta f(x)\} + \frac{u^2}{2}\{\Delta^2 f(x - h)\},$$

where $\Delta f(x) = f(x + h) - f(x)$.

This is the Newton–Sterling's interpolation formula up to second-order term.

Table 11.3. Brahmagupta's Tables of Sines.

k	Arc = kθ	Rsines	Rsine-differences	
			First order	Second order
1	15° = 900′	39	39	−3
2	30° = 1800′	75	36	−5
3	45° = 2700′	106	31	−7
4	60° = 3600′	130	24	−9
5	75° = 4500′	145	15	−10
6	90° = 5400′	150	5	

Table 11.3 constructed using the Brahmagupta's method at intervals of 15° (or 900 minutes) with the Rsin-differences of order up to two, is given below. The value of R used by Brahmagupta was 150.

11.8   The Work of Vatesvara (b. 880 CE)

In the first chapter of the *Vatesvara Siddhanta* the author announces both the year of his birth and the date of composition of his text.

> After the lapse of 802 years from the beginning of the *saka* era, I was born; and 24 (years) after my birth, this *Siddhanta* was composed by me by the with the blessings of all planets.

From this information we conclude that Vatesvara was born in 802 Saka era (or 880 CE) and composed his work 24 years later in 904.

Well versed in the work of the Aryabhathan School, Vatesvara showed considerable skill in trigonometric computations. In verses 2–51 of his *Siddhanta*, he gives a list of the values of 96 Rsines and 96 Rversed sines at intervals of 56.25 minutes from which tables of Rsines, Rversed sines and Rcosines can be constructed. This list is interposed with verses giving various relations between Rsines, Rcosines and Rversed sines in various quadrants, several methods for computing desired Rsines from given arc and tabular values of Rsines. Several forms of first- and second-order interpolation techniques and several inverse interpolation methods are also given in Verses 55–92 for finding the desired arc from the given Rsine and tabular values.

On the basis of this information, tables of 96 values of Rsines and Rcosines can be constructed and a comparison made of these values with corresponding modern values.[13] The value of R used by Vatesvara for computation of the tables was 3437′ 44″ and not 3438′. Earlier, the Kerala astronomer Govindasvami (800–850 CE) used the even more accurate value 3437′ 44″ 19‴ for R in sine computations. Note that the value of R used by Madhava (c. 1340–1425), the founder of the Kerala School of Mathematics and Astronomy, was 3437′ 44″ 48‴. Whether the Govindasvamin had any influence on Vatesvara or Madhava is a matter of conjecture and certainly worth further investigation.

After obtaining such highly accurate values of 96 Rsines and ninety six Rversed sines at interval of $56\frac{1}{4}$ minutes, Vatesvara proceeds to state several rules for interpolating the desired Rsines from a given arc and several rules for inverse interpolation of the arc from given Rsine using the tabular Rsines.

Vatesara's treatment of trigonometric relationships remains one of the most comprehensive and innovative achievements of Indian trigonometry.

11.9   The Work of Bhaskaracharya (b. 1114 CE)

Bhaskara II in the *Siddhantasiromani* gave the same rule for interpolating intermediate Rsine values with $h = 600'$ as Munisvara in his later commentary *Marici* (1635 CE) who modifies the rule in order to attain greater accuracy.

According to Bhaskaracharya,

"Multiply the difference of the crossed and to be crossed differences by the residual arc, divide by twenty and subtract it from or add to half the sum of the differences crossed and to be crossed. This will be the true difference to be crossed for getting *kramajya* or *utkramajya* respectively."

From this the true difference to be crossed over is given by

$$d = \frac{1}{2}(d_g + d_b) - (d_g - d_b)\frac{\theta}{20} \quad \text{for computing } R\text{sines}$$

and

$$d = \frac{1}{2}(d_g + d_b) + (d_g - d_b)\frac{\theta}{20} \quad \text{for computing } R\text{versine,}$$

where d_g is the tabular difference passed over (*gata*) and d_b is the tabular difference to be crossed (*bhogya*).[14]

The later commentator, Munisvara (1603), employs an iterative procedure numerically for attaining the functional difference with desired degree of accuracy using which *Rsines* can be interpolated with the desired degree of accuracy.

The interpolated functional value is given by $f(x + hu) = f(x) + u\delta$ where δ is the current functional difference with desired degree of accuracy and $u = \frac{\theta}{h}$.

Let Munisvara's successive approximations to δ be denoted by, $\delta^{(2)}, \delta^{(1)}\delta^{(3)}, \ldots$ If the number of iterations n becomes larger and larger, the desired degree of accuracy is attained when $\delta^{(n+1)} = \delta^{(n)}$. The common value is taken as the value δ of the current functional difference with the desired degree of accuracy.

The true tabular difference d for interpolating *Rsines* according to Bhaskara II is given by

$$d = \frac{1}{2}(d_g + d_b) - \frac{1}{2}(d_g - d_b)\frac{\theta}{h},$$

where

$$h = 600' = \frac{1}{2}(d_g + d_b) - \frac{\theta}{2h}d_g + \frac{\theta}{2h}d_b,$$

where d_g is the tabular difference crossed, d_b is the tabular difference to be crossed and θ is the residual arc.

The iteration process is initiated by taking as the first approximation to δ, $\delta^{(1)} = d$. Now using this, the other successive approximations to δ, viz; the second approximation $\delta^{(2)}$, third approximation $\delta^{(3)}$, fourth approximation $\delta^{(4)}$ and so on may be computed using the following formula

$$\delta^{(n+1)} = \frac{1}{2}(d_g + d_b) - \frac{\theta}{2h}d_g + \frac{\theta}{2h}\delta^{(n)}.$$

As n becomes larger, the desired accuracy will be attained at some stage when $\delta^{(n+1)} \approx \delta^{(n)}$ and this value is taken as the value of δ.

Hence when $n \to \infty$,

$$\delta = \frac{1}{2}(d_g + d_b) - \frac{\theta}{2h}d_g + \frac{\theta}{2h}\delta$$

or

$$\delta\left(1 - \frac{\theta}{2h}\right) = \frac{1}{2}(d_g + d_b) - \frac{\theta}{2h}d_g.$$

From this we get,

$$\delta = \frac{(h - \theta)d_g + hd_b}{2h - \theta}.$$

Substituting this value of δ in $f(x + hu) = f(x) + u\delta$, we get

$$f(x + hu) = f(x) + u\left\{\frac{(h - \theta)d_g + h \cdot d_b}{2h - \theta}\right\}.$$

Using this interpolated functional value one can obtain a desired degree of accuracy. In other words, Rsine of arc $x + hu$ can be computed with the desired degree of accuracy.

11.10   The *Golasara* of Nilakantha

Over the long period under discussion, Indian mathematicians made significant contributions to trigonometry under the caption '*jyotpatti*' (*jya* + *utpatti* = source of Rsines). This branch of mathematics evolved from astronomical needs, such as computation of latitudes or of position of planets, their movements etc. As Nilakantha pointed out in his *Golasara*,[15] while explaining the concept of *jya* (or Rsines), that he was computing sines and cosines because they were required

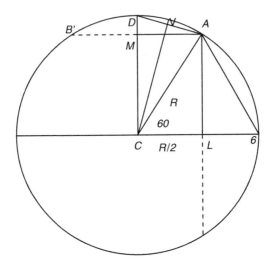

Figure 11.4. The *Golasara* method for computing the sine table.

for a discussion of the motion of planets in their respective orbits on the stellar sphere. As a case study of the advances made in trigonometry by the Kerala School, consider briefly Nilakantha's method of computing sine tables as explained in his *Golasara*.

The *Golasara* shows two methods of computing sine tables. The first method is intended to find the value of first Rsine, using sine tables of length 3×2^m (for $m = 0, 1, 2, 3, 4, 5, \ldots$). The relevant verses 6–14 have been translated by Mallayya (2004).

The method is to find the *R*sines geometrically using a circle of radius R. In Figure 11.4, R is to be taken as the base BC of an equilateral triangle ABC whose other two sides AC and AB equal to the radius R. Drop the perpendicular AL from the join A of the two sides AB and AC where C is the centre. Then the base segments CL and BL, each is equal to half the radius (R) i.e.

$$CL = BL == \frac{R}{2}.$$

But

$$CL = AM = R\sin 30° = \text{half the chord AB}'$$

or

$$R\sin 30° = \frac{R}{2} \quad \text{or } R\text{sine of one } rasi = \frac{R}{2}$$

$$\left(\text{Note that the modern } \sin 30° = \frac{1}{2}\right).$$

Now

$$AL = \text{half the chord } AA' = R\sin 60^0 \quad \text{and} \quad CM = AL = R\cos 30^0$$

Thus

$$R\sin 60^0 \quad \text{or} \quad R\cos 30° = AL \quad \text{or} \quad CM = \sqrt{R^2 - \left(\frac{R}{2}\right)^2} = R\sqrt{\frac{3}{4}}.$$

And Rversine of 30° arc is given by

$$MD = CD - CM = R - R\cos 30° = R(1 - \cos 30°).$$

Thus Radius $-$ Rcosine gives Rversine.

Now from the hypotenuse of these two (i.e. the Rsine of 30° and the Rversine of 30°), the Rsine of 15°, $7\frac{1}{2}°$, ..., etc. can be found by repeated application of the process on the circle.

Now

$$AM = R\sin 30° = \frac{R}{2}; \quad MD = R\text{versine } 30° = R(1 - \cos 30°)$$

$$AD^2 = AM^2 + MD^2$$

$$R\sin 15° = DN = \text{half the chord } AD = \frac{1}{2}AD = \frac{1}{2}\sqrt{AM^2 + MD^2}$$

$$= \frac{1}{2}\sqrt{\left(\frac{R}{2}\right)^2 + \{R(1 - \cos 30°)\}^2}$$

$$= \frac{1}{2}\sqrt{\left(\frac{R}{2}\right)^2 + \{R(1 - \sin 60°)\}^2} \text{ in other terms.}$$

Thus,

$$R\sin 15^0 = \frac{1}{2}\sqrt{\left(\frac{R}{2}\right)^2 + R^2\left(1 - \frac{\sqrt{3}}{2}\right)^2}$$

$$= \frac{1}{2}\sqrt{\frac{R^2}{4} + R^2\left(1 + \frac{3}{4} - \sqrt{3}\right)}$$

$$= \frac{R}{2}\sqrt{2 - \sqrt{3}}$$

or

$$\sin 15° = \frac{1}{2}\sqrt{2 - \sqrt{3}}.$$

Now

$$R\cos 15° = CN$$

$$= \sqrt{CD^2 - DN^2} = \sqrt{R^2 - \frac{R^2}{4}\left(2 - \sqrt{3}\right)} = \frac{R}{2}\sqrt{2 + \sqrt{3}}.$$

Similarly, Rsine and Rcosine of $7\frac{1}{2}°, 3\frac{3}{4}°$ ... etc. can be determined geometrically. The general formula embedded in this formulation is:

$$R\sin\left(\frac{\theta}{2}\right) = \frac{1}{2}\sqrt{(R\sin\theta)^2 + (R\text{versin}\,\theta)^2}$$

$$= \frac{1}{2}\sqrt{(R\sin\theta)^2 + \{R(1 - \cos\theta)\}^2}$$

which is true since the right side of this formula can be simplified as follows:

$$\frac{1}{2}\sqrt{(R\sin\theta)^2 + \{R(1 - \cos\theta)\}^2}$$

$$= \frac{1}{2}\sqrt{R^2\sin^2\theta + R^2 + R^2\cos^2\theta - 2R^2\cos\theta}$$

$$= \frac{1}{2}\sqrt{2R^2(1 - \cos\theta)}$$

$$= \frac{1}{2}\sqrt{2R^2 \times 2\sin^2\left(\frac{\theta}{2}\right)}$$

$$= \frac{1}{2}\left\{2R\sin\left(\frac{\theta}{2}\right)\right\}$$

$$= R\sin\left(\frac{\theta}{2}\right).$$

Next, to find the approximate length of small arcs, the following formula is suggested.

$$arc \approx \sqrt{\frac{4}{3}(R\text{versine})^2 + (R\text{sine})^2} \quad \text{for any small arc}$$

If the arc is $s = R\theta$, then the above formula can be expressed in the following form

$$s \approx \sqrt{\frac{4}{3}(R\text{versin}\,\theta)^2 + (R\sin\theta)^2} = \sqrt{\frac{4}{3}\{R(1-\cos\theta)\}^2 + (R\sin\theta)^2}$$

$$= \sqrt{\frac{4}{3}R^2\left\{2\sin^2\left(\frac{\theta}{2}\right)\right\}^2 + R^2\left\{2\sin\left(\frac{\theta}{2}\right)\cos\left(\frac{\theta}{2}\right)\right\}^2}$$

$$= \sqrt{\frac{4}{3}R^2\left\{4\sin^4\left(\frac{\theta}{2}\right)\right\} + R^2\left\{4\sin^2\left(\frac{\theta}{2}\right)\cos^2\left(\frac{\theta}{2}\right)\right\}}$$

$$= \sqrt{4R^2\sin^2\left(\frac{\theta}{2}\right)\left\{\frac{4}{3}\sin^2\left(\frac{\theta}{2}\right) + \cos^2\left(\frac{\theta}{2}\right)\right\}}$$

$$= \sqrt{4R^2\sin^2\left(\frac{\theta}{2}\right)\left\{1 + \frac{1}{3}\sin^2\left(\frac{\theta}{2}\right)\right\}}$$

$$= 2R\sin\left(\frac{\theta}{2}\right)\left\{1 + \frac{1}{3}\sin^2\left(\frac{\theta}{2}\right)\right\}^{\frac{1}{2}}$$

$$= 2R \sin\left(\frac{\theta}{2}\right) \left\{ 1 + \frac{1}{2} \times \frac{1}{3} \sin^2\left(\frac{\theta}{2}\right) + \frac{\frac{1}{2} \times \left(\frac{1}{2} - 1\right)}{1 \times 2} \right.$$

$$\left. \times \left\{ \frac{1}{3} \sin^2\left(\frac{\theta}{2}\right) \right\}^2 + \cdots \right\}$$

$$= 2R \sin\left(\frac{\theta}{2}\right) \left\{ 1 + \frac{1}{6} \left\{ \sin\left(\frac{\theta}{2}\right) \right\}^2 - \frac{1}{72} \left\{ \sin\left(\frac{\theta}{2}\right) \right\}^4 + \cdots \right\}.$$

If θ is small, then $R\sin\theta$ is taken as the corresponding arc itself and so

$$R \sin\left(\frac{\theta}{2}\right) \approx R\left(\frac{\theta}{2}\right).$$

Hence it follows that

$$\sqrt{\frac{4}{3}(R\text{versin }\theta)^2 + (R\sin\theta)^2}$$

$$\approx 2R\left(\frac{\theta}{2}\right)\left\{ 1 + \frac{1}{6}\left(\frac{\theta}{2}\right)^2 - \frac{1}{72}\left(\frac{\theta}{2}\right)^4 + \cdots \right\}$$

$$\approx R\theta\left\{ 1 + \frac{\theta^2}{24} - \frac{\theta^4}{1152} + \cdots \right\}.$$

Now, for sufficiently small values of θ such that θ^3 and higher powers of θ can be neglected, the series on the right side $\rightarrow R\theta$ or tends to the arc.

To compute Rsines and Rcosines we begin by using Nilakantha's circumference and diameter ratio with the diameter and circumference of a circle taken as 113 and 355 units respectively. This gives

$$\pi = \frac{355}{113} = 3.14159292\ldots.$$

Using this value, the value of R for computation of Rsines is given by

$$R = \frac{21600 \times 113}{2 \times 355} = 3437.746479 = 57°17'44.8''$$

or equal to 1 radian in modern terms.

Now the circumference is to be divided into a definite number of arcs of equal length for computing the Rsines and Rsine differences at equal intervals. The Rsine differences and the corresponding Rsines are interrelated and from the relation between them, the Rsines can be derived successively using the recursive formula mentioned earlier. For a detailed explanation of the derivation, see Mallayya (2004).

11.11   A Concluding Note: Spherical Trigonometry in India

We often tend to see trigonometry as applied to problems involving only planes with the mathematics required being results from similar triangles (or Rule of Three) and the Pythagorean Theorem. However, things get more complicated when the surface transforms from a plane to a sphere. From early times, astronomers recognised that arcs and distances relate to a celestial sphere and the concept of a spherical triangle therefore underpins trigonometric measurement. Before the sixteenth century, astronomy was based on the notion that the earth stood at the center of a series of nested spheres. To calculate the positions of stars or planets, we need a trigonometry modified for a spherical purpose.

Figure 11.5 shows two spherical triangles formed by the intersection of three great circles on the on the surface of a sphere. Triangle ABC is a spherical triangle with sides a, b and c, and angles A, B and C. The following cosine formulae, adapted from a plane surface, connects its sides and angles.

$$\cos a = \cos b \cos c + \sin b \sin c \cos A,$$

$$\cos b = \cos c \cos a + \sin c \sin a \cos B,$$

$$\cos c = \cos a \cos b + \sin a \sin b \cos C.$$

Also, the following sine formulae applies:

$$\frac{\sin a}{\sin A} = \frac{\sin b}{\sin B} = \frac{\sin c}{\sin C}.$$

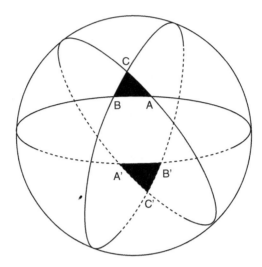

Figure 11.5. Spherical triangles.

And if sides a, b and c are small, the above formula reduces to

$$\frac{a}{\sin A} = \frac{b}{\sin B} = \frac{c}{\sin C}.$$

For an observer standing at the centre of the earth, all celestial objects seem to be situated on the surface of a sphere of large radius. It is this imaginary sphere which is the basic tool for analysing the daily and relative motion.[16] And just as we require the latitude and longitude to find the location of any point on the surface of the earth, we use three coordinate systems — horizontal, equatorial and the ecliptic — to specify the location of celestial objects. It is not our purpose to explain the Indian equivalents of these systems for it requires a thorough background of Indian astronomy. Methods found in Indian spherical astronomy may very well have their origins in Greek mathematical astronomy. But on the other hand the essential Greek tools such as analemma and the Menelaus's Theorem leave no direct traces in the methods of Indian astronomy.[17] We can only conclude that despite the early Greek influences, the application of Indian trigonometry to astronomy took its own direction both in content and method. For a useful survey of Indian spherical trigonometry and its applications to Indian astronomy, the reader should refer to Datta and Singh (1983).

Endnotes

[1] An alternative measure to an angle in degrees is the radian. A radian is the angle made at the centre of a circle by the arc whose length is equal to the radius of that circle. Each radian is approximately 57.296°.

[2] The value of radius $(R) = \frac{1}{2}[21600/3.1416] \approx 3438$ probably predates Aryabhata, but his is the first surviving text to record it.

[3] Another set of trigonometric functions, namely tangent and cotangent, developed from a different tradition, the study of the lengths of shadows cast by objects of various heights. Thales of Miletus was believed to have used shadow lengths to calculate the heights of the pyramids in around 600 BCE. Both Indian and Islamic mathematics developed a trigonometry based on shadow lengths, a tradition that subsequently influenced European work. The derived functions secant and cosecant were first used by navigators in the fifteenth century.

[4] An interesting application of this is in the calculation of the position of the earth. The length of the arc from winter solstice to summer solstice is approximately 176 degree and 18 minutes. Assuming that crd $176°\,18' = 6872$ minutes, it follows that half the chord is 3436'. Using the Pythagorean Theorem, the distance from the centre of the circle to this chord is $\sqrt{(3438^2 - 3436^2)} = 117'$. In other words, this chord is 117 minutes, almost two full degrees off-center. This arises from the fact that the lengths of the four seasons are not equal. Winter solstice to spring equinox is the shortest, 89 days long. Spring equinox to summer solstice is almost 90 days, with the summer solstice to autumnal equinox being the longest, just over 93 days and the autumnal equinox to the winter solstice being almost 93 days. Since the sun moves at a constant speed, the unequal lengths of the various seasons would indicate that the earth is off-centre. It was Hipparchus who first tackled the problem of calculating the position of the earth.

[5] Around the time of Aryabhata I or even earlier, other trigonometric functions were known and used. The *kotijya* or Cosine function is the length AB. *Utkrama-jya*, a less known function today, is the length BC and known as the versed (or reversed sine) such that $\text{vers}\,\theta = 1 - \cos\theta$. As in the case of the Indian Sine, these functions are R times as large as the ones we use today.

[6] There has been some discussion of the extent to which Indian sine tables owe a debt to earlier Greek work, notably that of Hipparchus's chord

table which precedes the work of Ptolemy. Recent studies have reignited the debate. The case for transmission rests primarily on circumstantial evidence of the observed similarities with regard to the structures of the tables and choice of parameters such as 360° circle, $3\frac{3}{4}^{\circ}$ step size interval and a trigonometric radius of 3438. However, it is possible to argue that once the 360° circle entered Indian astronomy through the transmission of what was essentially a Babylonian innovation, the Indians could have arrived independently to the idea of 24 *rasi* for each 90° quadrant that would lead to the subsequent calculation of sines.

[7] Other interpretations of the second sentence of the quotation include those by Paramesvara (*fl.* 1400), Nilakantha (*fl.* 1500) and the commentator Raganatha (1603)

(i) $d_{n+1} = d_n - \dfrac{d_1 + d_2 + \cdots + d_n}{R_1},$

(ii) $d_{n+1} = d_n - \dfrac{R_n}{R_1}(d_1 - d_2),$

(iii) $R_{n+1} = R_n + R_1 - \dfrac{R_1 + R_2 + \cdots + R_n}{R_1}.$

The equivalence of (i) to (iii) is easily established. Nilakantha provides a geometrical demonstration of (ii) that is interesting in the light of the work on infinite series in Kerala discussed in a later chapter.

[8] But this does not mean that d_{n+1} is equal to $225 - \frac{1}{2}n(n-1)$

[9] A *rasi* is literally a 'heap' or 'quantity' and in a more technical context denotes 'zodical sign' or 30°

[10] There are earlier instances, referred to earlier, of the use of $\sqrt{10}$ as the preferred ratio. In Jaina texts *Anuyoga Dwara Sutra* and the *Triloko Sutra* from around the beginning of the first millennium AD, the circumference of the *Jambo* island (a cosmographic representation of the earth) is 316227 *yojana*, 3 *krosa*, 122 *danda* and $13\frac{1}{2}$ *angula* where 1 *yojana* is about 10 kilometres, 4 *krosa* = 1 yojana, 2000 danda = 1 *krosa* and 96 *angula* (meaning literally a 'finger's breadth') = 1 *danda*. The result is consistent with taking the circumference to be given by $\sqrt{10 \times d}$ where the diameter (d) = 100,000 *yojanna*. The choice of the square root of 10 for π was convenient, since in Jaina cosmography islands and oceans always had diameters measured in powers of 10.

[11] It may be argued that this is the best rational approximation that can be devised.

[12] Note, that ever since Aryabhata devised a method to calculate square roots, Indian mathematicians could approximate Sine 60 by a rational number, i.e. $\sin 60 = 0.866$.

[13] For details, see Mallayya (2004).

[14] This is same as

$$d = \frac{1}{2}(d_g + d_b) - \frac{1}{2}(d_g - d_b)\frac{\theta}{h} \quad \text{for } R\text{sine}$$

and

$$d = \frac{1}{2}(d_g + d_b) + \frac{1}{2}(d_g - d_b)\frac{\theta}{h} \quad \text{for } R\text{versine,}$$

where $h = 10° = 600'$.

[15] This is a small text on spherical astronomy summarised in 56 verses in Sanskrit. For details, see *Golasara Siddhantadarpanamca* of *Gargya Kerala Nilakantha* Ms No. T 846. B, Transcript copy by Paramesvara Sastry, *C*. 1024.E (Mss Library, Trivandrum).

[16] One gets the impression from observing the sky that celestial objects are in continuous motion. The motion is in two parts. One is the apparent motion of all celestial objects from east to west which is due to the rotation of the earth. This is the diurnal motion. The other is due to the relative motion of a particular celestial body like the Moon or a planet in relation to the seemingly fixed background.

[17] Analema is the graduated scale in the shape of a figure eight, indicating the sun's declination and the equation of time for every day of the year, often found on sundials and globes. Menelaus' theorem, named after Menelaus of Alexandria (*fl.* 1st Century CE), is a theorem about triangles in plane geometry.

Chapter 12

Outside the Sanskrit Tradition: Texts, Translations and Dissemination

12.1 Introduction

From the beginning of the nineteenth century, the primary concern of historians of Indian mathematics was the exploration of mathematical sources in Sanskrit. This is not surprising since such sources have provided the main basis for the excavation of ancient Indian mathematics, even if it meant neglecting other sources. These included material in ancient languages such as the Prakrit with some of its more recent variants, in regional languages of South India and in imported languages such as those of Arabic–Persian origin. All these sources possessed their own rich and varied literature, and it would be surprising if they did not include works on mathematics as well. In earlier chapters we referred to the use of the Prakrit language in Jain and Buddhist mathematical literature, of Sarada (a North Indian script) in which the Bakshali Manuscript was recorded, and Malayalam in the case of notable texts of Kerala mathematics, such as the *Yuktibhasa*.[1] The scope for exchange between these vernaculars and Sanskrit offer a rich area for study, insufficiently explored as yet. Such exchanges were never one-sided and in a number of cases tended to complement one another. Often where the mathematical ideas were expressed in Sanskrit, the ideas took hold and were disseminated in regional languages.

A starting point would therefore be an exploration of mathematical literature in the regional languages derivative of Sanskrit. It should be remembered that popularisation of mathematics or any other knowledge systems in India was not always accompanied by a translation into a regional language. But there has been a long-standing tradition of producing popular texts in mathematics in Sanskrit. It may be remembered that Sridhara produced an abridged popular version of his exhaustive (but only partly extant) treatise *Patiganita* and named it *Trisatika*, which as the name suggests contained 300 verses. There has also been a tradition of writing for those with a limited knowledge of mathematics. Hayashi (1995) has argued convincingly that the Bakhshali Manuscript is a collection of such material from diverse sources, compiled for solving everyday problems. These and other publications in regional languages were aimed at numerate individuals such as accountants and merchants and give us a good glimpse of the contemporary society. In other words, while scholars generally wrote in Sanskrit, the practitioners wrote in vernaculars. It was in the writing of these local practitioners that we came across extensive problems of a computational nature involving short-cut methods of calculation and verification. Examples included 'casting off nines', monetary unit conversions, variations in the use of the 'Rule of Three', and simple approaches to solving the problems involving 'Rules of Five, Seven' etc.[2] Furthermore, although the Sanskrit texts in arithmetic employed units of measurement, weight and coinage, prevalent in earlier times texts, in regional languages, even though directly translated from Sanskrit, would update their metrological units, and thereby provide us with a glimpse of economic and social life of the area.

To illustrate, consider the work of Thakkura Pheru who composed the *Ganitasarakaumudi* ('Moonlight of the Essence of Mathematics') around 1315 CE in Apabhramsa (a mixture of Sanskrit, Prakrit and vernaculars of Assam.). He was a contemporary of Madhava of Kerala and predated Narayana Pandita's *Ganitakaumudi* by about four decades.

12.1.1 Thakkura Pheru: life and works

Thakkuru Pheru was born in the present-day Bhiwana district of the state of Haryana, a place of pilgrimage during the thirteenth and fourteenth centuries. He belonged to a group of Srimala Jains noted for their expertise in minting coins and banking. As a result of their long history of maritime contacts with the Persian Gulf and the Arabian Peninsula, the Srimala Jains had become important mediators between the Islamic and Sanskritic traditions of learning in the early medieval period. They were pioneers in learning Persian and disseminating such knowledge through the medium of Sanskrit. Pheru was one of these mediators who found ready employment in the court of Khalji Sultans of Delhi, and hence the title of 'Thakkura' ('Master') bestowed on him and his father.

Pheru wrote six texts on a variety of scientific and technical subjects in Apabhramsa, which in Sanskrit literally means 'corrupt' or 'ungrammatical' language. He probably chose to do so because his books would be accessible to a wider audience. Apart from texts on gems, architecture, coins and minerals, he wrote one on astronomy (*Jyotisasasaru*) and a textbook on arithmetic and geometry, entitled *Ganitasarakaumudi* (or in short *Ganitasara*). Composed before 1318, Pheru owed a great debt to two mathematicians whose works have been discussed in earlier chapters, namely Sridhara (the author of *Patiganita* and *Trisatika*) and Mahavira (the author of *Ganitasarasangraha*). Pheru's purpose was not to construct a formal mathematics text but produce a practical manual for a readership consisting of accountants, bankers, traders and architects. And hence, his emphasis on short-cut procedures in commercial arithmetic, entertaining mathematical riddles, magic squares and conversions from one era to another.[3]

Despite the fact that the *Ganitasara* is largely based in content and style on Sridhara's *Patiganita* it contains interesting information on the society of his day. This was not merely the units of measurement prevalent then but also practical methods in such occupations as shopkeepers, builders, carpenters, accountants and masons. A section on geometry contains rules for calculating volume of bridges and constructing staircases, domes and towers of various shapes. Some of these architectural designs and structures were introduced from outside India.[4] Pheru also touches upon daily aspects of life that are quantifiable, from the average yield of different crops to the quantity of ghee that can be extracted from cow's milk and buffalo's milk.

12.1.2   The contents of *Ganitasarakaumudi*

The text contains five chapters consisting of: (i) rules for 25 fundamental operations; (ii) eight classes of (reduction of) fractions; (iii) eight types of procedures; (iv) four topics useful in daily life, and (v) five topics of practical importance.

The first two chapters follow along the lines indicated in Sridhara's *Trisataki* and *Patiganita*. There are similarities and differences between the location and manner in which some of the 'fundamental operations' and 'procedures' are treated by Pheru compared to Sridhara. For example, Pheru discusses the rules of operating with zero (*sunyaganita*) between the rules for multiplication and division, as does Sridhara but unlike Brahmagupta or Bhaskara II. Pheru's algorithm for calculating the area of the segment of a circle is virtually identical to that found in the *Trisataki*.[5] And so are the rules for sums and differences of arithmetic series.[6] A recent introduction and translation of *Ganitasarakaumudi* by four scholars Sarma, Kasuba, Hayashi and Yano under the pseudonym SANKHYA contains an extensive discussion of the contents and historical background of this important text.

12.1.3   Other mathematics in regional languages of the north

With respect to mathematical activities in other part of North India, information is available from Assam on a group who kept the land records over a period of time. They developed a form of arithmetic expressed in verses. The verses covered not only weights and measures but also computational procedures prevalent at that time. In the sixteenth century, a number of translations of

Bhaskaracharya's *Lilavati* occurred. An early Telugu translation of Mahavira's *Ganitasarasamgraha* composed by Pavuluri Mallana in the eleventh century entitled *Sara Sangraha Ganitam*, served as a model for future translations because of its clarity and innovation. This was true in his choice of appropriate rules, his avoidance of algebraic methods and his inclusion of additional examples, including an introduction to an aesthetic appeal of patterns of numbers. The example of "necklace numbers" is a case in point.

Consider the following patterns formed by multiplication of two numbers:

$$37 \times 3 = 111,$$

$$101 \times 11 = 1111,$$

$$271 \times 41 = 11111,$$

$$37 \times 3003003 = 111111111,$$

$$37 \times 3003003003003 = 111111111111111.$$

There are also other 'necklaces' with '1's and '0's:

$$14287143 \times 7 = 100010001,$$

$$157158573 \times 7 = 1100110011,$$

$$142857143 \times 7 = 1000000001,$$

$$777000777 \times 13 = 10101010101.$$

One of the largest 'pearl necklace' referred to by translator Mallana who lived in the 11th century:

$$20408163265306122449 \times 49 = 10\ 000\ 000\ 000\ 000\ 000\ 000\ 01.$$

Finally, there are necklaces which are said to plot the pattern of the phases of the moon with the moon gradually increases to its fullest extent and then decreases to new moon.

$$111111 \times 111111 = 12345654321,$$

$$146053847 \times 448 = 65432123456.$$

This takes us to the role of recreational mathematics in mathematics education. It is indeed likely that problems such as these which produce startling results attracted the attention, not just of serious mathematicians who invented more problems like these, but also of laypersons who posed these problems as puzzles or riddles under the village tree. A large corpus of such mathematical riddles exists from different parts of world mainly as oral literature.[7]

During the seventeenth century, European travellers were impressed by the ordinary Indian traders' ability to perform speedy mental calculations without writing down the steps. In 1665, a French Jeweller, Jean-Baptiste Tavernier, remarked that the Indians had learned arithmetic "perfectly, using for it neither pen nor counters, but the memory alone, so that in a moment they will do a sum, however difficult it might be" [p. 124]. The answer to the existence of such feats may have been the way that arithmetic was taught where memorisation played a critical role. The secret lay in the number of multiplication and other tables the merchant had to commit to memory in childhood.[8] This was helped by also learning rhyming rules as an aid to easier memorisation.[9]

An interesting problem that occurred across both linguistic and geographical boundaries is the so-called Josephus problem.[10] The solution to this problem consists in arranging in a circle two groups of an equal number of persons or objects in such a manner that each nth person or object belongs to the same group. Though named after the Jewish historian, Flavius Josephus (37–100 CE), this problem was not known in Europe before the tenth century. There the problem is as follows: fifteen Jews and fifteen Christians were traveling in a boat when the boat developed a leak. So the Christian captain arranged all the 30 persons to stand in a circle and kicked out each ninth person to exclude all Jews. In a Japanese version, a man had 15 sons by his first wife. After her death, he married another woman who already had 15 sons of her own. The second wife got all 30 sons and stepsons to stand in a circle, explaining that she would count and take out each 10th one from the circle and that the last one in the circle would inherit the father's property. After eliminating 14 stepsons one after the other, the 15th stepson realised the trick and insisted that the counting should begin from him. She agreed to do so, but the consequence was that all her 15 sons were eliminated. The last one to remain in the circle and thus to inherit the patrimony was the 15th stepson, who cleverly saw through the stepmother's game.

An Indian version has its own cultural ambience. Fifteen Brahmins and 15 thieves found themselves on a dark night in an isolated temple of Durga. The goddess appeared in person at midnight and wanted to devour exactly 15 persons to satisfy her hunger. The plump Brahmins who were her primary targets suggested that all 30 of them would form a circle and Durga eat each ninth person. The proposal was accepted by Durga and the thieves. Durga then counted out each ninth person and devoured him, and then fourteen others leaving only Brahmins in the circle. The question is, how did the Brahmins arrange themselves and the thieves in the circle? A variant of the problem, involved arranging 30 Brahmins and 30 thieves in a circle in such a way that each 12th person would be a thief. Both the problem and solution was aimed at a wide audience.

A strong pedigree of mathematicians and astronomers arose in Maharashtra in the sixteenth century and extended well into the late seventeenth century. It commenced with Ganesa Daivajna of Nandigrama, a coastal village near Mumbai. He wrote the *Buddhivilasini*, a gloss on the *Lilavati*. His *Grahalaghava* was an immensely popular work on astronomy. Other works of his include *Siddhantasiromanivyakhya*, *Brihattithichintamani* and *Laghutithichintamani*. The grating method (*gelosia* method) of multiplication was developed by him.[11] He also substituted complex trigonometric calculations with simpler arithmetic formulations. Ganesa seems to have learnt mathematics and astronomy from his father Kesava, and in all likelihood trained Lakshmidasa and Nrisimha, his cousin and nephew respectively. He also taught Divakara Daivajna from a village on the Godavari, which bore the title of Golagrama. Divakara's sons Kesava and Visvanatha wrote glosses on a number of works. Visvanatha's commentaries included those on *Grahalaghava*, *Suryasiddhanta*, *Tithipatra*, *Siddhantasiromani* and *Karanakutuhala*. Three other sons of his, Vishnu, Krishna and Mallari, were also renowned astronomers of their times. Vishnu was the author of the

Saurapakshaganita, while Mallari commented on the *Grahalaghava* in his *Siddhantarahasya*. Krishna's son Nrisimha studied under Mallari and Vishnu, and went on to write the *Saurabhashya*, a commentary on the *Suryasiddhanta*, and the *Vasanavarttika* on the *Siddhantasiromani*. He trained four of his sons, Divakara, Kamalakara, Gopinatha and Ranganatha. Kamalakara studied Arabic and Persian sources and was acquainted with Euclid's *Elements*. Written in 1658, his *Siddhantatattvaviveka* shows his familiarity with Euclid. Kamalakara excelled in trigonometry and made interesting formulations in it. His brother Ranganatha was the author of two major works, the *Laghubhangivibhangi* and the *Siddhantachudamani*, a gloss on the *Suryasiddhanta*. He also wrote the *Mitabhashini* on the *Lilavati*. The influence of Ganesa's pedigree was not restricted to Maharashtra alone. Nrisimha spent many years of his life in Varanasi, where he might have engaged himself in teaching and observation. Vishnu trained Krishna Daivajna and Ranganatha, brothers who hailed from Elachpur near Dewas in Madhya Pradesh. Krishna went on to serve the Mughal ruler Jahangir who evinced deep interest in astronomy. He commented on the *Bijaganita* in his *Navankura* and on the *Lilavati* in his *Kalpalatavatara*. His brother Ranganatha produced *Gudharthaprakasika*, a gloss on the *Suryasiddhanta*. Ranganatha's son Munisvara was a leading astronomer of the seventeenth century. He wrote *Patisara* and *Siddhantasarvabhauma*. The *Marichi*, a gloss on the *Siddhantasiromani*, is another work of Munisvara's. The *Nisrishtharthaduti* was his commentary on the *Lilavati*. We know of at least one other pedigree from sixteenth century Maharashtra. It belonged to Parthapura on the Godavari in the Vidarbha region. The first known name in this pedigree is that of Naganatha, of whom we do not have any historical information. His son Jnanaraja was the author of *Siddhantasundara*, an extensive treatise on astronomy. His son Suryadasa wrote *Ganitamritakupaka* and *Suryaprakasa*, glosses respectively on the *Lilavati* and the *Bijaganita*. His disciple Dhundiraja was an accomplished astronomer of his time.

It is not easy to discuss the situation in other parts of India due to a paucity of adequate sources. But the evidence on hand suggests that mathematics and astronomy were taught and learnt in many regions. In Madhya Pradesh, we know of Gangadhara, who composed an astronomical treatise called *Chandramana* in the early half of the fifteenth century (1434). In the latter half of the century (1478), Makaranda, an astronomer from Varanasi, wrote the *Tithipatra*, which was based on the *Suryasiddhanta*. Earlier in the century, another Gangadhara produced a gloss on the *Lilavati*. He came from Gujarat. His brother Vishnu was the author of *Ganitasara*, which was deeply indebted to Sridhara. In 1639, Nityananda wrote the *Siddhantaraja*. Unlike Kerala, many of the works, which appeared in Maharashtra and Madhya Pradesh,were influenced by Persian and Arabic sources, to the extent that Kamilakara whose knowledge of such sources helped him to cast a critical eye on traditional Indian astronomy. However, interest in Indian mathematics in certain courts led to extensive projects of translations. For example, as mentioned earlier, Fyzi translated the *Lilavati* into Persian in 1587 in the court of Akbar; and the *Bijaganita* was made available in Persian in 1634 by Ataullah Rashidi at the court of Shah Jahan. Fyzi also had the *Brihatsamhita* of Varahamihira translated into Persian. There were also translations in the other direction, of which a notable one was Jagannatha in 1718 rendering into Sanskrit Euclid's *Elements* under the title *Rekhaganita*. Jagannatha was an interesting figure in the court of Savai Jai Singh II, He knew Indian and Arabic astronomy and was acquainted with the works of Islamic astronomers such as al Kashi, al Tusi and Ulugh Beg. He had studied Ptolemy's *Almagest* of which he produced a Sanskrit translation entitled *Siddhantasamrat* in 1732. He was also familiar with the methods of European mathematics and astronomy and helped to set up observatories at Jaipur, Mathura, Delhi, Varanasi and Ujjain.[12]

12.2   Indian Mathematics from Persian and Arabic Sources

The arrival of the Islamic scientist al-Biruni in India in 1018 CE as a prisoner in the invading army of Muhammad Ghaznavi was an important landmark in the scientific contact between the two cultures. There is the view that during his enforced stay in India, he profitably occupied the time by learning the language and culture of the country, and translating from Arabic into Sanskrit Euclid's *Elements*, Ptolemy's *Almagest* and his own work on the construction of the astrolabe. None of these Sanskrit translations are extant today. However, al-Biruni's example was followed by Sultan Tughlaq, who occupied the throne of Delhi between 1351–1388. He ordered the astronomical text *Brhatsamhita* of Varahmihira to be translated into Persian, while his court astronomer, Mahendra Suri, wrote *Yantraraja*, which introduced astronomical principles and practices from Central and Western Asia into India. Unfortunately, the ideas proposed in the book welding together principles from the *Siddhantas* and the Persian–Arabic system did not get a favourable reception. However, ideas from the West continued to flow into India, including those of Nasir al-Din al-Tusi (1201–1274). His *Kitab Zij-I1kani*, based on the observations made at the Marghah Observatory, became the model for future *zij* (astronomical tables) in the Islamic world. Two of his texts, one each in Arabic and Persian, became compulsory reading for students of astronomy in Indian *madrassahs* (mosque schools) and appears to have inspired an astronomer Muhammad Jaunpuri to question the validity of Ptolemaic planetary model.

Geometry has always been an integral part of astronomy. However, it has developed on different lines in different cultural areas. In India, as we saw in an earlier chapter, it originated to serve the needs of rituals and cosmographic speculations. In Central and Western Asia, under Islamic rule, Euclidean geometry, which developed as a deductive science, became dominant. As a result, students of mathematics and astronomy, taught in Arabic–Persian language schools in India, acquired a thorough knowledge of Euclidean geometry before they took up the study of Ptolemy's *Almagest*, Archimedes' *On the Sphere and Cylinder*, or Apollonius' *Conics*. The most widely used Arabic translation of Euclid's *Elements* was Nasir al-Din al-Tusi's *Tahrir Uqlidis*, which in turn went into a number of Persian translations. In 1732, seven centuries after al-Biruni's attempt to introduce Euclid to India, Jagannatha Samrata translated the Persian text into a Sanskrit version of the *Elements* entitled *Rekhaganita*. Only 100 years later, five chapters of Hutton's

Euclidean Geometry were rendered from English into Sanskrit by Yogadyana Misra in Calcutta. By then, English-medium education had become the norm in a number of Schools in major cities of India.

The major difference between the two streams of mathematical activity, namely those working within the Sanskritic tradition, and the other within the Arabic–Persian tradition, is well brought out in the research preoccupations of the two groups. The Kerala mathematicians, with their work on infinite series, were inspired by Aryabhata and his School. The Indian mathematicians working within the Greek–Arabic–Persian tradition whose interests were primarily in Greek geometry and Ptolemaic astronomy offered another model. An interesting illustration of the difference between the preoccupations of the two traditions is shown by the work of Ghulam-Hussain Jaunpuri (b. 1790), one of the notable Indian mathematicians from the Arabic–Persian tradition. In *Jame-i-Bahadur Khani*, composed in 1833, Juanpuri tackles the problem of trisecting an angle,[13] one that has engaged mathematicians from the Greek–Arabic–European tradition over a long period of time, including notable names such as Archimedes, Al-Biruni, Thabit ibn Qurra and François Viete. Various approaches were tried out, including those involving conic sections, transcendental curves, circles and *neusis* (insertion of a static line). Jaunpuri used the last method to achieve a construction that is not exact but sound and practicable. The details of his method are found in Rizvi (1983) and are of little importance within the context of this discussion. What would have been very unlikely is that such a problem would have engaged the interest of an Indian mathematician from the Sanskrit tradition.

The two parallel traditions met in a few cases involving astronomy in the courts of the Tughluq and Mughal emperors at Delhi and later in the court of Jai Singh in Jaipur, but hardly ever on matters relating to pure mathematics. The Islamic astronomical table (*zij*) and astronomical instruments (particularly the astrolabe, known in Sanskrit as the *yantraraja* or 'king of instruments') had a significant impact on mathematical astronomy in India. The practical advantage of referring to a table rather than calculating planetary positions using the rules from a *siddhanta* became immediately obvious. Indian

astronomers soon took to these tables and apart from using them for
working out planetary positions, became indispensable aids for syn-
chronising time units in calendars.[14] There were translations of texts
such as *Lilavati* and *Bijaganita* into Persian under the patronage of
Mughal emperors Akbar and Shah Jahan and dictionaries compiled
to help this process. However, this lack of contact between the two
traditions was a missed opportunity which has had considerable
repercussions for the development of Indian mathematics. But that
is another story.

Endnotes

[1] These were all regional languages or vernaculars, with Prakrit being
the ancient name for several Middle Indo-Aryan vernacular languages,
derived from dialects of Old Indo-Aryan languages. While they were
perceived as a "lower" medium of scholarship, they were more accessible
and over time the gradual "Sanskritisation" of these languages gave them
higher cultural cache.

[2] For example, in the case of the 'Rule of Five', the product of the last
three terms is divided by the product of the first two terms; or in the
case of the Rule of Seven, the product of the last four terms is divided
by the product of the first three terms and so on.

[3] It is interesting that Pheru was the first to give a rule for converting
dates from the Vikram era to the Islamic *Hijra* era. From a historical
point of view, he is also the first Indian mathematician to study magic
squares, although they appear in earlier writings, notably in the works of
Varamahira and Nagarjuna and of course Narayana Pandita. However,
as we saw in an earlier chapter, Narayana's discussion contains the first
systematic treatment of the subject.

[4] For example, the following definition is given for one of the constructions:
A Persian *minar* is like a "circular tower with a spiral stairway in
the middle, as far as the inside is concerned. But outside there is this
difference. The outer wall consists of half triangles and half circles." The
cryptic last sentence may be explained as: the horizontal cross-section of
the outer circumference consisting of alternate triangles and semicircle.

[5] Both texts provide an identical rule. Expressed in modern notation: Let
A = Area, h = arrow (or *sara*) and a = chord and b = arc

$$A = \sqrt{\left(\frac{a+h}{2} \cdot h\right)^2 \cdot 10 \cdot \frac{1}{9}}.$$

Pheru also provides a crude approximation for the length of the arc
$(a + h = b)$ and mentions an alternative one of $a^2 + 6h^2 = b^2$

[6] Expressed in modern notations, the rules found in *Ganitasarakaumudi*
were known from the time of *Aryabhatiya*:
Thus the sum $S(n)$ is given by

$$S(n) = \frac{n}{2}(n + 1),$$

where n is even

$$S(n) = n \cdot \frac{n + 1}{2},$$

where n is odd

$$S(n) = n \cdot \frac{(nx + x)}{2x},$$

where x is an optional number and Difference $S(n, m)$
Given two integers m and $n(n > m)$

$$S(n, m) = S_n - S_m = \frac{(m + 1 + n)(n - m)}{2}.$$

[7] David Singmaster (1996) has compiled an annotated source book on
Recreational Mathematics which has examples from all over the world.

[8] John Taylor records in the first quarter of the nineteenth century that "in
the Mahratta schools, tables [of multiplication] consists in multiplying
10 numbers as far as 30, and in Gujarati schools, in multiplying
10 numbers as far as 100." And this trend was found later for in
Writing in the first quarter of the nineteenth century, John Taylor records
that "in the Mahratta schools, this table [of multiplication] consists
in multiplying 10 numbers as far as 30, and in Gujarati schools, in
multiplying ten numbers as far as 100." And in the Gazetteer of the
Bombay Presidency at the beginning of the twentieth century there were
reports of sons of merchants memorising no fewer than 20 types of tables,
including tables of multiplication, fractions, squares and interest.

[9] This discussion owes a considerable debt to the work of S. R. Sarma
(2011) who has a number of other examples of approaches to numeracy
in Indian schools before the arrival of Westerners.

[10] The Josephus problem, in its multicultural variants, involves a simple
principle. A group of n people are asked to stand in a circle, numbered
clockwise from 1 to n. If we start with person numbered 2, and remove
every other person, proceeding clockwise. For example, if $n = 8$ the
persons removed in order $2, 4, 6, 8, 3, 5, 7$ leaves a single person number 1.
Denote the last person remaining as $j(n)$. We want to find a rule to

compute $j(n)$ for any positive integer $n > 1$. A search for a mathematical rule is not practical in this instance. Instead we use the binary notation where $j(n)$ is the left rotation of the binary digits of of n.

Thus if $n = x_1x_2x_3 \ldots x_n$, where x_i's are the digits of the binary representations of n and $x_1 \neq 0$, then $j(n) = x_2x_3 \ldots x_nx_1$. For example, if $n = 6$ (base 10) $\equiv 110$ (base 2), then the last person $j(n = 6) = 101$ (base 2) $= 5$.

[11] Gelosia multiplication owes its name to the arrangement of multiplication to resemble a medieval grating or lattice which custom dictated should be placed at windows or houses to shield women from public gaze. The method apparently makes its first appearance in India during the tenth century from where it spread to the Islamic world and soon found its way to Italy probably through Sicily, to finally the Italian Luca Pacioli (c. 1445–1517) who in his *Summa* includes it among eight methods of carrying out multiplication. It is likely that the Gelosia method may have still been popular, were it not for some of the technical problems of printing the lattice lines as well as the numbers in the grid.

[12] Different regions of the South produced their own versions of mathematics. A survey with some historical content by Rajagopalan discusses the mathematics from Karnataka, Tamil Nadu, Andhra and Kerala in various issues of the Bhavan's Journal between 1958–1959. Also Gupta (1978, pp. 25–28) discussed works in Telugu at the 44th Conference of the Indian Mathematical Society. Senthil Babu (2007), following Dharampal (1983), examines school mathematics in the Tamil region of South India during the 18th and 19th centuries where the system of pedagogy places more emphasis on memory rather than technique and suggests that this may explain the greater numeracy evident among indigenous school children compared to their English counterparts.

[13] Bisecting a given angle using only a pair of compasses and a straight edge is relatively easy and forms part of the arsenal of elementary geometry. However, it took mathematicians centuries to realise that dividing into three equal angles is in most cases impossible. The reason for this, as shown by Pierre Wantzel in 1837, is found in the equivalence between the problems of trisecting an angle and of solving a cubic equation. Only a few cubic equations can be solved using the straight-edge-and-compass method and thus deduced that most angles cannot be trisected.

[14] Since the Indian calendar is a lunisolar calendar, there is need for true lunar months to be synchronised with true solar years and this is most efficiently achieved with the help of tables. One of the best known examples of such a text is *Tithicintamani* by Ganesa Daivajna (1525). For further details see Ikeyama and Plofker (2001).

Chapter 13

Battle for the Mind: The Rise of Western Mathematics

13.1 Introduction

The British presence in India set the scene for a number of cognitive encounters, the nature and scale of which varied considerably over time. Five of these encounters, three of them initiated by Britons from 1790 over a period of 50 years and the others by two Indians one writing in 1859 and the other over 50 years later, set the scene. The framework within which these encounters took place was the changing nature of British rule over India and the growing self-confidence of Western scholarship that went hand in hand with the spread of the imperialist ideology of dominance. An investigation of these encounters would show how the naive wonder and respect for the antiquity and achievement of Indian exact sciences (notably astronomy), exemplified in the writings of Burrow (1790), was soon replaced by a more measured though condescending assessment of 'Hindu Astronomy' by John Playfair. This was then followed by silence and indifference that met the startling disclosures of Charles Whish on the beginnings of mathematical analysis in Kerala, 200 years before its appearance in Europe. And in the next encounter, there was the case of an unsuccessful attempt by Yesudas Ramchandra (1821–1880) in the post-Macaulay period to graft traditional Indian mathematics on to modern mathematics education. Fifty years later emerged the remarkable figure of Srinivas Ramanujan (1887–1920) whose mathematics had an indigenous resonance, rarely recognised, both in the content as well as the manner in which it was presented.

13.2   Early European Accounts of Indian Astronomy and Mathematics

While waiting in Pondicherry (in South India) for the 'Transit of Venus'[1] in 1769, the French astronomer, Le Gentil, tried to gather information about astronomical methods of the local people. With the help of a Tamil interpreter he succeeded in computing the length of the lunar eclipse of August 1765 which he then checked against the lunar tables of Tobias Mayer.[2] The 'Tamil' method underestimated the duration of the eclipse by 41 seconds while the Mayer table's estimate was a minute and 41 seconds too long. Le Gentil expressed amazement having observed the Tamil astronomers carrying out their calculations only with the aid of cowrie shells. It turned out that the method used by the astronomers was the *Vakya* method, prevalent in South India. Neugebauer (1952) referred to this method as the 'Tamil' method and noted that John Warren had mentioned it in his book, *Kalasankalita*, to calculate the lunar eclipse that occurred on 1st June 1825, which had predicted the midpoint of the eclipse with an error of about 23 minutes.

Even earlier in 1688, De la Loubère, the French Ambassador to Siam, brought to Paris manuscripts describing Indian methods of astronomical computations.[3] The methods were commented on by the Italian astronomer Cassini and then by the German astronomer Bayer (1738) and finally in a note later by the great German mathematician Leonard Euler on the 'On the Solar Astronomical Year of the Indians' (*De Indorum Anno*). This increasing interest in Indian astronomy was helped by Le Gentil's detailed account of Indian astronomy based on the information that he had gathered in Pondicherry.

In 1787, Bailly brought out his treatise entitled *Traite de l'Astronomie Indian et Orientalle* which was reviewed by Playfair in 1790 in the *Transaction of the Royal Society*. Around this period, the Asiatic Society was founded by William Jones and in the early years of the Society, there were articles by himself, Samuel Davis and John Bentley. Translations of Indian texts on Astronomy and Mathematics gained apace with *Bijaganita* of Bhaskaracharya, translated from a Persian text by Edward Strachey, and *Lilavati* by John Taylor. In 1817, Colebrooke brought out the seminal work, *Algebra with Arithmetic and Mensuration from the Sanskrit of Brahmagupta and Bhascara.*

Interest in Indian mathematics and astronomy continued to grow. John Warren (1769–1830) wrote on Indian calendrical computations, based on both the *siddhanta* and *vakya* methods, in *Kalasankalita* (1825). In an earlier chapter, we referred to the exchanges between Warren, Heyne and Whish on the provenance of various infinite series discovered in Kerala. Whish (1794–1833) collected and studied several important manuscripts of the Kerala School (archived today in the Whish Collection of the Royal Asiatic Society in London) and made extensive notes on them. His seminal article, already referred to, was published in 1835 in the Royal Asiatic Society Journal. In the 1840's Lancelot Wilkinson, political agent at Bhopal, edited the *Siddhantasiromani* of Bhaskaracarya and *Grahalaghava* of Ganesa Daivajni. He also translated the *Goladhyaya* from *Siddhantasiromani* in 1861 and brought out an edition of *Suryasiddhanta* which translated into English by Burgess in 1860. There appeared editions of *Vedanga Jyotisa* by Weber, of *Brhatsamhita* of Varahamihira by Kern who also brought out *Aryabhatiya* with the commentary of Paramesvara in 1875. Also notable were Thibaut's edition of the Baudhayana *Sulbasutra* in 1874 and his translation of the *Pancasiddhantika* of Varahamihira (1884). There was therefore a substantial corpus of Indian texts available in Europe by the beginning of the 20th century.[4]

13.2.1 Encounter of wonder and respect: the case of Reuben Burrow

In June 1783, Reuben Burrow[5] wrote to Governor General Warren Hastings his views on the reasons for renovating the Observatory at Benares. Making a case for studying the 'Asian Sciences' and bemoaning the lost opportunities in the past for doing so, he argued that:

> [If we were] in possession of the most finished productions of Asia as well as of Europe, the sciences might, in consequence, have been carried to a much higher degree of perfection with us than they are at present. The elegance and superiority of the Asiatic model might have prevented that neglect and depravity of Geometry, and that inundation of Algebraic barbarism which has ever since the time of Descartes both vitiated taste and overrun the publications of most of the philosophical societies in Europe.

In 1790, Burrow published an article entitled 'A Proof that the Hindoos had the Binomial Theorem'. A notable feature of Burrow's writings is his conviction that India was the land from which all knowledge originated some time in the distant past but of which there is at present, only the most imperfect of information. While traces of such views may be found today only among a few who continue to subscribe to a highly romantic vision of the Indian past, these views were not uncommon then among the soldiers and administrators of the East India Company who returned home and were often elevated as giants of scholarship. When the reaction set in at the same time as the British hold over India became firmer, the pendulum swung the other way. And the burning issue became the question of the originality and antiquity of Indian astronomy.

13.2.2   Encounter of scrutiny and creeping condescension: the case of John Playfair

In 1789, John Playfair[6] published a long article in the Transactions of the Royal Society of Edinburgh entitled 'Remarks on the astronomy of the Brahmins'. He specified the following three objectives for his study: (i) ... to give a short account of the Indian astronomy, so far as it is known to us, from the four sets of tables (i.e. the Siamese and South Indian tables) ... ; (ii) to state the principal arguments, that can be deduced from these tables, with respect to their antiquity ... ; (iii) to form some estimate of the geometrical skill with which this astronomical system is constructed.

We will only discuss (ii) and (iii) here. Playfair begins with the assertion that the antiquity of the epoch (the beginning of *Kaliyuga*)[7] offers a criterion "by which we are able to judge of the pretensions of the Indian astronomy":

> It is for astronomy, in its most perfect state, to go back to the distance of forty-six centuries, and to ascertain the situation of the heavenly bodies at so remote a period. The modern astronomy of Europe, with all the accuracy that it derives from the telescope and the pendulum, could not venture on so difficult a task, were it not assisted by the theory of gravitation, and had not the integral calculus, after a hundred years of almost continual improvement been able, at last, to determine the disturbances in our system which arise from the action of the planets on one another... Unless the corrections for these disturbances be taken into account, any set of astronomical tables, however accurate at the time of its formation, and however diligently copied from the heavens, will be found less exact for every instant, either before or after that time, and will continually diverge more and more from the truth, both for future and past ages. Indeed, this will happen, not only from the neglect of these corrections, but also from the small errors unavoidably committed, in determining the mean motions, which must accumulate with time and produce an effect that becomes every day more serious, as we retire on either side, from the instant of observation. For both these reasons, it may be established as a maxim, that, if there be given a system of astronomical tables, founded on observation of an unknown date, that date may be found by taking the time when the tables represent the celestial motions most exactly.

Playfair's final conclusion gives a strong endorsement to the advanced state of Indian astronomy. The position of the planets as given in the tables at the beginning of the *Kaliyuga* is very close to the position as calculated by astronomical methods of his time which involved knowledge of modern integral calculus and theory of gravitation. He tests out the possibility that the Indians may have used other systems of calculation, whether Babylonian or Greek, and finds that they give results which do not correspond to those of the tables.

His conclusions with respect to objectives (ii) and (iii) are summarised in the following quotations.

> Thus we have enumerated no less than nine astronomical elements to which the tables of India assign such values as they do . . . [which] the theory of gravity proves to have belonged to them 3,000 years before the Christian era. At that time, therefore, or in the ages preceding it, the observations must have been made from which these elements were deduced [correctly].

And:

> Of such high antiquity, therefore, must we suppose the origin of this astronomy, unless we can believe, that all the coincidences which have been enumerated, are but the effects of chance: or indeed still more wonderful that some ages ago there had arisen a Newton among the Brahmins. to discover that universal principle which connects, not only the most distant regions of space, but the most remote periods of duration; and a De La Grange, to trace, through the immensity of both, most subtle and complicated operations.

What Playfair in effect is arguing is that the remarkable accuracy of the Indian astronomical observations for the year 3012 BCE can be explained *either* through meticulous direct observations in that year, using quite sophisticated instruments, *or* through analytical methods, including integral calculus, to extrapolate back into time.

He chose the first explanation, since the second one, invoking astronomical calculations, would have meant admitting the appearance of an Indian Newton with knowledge of the theory of gravity, and of an Indian La Grange who had discovered the procedure for correcting the inequality in the precession of the equinoxes. Playfair would rather concede antiquity of accurate observation to Indian astronomy rather than such knowledge of advanced methods and underlying theories needed for backward extrapolation.

But even this concession was only of short duration. In the eighth volume of the *Asiatic Researches*, published in Calcutta in 1805 appeared an article by John Bentley entitled "On the Hindoo Systems of Astronomy and their Connection with History in Ancient and Modern Times". It contains one of the first expositions of the principles of Indian astronomy in English. In it, Bentley pours scorn on the view that Indian astronomy is very old. He bases his argument mainly on a study of the astronomical text, *Suryasiddantha* — an early version of which has been dated around the fourth or fifth century CE — of which excessive claims of antiquity had been made, including one which suggested that the text was received by divine revelation about 2 million years ago! Bentley criticises the 'delusions' of Bailly and Playfair for supposing that Indian astronomy could be traced back to the year 3012 BCE. He was probably referring to Playfair's inference that actual astronomical observations must have been made at that time to explain the high degree of accuracy attained. Bentley also criticised the 'artificial systems' of Indian astronomy as hindering the development of progressive accuracy in astronomical calculations.[8]

Irrespective of the relative merits of Bentley's assessment of Indian astronomy, what is evident is the sea change in British attitudes to the culture and history of the Indian people, culminating in Macaulay's verdict on Indian science and learning in 1835. The relevant section of Macaulay's famous minute reads:

> But when we pass from works of imagination to works in which facts are recorded and general principles investigated, the superiority of the Europeans becomes absolutely immeasurable. It is, I believe, no exaggeration to say that all the historical information which has been collected from all the books written in the Sanskrit language is less valuable than what may be found in most paltry abridgement used at preparatory schools in England. In every branch of physical or moral philosophy the relative position of the two nations is nearly the same.

13.2.3 Encounter of silence and indifference: the case of Charles Whish

In 1832, Charles Whish read a paper to a joint meeting of the Madras Literary Society and the Royal Asiatic Society in which he referred to four works of the period 1450 to 1830 — Nilakantha's *Tantrasangraha*, Jyesthadeva's *Yuktibhasa*, Putumana Somayaji's

Karanapaddhati and Sankara Varman's *Sadratnamala* — as being significant astronomical and mathematical texts from Kerala. Writing about Nilakantha's work, he claimed that this "laid the foundation for a complete system of fluxions." As we saw in an earlier chapter, the Kerala discoveries include the Gregory series for the inverse tangent, the Leibniz series for π, and the Newton power series for sine and cosine as well as certain remarkable rational approximations for circular and trigonometric functions, including the well-known Taylor series approximations for the sine and cosine functions. And these results had been obtained without the use of infinitesimal calculus.

The question remains why a discovery of such importance was ignored then and, at best, only noted in a cursory fashion even till recently.[9] The following footnote, buried in a book by John Warren, entitled *Kala Sankalita* (1825), may give a clue to the answer:

> I owe the following note to Mr Hyne's favour: 'The Hindus never invented this series [i.e. the arctan θ series where $\theta = \pi/4$ radians]; it was communicated with many others, by Europeans to some learned Natives in modern times. Mr. Whish sent a list of the various methods of demonstrating the ratio of the diameter and circumference of a circle employed by the Hindus to the Literary Society, being impressed with the notion that they were the inventors. I requested him to make further inquiries, and his reply was that he had reasons to believe them entirely modern and derived from Europeans, observing not one of those who used the Rules could demonstrate them. Indeed, the pretensions of the Hindus to such a knowledge of Geometry is too ridiculous to deserve refutation.

There are some interesting issues raised by the above quotation. All we know about Hyne was that he was a junior employee in the East India Company's Medical Service, Madras. He joined the service at the age of 20 in 1820 and died six years later. There is little reason to suspect that his knowledge of Indian mathematics and astronomy was considerable. Also, his statement regarding Whish's change of views about the origins of the series seems odd if it should have taken place at least seven years before Whish addressed the

Madras Literary Society and 10 years before the paper was published in the *Transactions of the Royal Asiatic Society of Great Britain and Ireland*. Instead, Whish announced in his published paper that he had 'ascertained beyond a doubt, that the invention of infinite series of these forms had originated in Malabar' and, in describing one of the books, he claimed that it 'abounds with fluxional forms and series, to be found in no work of foreign or other Indian countries.

The momentous work of Orientalists such as William Jones (1746–1794) and Max Mueller (1823–1900) brought to the forefront the rich linguistic, religious and philosophical heritage of India which many Indians were unaware of. But there was a negative side to their promotion of Indian heritage. The work of the Orientalists has developed the view that Indian mathematics originated in the service of religion. Thibaut (1848–1914) was responsible for bringing to a wider public notice the geometry contained in the *Sulbasutras* (*c.* 800–500 BCE). Such an outcome was not to be the fate of Whish's work.

13.2.4   Encounter and an 'aborted pedagogy': the case of Ramchandra[10]

Yesudas Ramchandra was born in Panipat, a historic town near Delhi, in 1821. Admitted to an 'English' school in 1833, his ability in mathematics soon became evident, though there was no facility to learn the subject at school. So he taught himself, which turned out to be a blessing in disguise, since he became acquainted early on with texts such as Bhaskaracharya's *Lilavati* and *Bijaganita*, important influences on his later work. In 1850 he published his first book on mathematics called A Treatise on the *Problem of Maxima and Minima Solved by Algebra* (hereafter referred to as *The Treatise*) in Calcutta. Nine years later and as a result of the efforts of Augustus De Morgan, an English logician and mathematician, the book was reprinted in London. It is not the purpose here to investigate the mathematical content of this book, nor to locate Ramchandra's efforts as a teacher, translator and mathematician within the context of an emerging colonial science. This has already been done by Irfan Habib and Dhruv Raina (1989, 1990) in various publications. Instead, we use the case of Ramchandra to illustrate how once the dominant paradigm is *in situ* (in this case Western mathematics), it is very difficult to slot in another traditional mode, even if there are strong arguments for doing so.

By the middle of the nineteenth century, the discontinuity between traditional and modern knowledge systems had widened to such an extent that a number of Indian intellectuals were seeking solutions to this dilemma. Apart from a few who succumbed to the Macaulay's prescription of discarding the old, others were interested in attempting to bridge this growing discontinuity. This was to be done in two ways: (i) by establishing historically the Indian roots of modern science which would dent British cultural superiority at one stroke, and (ii) by reviving Indian traditions to see whether they could be merged with modern science. Ramchandra's efforts should be seen as falling into the second category. He dedicated his book, *The Treatise*, to reviving the spirit of algebra so as to "resuscitate the native disposition of these people" that had been eroded over the centuries.

Ramchandra's essential contribution was the application of the theory of equations as found in Bhaskaracharya's *Bijaganita* to the solution of an elementary problem in calculus — the obtaining of the maxima and minima of a function. The function could be quadratic or of higher order. And he did this without the help of differential calculus and made the theory of equations as the starting point. This attempt should be seen in the context of a widespread perception even today that may be traced back to Colebrook's book on Indian algebra, first published in 1817. It is a perception well summed up by his remark that the Indians had "cultivated algebra much more, and with greater success than geometry, as is evident from the comparatively low state of knowledge in the one and the high pitch of attainments in the other."

So one should see Ramchandra's effort as bridging the dis-continuity between an Indian mathematical tradition, which was perceived as algebraically strong but geometrically weak, and modern calculus — with Bhaskaracharya's theory of equations serving as the bridge. Ramchandra certainly subscribed to this view, while De Morgan believed that the strength of the book lay in drawing upon the native resources and not on the 'imported science of his teachers'. But neither Ramchandra nor De Morgan saw the book as being useful only to the Indians.

Ramchandra wrote the book in the hope that his "labours will be of some use to those mathematical students who are not advanced in their study of the differential calculus, and that the lovers of science, both in India and Europe, will give support to my undertakings." And De Morgan, whose interest in the pedagogy of mathematics teaching was well known, recommended that "selections from Ramchandra's work might advantageously be introduced into elementary instruction in this country. The exercise in quadratic equation which it would afford, applied as it is to real problems, would advantageously supersede some of the conundrums which are manufactured under the name of problems producing equations."

Yet this interesting attempt at building a pedagogic bridge between two mathematical traditions was a failure. The Treatise did not gain acceptance in any Indian school and, while there is the intriguing suggestion by Mary Boole, the widow of the renowned algebraist, George Boole, that English students were being taught to solve problems in maxima and minima by "other simple devices similar in essence to Ramchandra's and probably superior in efficiency," the interest petered out there as well. The book was reviewed poorly in India when it was first published, though it picked up well after De Morgan's endorsement — a characteristic common to many other Indian endeavours which gain in value only after western endorsement.

It is necessary to investigate this encounter within the colonial context in which it took place. By 1850, the last vestiges of an education programme which tried to blend traditional learning with modern science, advocated by the so-called Orientalists like Prinsep, Lobb and Colebrook, had disappeared. It was not that there was a conscious attempt to keep texts such as *The Treatise* out of schools and colleges, but that they were seen as being irrelevant to the needs of modernising India. The process of modernisation undertaken by the Indian intelligentsia, in cognitive encounters with the British, initiated a number of developments in the fields of science and education. But none included Ramchandra's mathematical project, which was therefore abandoned. But that is not the end of story. For about 50 years later emerged in the firmament the extraordinary figure of Srinivasa Ramanujan.

13.3   Srinivasa Ramanujan (1887–1920)

13.3.1   Life and work

Srinivasa Ramanujan, was born in Erode on December 22, 1887. His father, Srinivasa Iyengar, hailing from the district of Tanjavur worked as a clerk in a shop selling saris. His mother, Komalathammal, a dominating figure in his life, was a housewife who also sang at a local temple and had gained a reputation as a folk astrologer. In 1892, Ramanujan was enrolled in a primary school in Kumbakonam. In 1898, having passed his primary school examinations with distinction, joined the Town High School. He passed out of that school as an outstanding student and received a scholarship to study at the Government College, Kumbakonam.

While at high school, Ramanujan got a copy of Loney's *Plane Trigonometry* which he soon mastered, but was surprised to find in it some of the results that he had obtained himself. Around 1903, Ramanujan acquired a copy of Carr's *Synopsis of Pure and Applied Mathematics* (1880), a collection of 5000 results, and worked through painstakingly using his own approach, but adopting the style in which this text presented results. By 1904 he began recording his discoveries in a Notebook However, at the college, his concentration on mathematics to the exclusion of other subjects, resulted in his failure in examinations and as a result lost his scholarship. He then joined another College in Madras, but had to discontinue due to bad health. In 1909, Ramanujan married Janaki and with the effort and help of various individuals impressed by his mathematical ability, found a job in Madras Port Trust as a clerk in 1912. This period of poverty and instability did not seem to have deterred his mathematical work for he published not only five papers in the *Journal of Indian Mathematical Society* and also posed 30 problems in the same journal for which he himself offered 20 solutions. An early paper on which he began to work on during this period, published later in London in 1914, was the "Modular equations and approximations to π", a subject that had fascinated the Kerala mathematicians as we found in an earlier chapter. Incidentally one of the results was used by R.W. Gosper in 1985 to compute the value of π, correct to 17,526,100 digits, which was a record at that time.[11]

In 1913 Ramanujan wrote to Prof G. H. Hardy, a distinguished number theorist at the University of Cambridge, enclosing a 11 page list of over 100 results.[12] In reply, Hardy wrote back expressing his interest in Ramanujan's work and inviting him to Cambridge. Ramanujan arrived in London in April 1914 and returned to India in February 1919. Of the nearly five years he spent there, he was ill for more than two.

During his stay in England, Ramanujan wrote about 30 papers, 7 of them in collaboration with Hardy. His work was highly acclaimed. In March 1916 he was awarded Bachelor of Science degree by Research from the Cambridge University. He was elected a Fellow of Royal Society in February 1918 and in October of the same year he was elected a Fellow of the Trinity College. Later that year, the University of Madras offered a matching grant of 250 pounds sterling a year. In response, Ramanujan wrote back to the Registrar of the University of Madras that after deducting certain basic expenses incurred for his family, the money that remained should "should be used for some educational purpose, such in particular as the reduction of school-fees for poor boys and orphans and provision of books in schools."

Finding that his health wasn't improving, on March 27, 1919 Ramanujan returned to India. Though seriously ill, he continued his work all the while. On January 12, 1920, just before his death, Ramanujan wrote to Hardy:

I discovered very interesting functions recently which I call 'Mock-functions'. Unlike the 'False–functions' (studied partially by Professor Rogers in his interesting paper) they enter into mathematics as beautifully as the ordinary functions. I am sending you with this letter some examples...

The so-called "Lost Notebook" of Ramanujan, being a sheaf of over hundred sheets, contains about 600 results that Ramanujan had discovered during the last year of his life. This had been sent to Hardy along with all other papers of Ramanujan in 1923. It was finally discovered by George Andrews in 1976 in Trinity College at the University of Cambridge.

13.3.2   Ramanujan's legacy

Soon after Ramanujan's death, Hardy wrote an Obituary Notice in the Proceedings of London Mathematical Society which was later reproduced in the Collected Papers of Ramanujan (Cambridge 1927). Hardy first gives his impression of Ramanujan on his arrival in England in 1914:

> The limitations of his knowledge were as startling as its profundity. Here was a man who could work out modular equations, and theorems of complex multiplication, to orders unheard of, whose mastery of continued fractions was, on the formal side at any rate, beyond that of any mathematician in the world, who had found for himself the functional equation of the Zeta-function, and the dominant terms of many of the most famous problems in the analytic theory of numbers; and he had never heard of a doubly periodic function or of Cauchy's theorem, and had indeed but the vaguest idea of what a function of a complex variable was. His ideas as to what constituted mathematical proof were of the most shadowy description. *All his results, new or old, right or wrong had been arrived at by a process of mingled argument, intuition and induction, of which he was entirely unable to give any coherent account* (My emphasis).

After listing the unique strengths of Ramanujan Hardy concludes with this assessment:

> Opinions may differ as to the importance of Ramanujan's work, the kind of standard by which it should be judged, and the influence which it is likely to have on the mathematics of the future. It has

not the simplicity and the inevitability of the very greatest work; it would be greater if it were less strange. One gift it has which no one can deny, profound and invincible originality. *He would probably have been a greater mathematician if he had been caught and tamed a little in his youth; he would have discovered more that was new, and that, no doubt of greater importance...* (My emphasis)

Hardy then offers reasons for Ramanujan not achieving full potential apart from his early death:

It was inevitable that a very large part of Ramanujan's work should prove on examination to have been anticipated. *He had been carrying an impossible handicap, a poor and solitary Hindu pitting his brains against the accumulated wisdom of Europe. He had had no real teaching at all;* there was no one in India from whom he had anything to learn...

In 1988, during the centenary of Ramanujan's birth, Selberg (1917–2007), a leading number theorist, gave his reflections on Ramanujan's work and collaboration with Hardy:

Srinivasa Ramanujan's *particular talent will seem to be primarily of an algebraic and combinatorial nature.* He developed it, for a long time in complete isolation really without any contact with other mathematicians. *He had on his own acquired an extraordinary skill of manipulation of algorithms, series, continued fractions and so forth, which certainly is completely unequalled in modern times...* (my emphasis)

(It) seems clear to me that it must have been, in a way, Hardy who did not fully trust Ramanujan's insight and intuition, when he chose the other form of the terms in their expression, for a purely technical reason, which one analyses as not very relevant. I think that if Hardy had trusted Ramanujan more, they should have inevitably ended with the Radmacher series. There is little doubt about that. Littlewood and Hardy were primarily working with hard analysis and they did not have a strong feeling for modular forms and such things...

It is at this point that Selberg adds an important rider regarding mathematical discovery. "A felicitous but unproved conjecture may be of much more consequence for mathematics than the proof of many a respectable theorem." Selberg implicitly disagreed with Hardy's opinion of the extent to which the Ramanujan's legacy would last. Ramanujan's stature in mathematics has grown over the period since his death and his contributions are increasingly pivotal in the development of new areas of mathematics.

> Ramanujan's recognition of the multiplicative properties of the coefficients of modular forms that we now refer to as cusp forms and his conjectures formulated in this connection, and their later generalisation have come to play a more central role in the *mathematics of today, serving as a kind of focus for the attention of quite a large group* of the best mathematicians of our time. Other discoveries like the mock theta functions are only in the very early stages of being understood and no one can yet assess their real importance. So the final verdict is not really in, and it may not be in for a long time, but the estimates of Ramanujan's stature in mathematics certainly have been growing over the years. There is no doubt about it. (My emphasis)

Until the second half of the twentieth century, the only work generally available of Ramanajuan consisted to the 37 odd papers and 57 questions and solutions available in various Journals published between 1911 and 1920. These together with the two letters to Hardy in 1913 constituted the corpus of Ramanujan's work, gathered and published in 1927 by Cambridge University Press as *Collected Works of Srinivasa Ramanujan,* This of course excluded most of Ramanujan's work done both before he left for England and after his return to India. The corpus of work done before leaving for England is now found in two Notebooks and the work after return from England is found mainly in the so-called "Lost Notebook" which is said to have about 600 results. The study of these notebooks that began only in the last quarter of the 20th century and continuing even today has led to a better appreciation of the genius of Ramanujan.

It is believed that Ramanujan began recording his results in a Notebook around the time he was admitted to the Government College of Kumbakonam in 1904, but copied these results with some amendments and extensions during the period of 1911–1913 into a second notebook. This notebook contained 21 sections in 256 pages and an additional 100 pages of miscellaneous material. The third notebook (usually referred to as 'Lost Notebook') was found in the Trinity College Library by G.G. Andrews in 1976 and consisted of work done by Ramanujan during 1919–1920 after his return to India. It is a manuscript of 100 pages with 138 sides of around 600 results. They had originally been sent to Hardy along with other papers of Ramanujan in 1923. In 1987 this notebook ('The Lost Notebook') was published with the other unpublished papers of Ramanujan on his birth centenary in 1987. Andrews and Berndt then embarked on an edition of all this material in five volumes; of which the first three have appeared in 2005, 2009 and 2013. They note in the first volume of their edition of the 'Lost Notebook': "... (Only) a fraction (perhaps 5%) of the notebook is devoted to the mock theta functions themselves for which Ramanujan is best known.[13] Berndt also gave the following overall assessment of Ramanujan's notebooks:

> Altogether, the notebooks contain over 3,000 claims, almost all without proof. Hardy surmised that over two-thirds of these results were rediscoveries. This estimate is much too high; on the contrary, at least two-thirds of Ramanujan's claims were new at the time that he wrote them, and two-thirds more likely should be replaced by a larger fraction. Almost all the results are correct; perhaps no more than five to ten are incorrect.[14]

13.3.3   The enigma of Ramanujan's mathematics

For the past hundred years, the problem of comprehending and assessing Ramanujan's mathematics has centred around the thorny issue of 'proof'. In 1913, in response to Ramanujan's letter, Hardy asked for 'proofs' of his results. Ramanujan responded by stating that he had a systematic method for deriving all his results but could not communicate them in letters. In his earlier published work in India and a few of the results contained in the unpublished notebooks there are outlines of 'proofs', described as sketchy, incomplete and even faulty. Ramanujan had no doubts about the validity of his results and was willing to wait for proofs to be added in the necessary format before his papers could be considered for publication.

The Graeco-Western tradition of mathematics has equated mathematics with formal proof, so that the process of discovery of mathematical results on the part of Ramanujan has been characterised vaguely as "intuition", "natural genius", etc. Since mathematical truths are believed to be non-empirical, there are no systematic way of arriving at them except by pure logical reason, an essentially Platonic position. There are some philosophers who argue that such a "philosophy of mathematics" is indeed sterile constituting little more than mind games, and has therefore little validity when viewed in terms of mathematical practice, either in history or in our times.

13.4   The Indian 'Way of Knowing'*

In the Indian mathematical tradition, known from the texts of the last two to three millennia, mathematics was not equated with proof. Mathematical results were not perceived as being non-empirical and could therefore be validated in different ways. The process of discovery and the process of validation were not seen as separate entities. Proof or logical argumentation to demonstrate the results could be important. But the purpose of proof was mainly to establish credibility in a community of well-informed individuals.[15] However, the Indian mathematicians were clear that results in mathematics, even those stated in authoritative texts, cannot be considered as valid unless they are supported by reason (*yukti*) or demonstration (*upapatti*).

For example, in Ganesa's commentary on Bhaskaracharya's *Lilavati*, the role of Indian Proof (*Upapatti*) is explained thus:

> Whatever is discussed in *vyakta* (computation) or *avyakta* (analytic) branches of mathematics, without *upapatti*, it will not be *nirbharanta* (free from misunderstanding). It will not acquire any standing in an assembly of scholars. The *upapatti* is directly perceivable, like looking into a hand mirror. It is there to elevate the intellect (*buddhi vriddhi*) that I proceed to state the *upapatti*.

---

*This section is based on the work of Srinivas (2008, pp. 267–293).

The multiple functions of the Indian 'proof' may be summarised thus:

(i) *Psychological*, where the purpose is to convince. Success in this case depends on how the argument is formulated, organised and presented;

(ii) *Logical*, where the purpose is to minimise error so that claims made about mathematical logic are coherent and consistent; and

(iii) *Social and Cultural* where claims are made about mathematical objects *or* about the degree to which the preparedness of the audience is recognised.

To highlight some of the important characteristics of *upapattis* in Indian mathematics:

1. Mathematical results, even those stated in authoritative texts, should not to be accepted as valid unless they were supported by *yukti* (reason). It was not enough that one observes a result in a large number of cases to conclude that it is valid.

2. Generally, the demonstration (or *upapatti*) is found in the commentaries on major texts in Indian mathematics and astronomy.

3. A typical *upapatti* is presented in an ordered sequence proceeding systematically from a known or self-evident premise to arrive at a final answer.

4. An *upapatti* should be expressed in a way so as to eliminate doubts and obtain approval from a community of scholars.

5. An *upapatti could be* sought from observation and/or experimentation, depending on the nature of the mathematical object or relationship between mathematical objects involved.

6. Unlike Greek or Western mathematics, the method of 'indirect proof' (or proof by contradiction) is rarely found in *upapattis*, although a method, known *as tarka*, was used to show the *non-existence* of certain entities.[16] In rejecting the method of indirect proof as a valid means for establishing the existence of an entity (whose existence is not even in principle 'establishable'), Indian mathematics adopted what would be known today as the 'constructivist' approach to the issue of mathematical existence.[17]

7. The Indian mathematical tradition did not subscribe to the ideal prevalent in Greek (and Western) traditions that *upapattis* should seek irrefutable demonstrations to establish the absolute truth of mathematical results.

8. There was no attempt in Indian mathematical tradition to present the *upapattis* in an axiomatic framework based on a set of self-evident (or arbitrarily postulated) axioms which are fixed at the outset.

9. Consequently, while in Indian mathematics there was a veritable explosion of algorithms, in the invention and manipulation of symbols for presenting mathematical results and in facilitating mathematical processes, this led to a computational revolution without any attempt at formalisation of mathematics.

In a wider context of understanding different ways of 'knowing' (or 'proving'), consider the history of Indian logic where we see that the earliest forms of logic, namely their syllogisms, make specific use of persuasion and appeal to open-mindedness. From early times commentators considered ten part syllogism, of which five parts involved issues like 'Do we actually have any doubt regarding this proposition?' or 'Why do we seek to prove or disprove this proposition?' and so on. Only then does the syllogism lay out a major premise, a minor premise, and draw a conclusion. As time went on, the ten part syllogism became five when the Indian philosophers and logicians assumed that the other five could be taken for granted or had become redundant, for why would one discuss something that one would be certain about, or didn't believe could be answered one way or another. Later some Indian logicians such as the Buddhist *Dignaga*[18] thought that even five was excessive, and that a further reduction on grounds of redundancy left two premises, which seem to be almost identical to the Aristotelian syllogism.

More recently, Wittgenstein has argued for changing the way of demonstrating proof in mathematics. "Yes, that is true" or "Behold!" (Bhaskara's demonstration of the Pythagorean result) could be a valid geometrical demonstration. Here, Wittgenstein does not equate "proof" with "formal derivation". The question then arises: If "no proof convinces solely in virtue of its capacity to be formalised in a logical structure", then what is proof? We will not try to explore this question any further except to add that Wittgenstein is arguing that we often follow rules in mathematical reasoning because of well-tried custom, not because of logical necessity. So Wittgenstein's contribution is to indicate that what mathematicians do in practice, and not what logical theories tell us, becomes the engine driving the development of mathematical knowledge. This is a view that resonates in the Indian approach to mathematics.

In 1913, Bertrand Russell jocularly remarked that Hardy and Littlewood had discovered a "second Newton" in a "Hindu clerk". If parallels are to be drawn, Ramanujan's intellectual ancestor should be Madhava rather than Newton. It is not merely in terms of his methodology and philosophy that Ramanujan was closer to the earlier Indian tradition of mathematics. Even in his extraordinary felicity in handling iterations, infinite series, continued fractions and transformations of them, Ramanujan is indeed a worthy successor of Madhava. Ramanujan represents an example of how mathematics in modern times in India involved a fruitful fusion of Western mathematics and Indian traditions. This implies that the influence of Indian mathematics on the world and world mathematics on India has continued into recent times.

To conclude, the roots of the 'unreasonable effectiveness of mathematics' that provided the engine for the Scientific Revolution are found in the combination of Greek rational and deductive traditions and Indian (Chinese) algorithmic/empirical traditions. Such a combination was mediated through the agency of the Islamic conduit.

Endnotes

[1] The planet Venus appears to pass in front of the Sun seen from the surface of the Earth. This happens only rarely and observing the transits of Venus became important when in 1716 Edmund Halley proposed that observations of this phenomenon from different locations on the Earth could be used to determine the distance between the Sun and the Earth, and the scale of the solar system could be subsequently determined by

applying Kepler's third law of planetary motion. Gentil's expedition was part of an international expedition that failed because war broke out between England and France.

2 Tobias Mayer (1723–1762) was a German astronomer known for his studies of the Moon. In 1752, he produced a set lunar tables to determine the position of the moon accurate to five seconds and therefore the longitude at sea to half a degree. This was of great interest to various governments (notably the British) who were in the midst of their voyages of discovery.

3 De la Loubère also brought to France from his Siamese travels a very simple method for creating an odd magic square, known as the 'Siamese method'. Originally discovered in Surat (India) by another Frenchman, M. Vincent, who sailed on his return journey with de la Loubère, Vincent showed it to Le Loubère who then amused himself by constructing magic squares with this faster procedure which he named the 'Hindu continuous method'. Resemblances between Narayana's treatment of magic squares (discussed in Chapter 9) and the Loubère method raise interesting questions regarding transmission. On his return to Paris, Loubère devoted a whole chapter to the subject of magic squares in his travel account *Du Royaume de Siam*.

4 What has been ignored so far in this discussion of the role that the Westerners played in the rediscovery of Indian mathematics and astronomy and a growing recognition by the Indians themselves of their legacy. Apart from some of the names already mentioned, the part played by the translations of texts into regional languages, notably *Lilavati*, *Bijaganita* and *Siddhantasiromani* has been noted in the last chapter. One name stands out. Sudhakar Dwivedi (1855–1910) was not only a custodian and disseminator of traditional knowledge but also wrote on history of mathematics, on differential calculus and the lives of Indian mathematicians and astronomers.

5 Reuben Burrow (1747–1792) was appointed assistant to Maskelyne, the Astronomer Royal at Greenwich in 1770. In 1782 he accepted an appointment as Chief Engineer in Bengal. He taught mathematics to the engineers and helped to establish the Trigonometrical Survey of Bengal. He was one of the first members of the Asiatic Society.

6 John Playfair (1748–1819) was born in Scotland and became a Presbyterian minister in 1770. In 1795, he became a professor of mathematics in the University of Edinburgh and later took up the chair of natural philosophy at the same university One of the foundation members of the Royal Society of Edinburgh, he subsequent became the Secretary and held the post until his death. He was elected a Fellow of the Royal Society in 1807. He is noted in the history of mathematics for the

Playfair Axiom, a proposal for an alternative version to Euclid's parallel postulate.

[7] According to the *Suryasiddhanta*, the *Kaliyuga* began at midnight on 18 February 3102 BCE.

[8] The introduction of a fixed epoch, the beginning of *Kaliyuga*, and the practice of starting with that epoch in all subsequent calculations is what Bentley describes as the 'artificial' system of Indian astronomy A fixed reference point in time does not build into the system improvements arising from more accurate observations and calculations after the tables are first constructed.

[9] See, for example, this opening statement: "Until the development of the calculus, all the methods of calculating π; depended on inscribed and circumscribed regular polygons within and about a circle...The new techniques of the calculus replaced the older geometric methods with analytical functions which are given by integrals of quadratic functions that characterise the curvature of a circle, both Newton and Leibniz calculated approximations for π; using series expansions of these functions. Newton used the inverse sine function while Leibniz preferred the inverse tangent function" (Abeles, 1993). One searches in vain for even a bare mention of the Kerala work on arctangent series approximation for π at least two hundred years earlier than European work in this area.

[10] Ramchandra worked as a science teacher and later a headmaster in Delhi. He translated a number of scientific texts from English to Urdu. He also wrote books in Urdu which attempted to popularise modern science, and founded an Urdu newspaper which contained incisive criticisms of traditional practices and thought. In 1852, he became a Christian, probably as a result of which his life was endangered during the Indian Uprising of 1857. An important member of the so-called Delhi Renaissance, he died in 1880 at the age of 59.

[11] The following modified algorithm of Gosper was used in obtaining this accurate estimate:

$$\frac{1}{2\pi\sqrt{2}} = \frac{1103}{99^2} + \frac{27493}{99^6} \cdot \frac{1}{2} \cdot \frac{1 \times 3}{4^2}$$
$$+ \frac{53883}{99^{10}} \cdot \frac{1 \times 3}{2 \times 4} \cdot \frac{1 \times 3 \times 5 \times 7}{4^2 \times 8^2} + \cdots.$$

[12] In his lectures on Ramanujan, published in 1940, Hardy gave an assessment of Ramanujan's early work before he went to England:

He [Ramanujan] published abundantly...but he also left a mass of unpublished work which has never been assessed properly until the last few years. This work includes a great deal that is new, but much more that is rediscovery, and often imperfect rediscovery; and it is sometimes still impossible to distinguish between what he must have rediscovered and what he may somehow have learnt.

[13] They then go on to list the range of subjects in which Ramanujan made major contributions as indicated by a study of the 'Lost Notebook'. "These include mock theta functions, theta functions, partial theta function expansions, false theta functions, identities connected with the Rogers-Fine identity, several results in the theory of partitions, Eisenstein series, modular equations, the Rogers-Ramanujan continued fraction, other q-continued fractions, asymptotic expansions of q-series and q-continued fractions, integrals of theta functions, integrals of q-products and incomplete elliptic integrals. Other continued fractions, other integrals, infinite series identities, Dirichlet series, approximations, arithmetic functions, numerical calculations, Diophantine equations. . ."

[14] Even a mistake by Ramanujan led to a new area of discovery. A result contained in the famous first letter from Ramanujan to Hardy in 1913 is given as:

$$\text{The coefficient of } x^n \text{ in } \frac{1}{1 - 2x + 2x^4 - 2x^8 + 2x^{16} - \cdots}$$

$$= \text{the nearest integer to } \frac{1}{4n} \left\{ \cosh\left(\pi\sqrt{n}\right) - \frac{\sinh(\pi\sqrt{n})}{\pi\sqrt{n}} \right\}.$$

Commenting on this result later, Hardy added: "The function in. . . [the right hand side] is a genuine approximation to the coefficient, though not at all so close as Ramanujan imagined, and Ramanujan's false statement was one of the most fruitful he ever made, since it ended by leading us to all our joint work on partitions."

[15] Briefly, the uses of proof are to: (i) verify a result, (ii) communicate and persuade others, (iii) interpret a result and (iv) systematise results into a Deductive System. The last became important after leading 19th century mathematicians such as Abel, Bolzano, Cauchy, Lagrange and Weierstrass laid the foundations of the present deductive proof tradition in mathematics.

[16] Thus in showing that a negative number cannot have a square root a commentator, Krishna Daivajna (*fl.* 1600), used the method of indirect proof. For a further discussion of Daivajna's mode of argument, see Joseph (1994).

[17] But the Indian logician did more than merely reject certain existence proofs. The general Indian philosophical position is one of completely

eliminating from logical discourse all reference to those entities whose existence are not even in principle accessible to direct means of verification. This appears to be the position adopted by Indian mathematicians for whom a method of recursive reasoning called *samskaram* (or 'refining') took the place of indirect proof And it is for this reason that many an 'existence theorem' (where all that has been proved is that the non-existence of a hypothetical entity is incompatible with the accepted set of postulates) of Greek mathematics would not be considered as significant or even meaningful in Indian mathematics.

[18] Dignaga (ca. 480–540 CE) was one of the Buddhist founders of Indian logic. His first work on formal logic, *Hetucakra* ('The wheel of reason'), proposed a new form of deductive reasoning, a concept resonant of the Western notion of 'implication' (i.e. philosophical analysis of logical consequence or in what sense does a conclusion follow from its premises?).

Bibliography

Abeles, F.F. (1993). Charles Dodgson's geometric approach to arctangent relations for π. *Historia Mathematica*, 20(2), pp. 151–159.

Almeida, D.F. and Joseph, G.G. (2009). Kerala mathematics and its possible transmission to Europe. In P. Ernest *et al.* (eds.). *Critical Issues in Mathematics Education*, Information Age Publishing, Charlotte, NC.

Ayyangar, K. (1938). The earliest solution of the biquadratic, *Current Science*, 7(4).

Baldini, U. (1992). *Studi su filosofia e scienza dei gesuiti in Italia 1540–1632*, Bulzoni Editore, Firenze.

Baron, M.E. (1969). *The Origins of the Infinitesimal Calculus*, Pergamom, Oxford.

Barrett, J.L. and Nyhof, M. (2001). Spreading non-natural concepts: the role of intuitive conceptual structures in memory and transmission of cultural materials, *Journal of Cognition and Culture*, 1, pp. 69–100.

Baumgardt, C. (1951). *Johannes Kepler: Life and Letters*, Philosophical Library, New York.

Beggren, J. (1986). *Episodes in the Mathematics of Medieval Islam*, Springer, New York.

Beggren, J. (2007). Mathematics in Medieval Islam, In V. Katz (ed.) *The Mathematics of Egypt, Mesopotamia, China, India and Islam: A Source Book*, Princeton University Press.

Boyer, P. and Ramble, C. (2001). Cognitive templates for religious concepts: cross-cultural evidence for recall of counter-intuitive representations, *Cognitive Science*, 25, pp. 535–564.

Brahmasphutasiddhanta of Brahmagupta, edited with Dvivedi commentary Vasana, and Hindi Translation by R.S. Sharma in 4 Volumes, Indian Institute of Astronomical and Sanskrit research, New Delhi (1966).

Bressoud, D. (2002). Was calculus invented in India? *The College Mathematics Journal*, 33(1), pp. 2–13.

Briggs, R. (1985). Knowledge representation in sanskrit and artificial intelligence, *AI Magazine*, 6, pp. 22–38.

Burgess, E. (1860). *The Surya Siddhanta: A Text-Book Of Hindu Astronomy*, Motilal Banarsidass Publishers Private Limited, New Delhi, Reprinted 1997.

Burnett, C. (2002). Indian numerals in the Mediterranean Basin in the twelfth century, with special reference to the 'Eastern Forms', In Y. Dold-Samplonius *et al.* (eds.). *From China to Paris: 2000 Years Transmission of Mathematical Ideas*, Franz Steiner Verlag, Stuttgart, pp. 237–288.

Burrow, R. (1790). A proof that the Hindoos had the binomial theorem, *Asiatik Researches*, 2, pp. 487–497.

Calinger, R. (1999). *A Contextual History of Mathematics to Euler*, Prentice-Hall, New Jersey.

Coedès, G. (1930–1932). A propos de l'orgine des chiffres arabes, *Bulletin of the School of Oriental and African Studies*, 6, pp. 323–328.

Cohen, H.F. (1994). *The Scientific Revolution: A Historiographical Inquiry*, University of Chicago Press, Chicago.

Correia-Afonso, J. (1969). *Jesuit Letters and Indian History: 1542–1773*. Oxford University Press, Bombay.

Correia-Afonso, J. (1997). *The Jesuits in India, 1542–1773: A Short History*, Gujarat Sahitya Prakash, Gujarat.

Cronin, V. (1984). *The Wise Man From The West — Matteo Ricci and His Mission To China*, Collins, London.

Datta, B. (1927). Early history of the arithmetic of zero and infinity in India, *Bulletin of the Calcutta Mathematical Society*, 18, pp. 165–176.

Datta, B. (1932). *The Science of the Sulbas. A Study in Early Hindu Geometry*, Calcutta University Press, Calcutta.

Datta, B. and Singh, A.N. (1938). *History of Hindu Mathematics, Vol. 2: Algebra*, Lahore; Reprint 1962, Asia Publishing House, Mumbai.

Datta, B. and Singh, A.N. (1983). Hindu trigonometry, revised by K.S. Shukla, *Indian Journal of History of Science*, 18(1), pp. 39–108.

Datta, B. and Singh, A.N. (1993). Approximate values of surds in Hindu mathematics, *Indian Journal of History of Science*, 28(3), pp. 265–275.

Datta, B. and Singh, A.N. (with revision by K. S. Shukla) (1993). Surds in Hindu mathematics, *Indian Journal of the History of Science*, 28(3), pp. 253–264.

De Cruz, H. and De Smedt, J. (2013). Mathematical symbols as epistemic actions, *Synthese*, 190, pp. 3–19.

D'Elia, P. (1960). *Galileo In China Relations through the Roman College Between Galileo and the Jesuit Scientist-Missionaries* (1610–1640). Harvard University Press, Cambridge.

Dharampal (1983). *The Beautiful Tree: Indigenous Indian Education in the Eighteenth Century*, Biblia Impex Private Limited, New Delhi.

Duan Yaoyong (1996). *The Discussion that Indian Trigonometry affected the Chinese Calendar Calculation in Tang Dynasty* (*A.D.* 618–907), Institute for the History of Science, Inner Mongolia Normal University.

Edwards, C.A.H. (1979). *The Historical Development of the Calculus*, Springer, New York.

Elfering, K. (1977). The area of a triangle and the volume of a pyramid as well as the area of a circle and the surface of the hemisphere in the mathematics of Aryabhata I, *Indian Journal of the History of Science*, 12(2), pp. 232–236.

Eves, H. (1983). *An Introduction to the History of Mathematics*, Saundeers, Philadelphia.

Ferroli, D. (1939). *The Jesuits in Malabar*, 2 Vols, Bangalore Press, Bangalore.

Fiegenbaum, L. (1986). Brook Taylor and the method of increments, *Archive for the History of Exact Sciences*, 34(1), pp. 1–140.

Filliozat, P.-S. (2002). *The Sanskrit Language: An Overview — History and Structure, Linguistic and Philosophical Representations, Uses and Users*, Indica Books, New Delhi.

Foucault, M. (1972). *The Archaeology of Knowledge and the Dioscourse on Language*, Pantheon Books, New York.

Friberg, J. (2005). *Unexpected Links between Egyptian and Babylonian Mathematics*, World Scientific, Singapore.

Friberg, J. (2007). *Amazing Traces of a Babylonian Origin in Greek Mathematics*, World Scientific, Singapore.

Ganitakaumudi of Narayana Pandita (1998–2002). Edited and Translated with Notes by Paramanand Singh, Ganita Bharati 20, 1998, pp. 25–82; *ibid.* 21, 1999, pp. 10–73; *ibid.* 22, 2000, pp. 19–85; *ibid.* 23, 2001, pp. 18–82; *ibid.* 24, 2002, pp. 34–98.

Gironi, F. (2012). Sunyata and the zeroing of being: a reworking of empty concepts. *Journal of Indian History and Philosophy*, 15, pp. 1–42.

Gordon Childe, V. (1935). *New Light on the Most Ancient East*, Kegal Paul, London.

Gosper, R.W. (1978). Decision procedure for indefinite hypergeometric summation, *Proc. Natl. Acad. Sci. USA* 75, pp. 40–42.

Gupta, R.C. (1969). Second order interpolation of Indian mathematics up to the fifteenth century, *Indian Journal of History of Science*, 4(1–2), pp. 86–98.

Gupta, R.C. (1977). Parameśvara's rule for the circumradius of a cyclic quadrilateral, *Historia Matematica*, 4(1), pp. 67–74.

Gupta, R.C. (1978). Some telugu authors and works on ancient Indian mathematics, *Souvenir of the 44th Conference of the Indian Mathematical Society*, Hyderabad, pp. 25–28.

Gupta, R.C. (1982). Indian mathematics abroad up to the tenth century A.D., *Ganita Bharati*, 4, pp. 10–16.

Gupta, R.C. (1986). On derivation of Bhaskara I's formula for the sine, *Ganita Bharati*, 8(1–4), pp. 39–41.

Gupta, R.C. (1986). India and the ancient world: transmission of scientific ideas, In P.R. Roy and S.N. Sen (eds.), *The Cultural Heritage of India. Volume VI: Science and Technology*, Calcutta, 1986 (reprint 1991), pp. 220–247.

Gupta, R.C. (1989). Sino-Indian interaction and the great Chinese buddhist astronomer-mathematician I-Hsing (A.D. 683–727)." *Ganita Bharati*, 11, pp. 38–49.

Gupta, R.C. (1990). The chronic problem of ancient Indian chronology, *Ganita-Bharati*, 12, 17–26.

Hardy, G.H. (1940). *Ramanujan: Twelve Lectures on Subjects Suggested by His Life and Work*, Cambridgr University Press, Cambridge.

Hayashi, T. (1995). *The Bakhshali Manuscript: An Ancient Indian Mathematical Treatise*, Egbert Forsten, Groningen.

Hayashi, T. (1998). Twenty-one algebraic normal forms of Citrabhanu, *Historia Mathematica*, 25, pp. 1–21.

Hayashi, T., Kusuba, T. and Yano, M. (1989). Indian values for π derived from Aryabhata's value, *Historia Scientiarum*, 37, 1–16.

Henderson, D.W. (2000). Square roots in the Sulba Sutra, In C.A. Gorini (ed.) *Geometry at Work: Papers in Applied Geometry*, MAA Notes Number 53, pp. 39–45.

Hettle, C. (2015). The Symbolic and Mathematical Influence of Diophantus's Arithmetica, *Journal of Humanistic Mathematics*, 5(1), pp. 139–166.

Hoemle, A.F.R. (1888). The Bakhshali manuscript, *Indian Antiquary*, 17, pp. 33–48 and pp. 275–279.

Hoyrup, J. (2010). Old Babylonian 'algebra', and what it teaches us about possible kinds of Mathematics", *Ganita Bharati*, 32(1–2), pp. 1–24.

Hunter, G.R. (1934). The *Script of Harappa and Mohenjodaro and Its Connection with Other Scripts*, Kegan Paul, London.

Iannaccone, I. (1998). *Johann Schreck Terrentius*, Instituto Universitario Orientale, Napoli.

Ifrah, G. (2000). *The Universal History of Numbers: From Prehistory to the invention of the Computer*, Wiley.

Ikeyama, S. and Plofker, K. (2001). The *Tithcitamani* of Ganesa, a medieval Indian treatise on astronomical tables, *SCIAMVS*, 2, pp. 251–289.

Ingalls, D. (1951). *Materials for the Study of Navya-Nyaya Logic*, Harvard University Press, Cambridge.

Jesseph, D.M. (1999). *Squaring the Circle: The War Between Hobbes And Wallis*, University of Chicago Press, Chicago.

Jha, N. and Rajaram, J.S. (2000). *The Deciphered Indus Script*, Aditya Prakashan, New Delhi.

Joseph, G.G. (1994). Different ways of knowing: contrasting styles of argument in Indian and Greek mathematical traditions, In P. Ernest (ed.) *Mathematics, Education and Philosophy: An International Perspective*, The Falmer Press, London, pp. 194–204.

Joseph, G.G. (1995). Cognitive encounters in India during the age of imperialism, *Race and Class*, 36(3), pp. 39–56.

Joseph, G.G. (2009a). A *Passage to Infinity:Medieval Indian Mathematics from Kerala and Its Impact*, Sage, New Delhi.

Joseph, G.G. (2009b). *Kerala Mathematics: Its History and Possible Transmission to Europe*, B.R. Publishing Corporation, New Delhi.

Joseph, G.G. (2011). *The Crest of the Peacock: Non-European Roots of Mathematics*, 3rd Edition, Princeton University Press, Princeton.

Joseph, G.G. (2016). From Zero to Infinity: The Indian Legacy of the Bright Dark Ages, in Bala, A. and Duara, P. (eds.) *The Bright Dark Ages: Comparative and Connective Perspectives*, Brill, Leiden, pp. 186–198.

Jushkevich, A.P. (1964). *Geschichte der Matematik im Mittelater*, German translation, Leipzig.

Katz, V.J. (1992). *A History of Mathematics: An Introduction*, Harper Collins, New York.

Katz, V.J. (1995). Ideas of calculus in Islam and India, *Math Magazine*, 68(3), pp. 163–174.

Kak, S. (1987). The Paninian approach to natural language processing, *International Journal of Approximate Reasoning*, 1, pp. 117–130.

Kak, S.C. (1993). Astronomy of the vedic altars, *Vistas in Astronomy*, 36, pp. 117–140.

Kaye, G.R. (1933). *The Bakhshālī Manuscript: A Study in Mediaeval Mathematics*, Archaeological Survey of India, New Imperial Series, Vol. 43, Parts 1–3, Government of India Central Publication Branch.

Keller, A. (2006). *Expounding the Mathematical Seed*, Vols. 1 and 2, Birkhäuser, Basel.

Kulkarni, R.P. (1971). Geometry as known to the people of Indus civilisation, *Indian Journal of History of Science*, 13, pp. 117–24.

Kusuba, T. (1993). *Combinatorics and Magic Squares in India: A Study of Narayan Pandita's Ganita Kaumudi*, Chapters 13–14, Brown University.

Lach, D.F. (1965). *Asia in the Making of Europe*, University of Chicago Press, Chicago.

Lam Lay Yong (1986). The conceptual origin of our numeral system and the symbolic form of algebra, *Archive for History of Exact Science* 36, pp. 183–199.

Lloyd, G.E. (1990). *Demystifying Mentalities*, Cambridge University Press, Cambridge.

Mackie, E.W. (1977). *The Megalithic Builders*, Phaidon, Oxford.

Madhavan, S. (1991). Origins of Katapayadi System of Numerals, *Sri Rama Varma Samskrta Grantavali Journal*, 18(2), pp. 35–48.

Mahadevan, I. (2002). Aryan or Dravidian or Neither? A Study of Recent Attempts to Decipher the Indus Script (1995–2000), *Electronic Journal of Vedic Studies (EVJS)*, 8(1).

Mainkar, V.B. (1984). Metrology in the Indus Civilisation, in B. Lal and S.P. Gupta (eds.) *Frontiers of the Indus Civilisation*, Motilal Banarsidass, New Delhi.

Mallayya, V.M. (2002). Geometrical Approach to Arithmetic Progressions from Nilakantha's Aryabhatiyabhasya and Sankara's Kriyakramari, proceedings of the *International Seminar and Colloquium on 1500 Years of Aryabhateeyamn*, Kerala Sastra Sahitya Parishad, Kochi, pp. 143–147.

Mallayya, V.M. (2004). An interesting algorithm for computation of sine tables of sine tables from the Golasara of Nilakantha, *Ganita Bharati*, 26, pp. 40–55.

Mallayya, V.M. (2009). *Trigonometric Sines and Sine Table: A Survey*, Preprint, AHRB, Project, University of Manchester.

Mallayya, V.M. (2011). Śankaras Geometrical Approach to Citrabhānu's Ekaviṁśati Praśnottara, In P.V. Arunachalam, C. Umashankar, and V. Ramesh Babu (eds.). *Proceedings of the National Workshop on Ancient Indian Mathematics with Special Reference to Vedic Mathematics and Astronomy*, Rashtriya Sanskrit Vidyapeetha, pp. 99–127.

Maor, E. (1987). *To Infinity and beyond: Cultural History of the Infinity*, Princeton University Press, Princeton.

Matilal, B.K. (1968). *The Navya-Nyaya Doctrine of Negation*, Harvard University Press, Cambridge.

Menninger, K.A. (1969). *Number Words and Number Symbols: A Cultural History of Numbers*, MIT Press, Cambridge, MA.

Menninger, K.A. (1977). *Number Words and Number Symbols: Cultural History of Numbers*, New Edition, MIT Press, Cambridge, MA.

Michels, A. (1978). *Beweisverfahren in der vedischen Sakralgeometrie: Ein Beitrag? ur Ent-stehungsgeschichte von Wissenschaft*, Steiner, Wiesbaden.

Montelle, C. (2013). Book Review of "The Crest of the Peacock" [3rd edition 2011] by George Gheverghese Joseph, Notices of the American Mathematical Society (AMS). December, 1459–1463.

Narayanan, M.G.S. (2014). Social Background of Science in Medieval Kerala, *Kerala School of Mathematics: Trajectories and Impact*. Calicut, pp. 5–20.

Needham, J. (1954). *Science and Civilisation in China*, Vol. 1, Cambridge University Press.

Neugebauer, O. (1952). Tamil astronomy, *Osiris*, 10, pp. 252–276.

Neugebauer, O. (1962). *The Exact Sciences in Antiquity*, Harper, New York.

Palat, R.A. (2002). Is India part of Asia? *Environment and Planning D: Society and Space*, 20(6), pp. 669–691.

Patiganita of Sridharacarya, with an ancient Sanskrit Commentary (1959). edited Kripa Shankar Shukla.

Peacock, G. (1849). Arithmetic — including a history of the science, In *Encylcopedia Metropolitana or Universal Dictionary of Knowledge*, Part 6 First Division, J J Griffin and Co., London.

Pingree, D. (1981). History of mathematical astronomy in India, In C. C. Gillespie (ed.). *Dictionary of Scientific Biography*, Vol. 15 (Suppl.), Charles Scribner's Sons, New York.

Playfair, J.J. (1789). Remarks on the astronomy of the brahmins, *Transactions of the Royal Society of Edinburgh*, 2, Part 1, pp. 135–192.

Plofker, K. (2009). *Mathematics in India*, Princeton University Press, Princeton.

Potache, D. (1989). The commercial relations between Basrah and Goa in the sixteenth century, *STUDIA*, Lisbon, 48.

Puttusamy, T. (2012). *Mathematical Achievements of Pre-modern Indian Mathematicians*, Elsevier, London.

Rajagopal, C.T. and Marar, M. (1944). On the Hindu quadrature of the circle, *Journal of the Royal Asiatic Society (Bombay Branch)*, 20, pp. 65–82.

Rajagopal, C.T. and Venkataraman, A. (1949). The sine and cosine power series in Hindu mathematics, *Journal of the Royal Asiatic Society of Bengal*, 15, pp. 1–13.

Rajagopal, C.T. and Aiyar, T.V.V. (1951). On the Hindu Proof of Gregory's Series, *Scripta Mathematica*, 17, pp. 65–74.

Rajagopal, C.T. and Vedamurthi, T.V. (1952). On the Hindu proof of Gregory's series, *Scripta Mathematica*, 18, pp. 65–74.

Rajagopal, C.T. and Rangachari, M.S. (1978). On an untapped source of medieval Keralese mathematics, *Archive for the History of Exact Sciences*, 18, pp. 89–102.

Rajagopal, C.T. and Rangachari, M.S. (1986). On Medieval Keralese Mathematics, *Archive for the History of Exact Sciences*, 35, pp. 91–99.

Raina, D. and Habib, S.I. (1989). The introduction of scientific rationality into India: a study of Master Ramchandra — Urdu journalist, mathematician and educationalist, *Annals of Science*, 6, pp. 597–610.

Raina, D. and Habib, S.I. (1990). Ramchandra's 'treatise through 'the haze of the golden sunset': An aborted pedagogy, *Historia Mathematica*, 1992, pp. 371–384.

Ramasubramanian, K., Srinivas, M.D. and Sriram, M.S. (1994). Modification of the earlier Indian planetary theory by the Kerala astronomers (*c.* 1500 AD) and the implied heliocentric picture of planetary motion, *Current Science*, 66, pp. 784–790.

Rao, T.R.N. and Kak, S. (1998). *Computing Science in Ancient India*, Center for Advanced Studies, University of Southwestern Louisiana, Lafayette, LA.

Reich, E. (2000). Ein Brief des Severus Sebokt, In M. Folkerts and R. Lorch (eds.), *Sic ad Astra: Studien zur Geschichte der Mathematik und Naturwissenschaften: Festschrift für den Arabisten Paul Kunitzsch zum 70.* Geburtstag, Weisbaden, 2000, pp. 478–489.

Rizvi, S.A.H. (1983). On trisection of an angle leading to the derivation of a cubic equation and computation of value of sine, *Indian Journal of History of Science*, 19(1), pp. 77–85.

Rosenfeld, B.A. (1990). Al-Khwārizmī and Indian Science, In W.H. Abdi et al. (ed.) *Interaction between Indian and Central Asian Science and Technology in Medieval Times*, Indian National Science Academy, New Delhi, Vol. I, pp. 132–139.

Sachau, E. (ed.) (1888) *Kitab fi tahqiqi ma li'l Hind* (Book on India): *Aberuni's India: An Account of the Religion, Philosophy, Literature, Geography, Chronology, Astronomy, Customs, Laws and Astrology of India about A.D.1030.* Trübner's oriental series, London.

Salem, S.I. and Kumar, A. (1996). Said al-Andalusi *Tabaqat al-Uman* (*Book of the Categories of Nation*), University of Texas Press, Austin.

Sarukkai, S. (2008). *Indian Philosophy and Philosophy of Science*, 2nd Edition, Centre for Studies in Civilizations, New Delhi.

Sarasvati, T.A. (1961/1962). The mathematics in the first four Mahad-hikara in the Trilokaprajnpari, *Journal of Ganganath Research Institute*, 18, pp. 27–51.

Sarasvati, T.A. (1963). Development of mathematical series in India, *Bulletin of the National Institute of Sciences (India)*, 21, pp. 320–343.

Sarasvati, T.A. (1979). *Geometry in Ancient and Medieval India*. Motilal Banarsidass, Delhi.

Sarma, K.V. (1972). *A History of the Kerala School of Hindu Astronomy*, Vishveshvaranad Instiute, Hoshiarpur.

Sarma, K.V. (2008). *Ganita-Yukti-Bhasa of Jyesthadeva*, Epilogue (by M.D. Srinivas), Vol. I, Hindustan Book Agency, Delhi.

Sarma, S.R. (2002). Rule of Three and its Variations in India, In Y. Dold-Samplonius *et al.* (eds.) *From China to Paris: 2000 Years Transmission of Mathematical Ideas*. Stuttgart, pp. 133–156.

Sarma, S.R. (2011). Mathematical Literature in Regional Languages of India, in *Ancient Indian Leaps into Mathematics*, (eds.) B.S. Yadav and Man Mohan, Springer, New York.

Sarma, U.K., Bhat, V., Pai, V. and Ramasubramanian, K. (2010). The discovery of madhava series by whish: an episode in the historiography of science, *Ganita Bharati*, 32(1–2), pp. 115–126.

Sastri, K. (1985). *Vedanga Jyotisa of Lagadha*, Cr. Edition by K.V. Sarma, INSA, New Edition .

Scott, J.F. (1981). *The Mathematical work of John Wallis* (1616–1703), Taylor & Francis, London (1938); Chelsea, New York.

Sedillot, L.A. (1875). The Great Autumnal execution, *Bulletin of the Bibliography and History of Mathematical & Physical Sciences* published by B. Boncompagni, Reprinted in *Sources of Science*, No. 10 (1964), New York and London.

Seidenberg, A. (1962). The ritual origins of geometry, *Archives for History of Exact Sciences*, 1, pp. 488–522.

Seidenberg, A. (1978). The origin of mathematics, *Archives of the History of Exact Sciences*, 18(4), pp. 301–342.

Seidenberg, A. (1983). The Geometry of Vedic Rituals, in F. Staal (ed.). *Agni, the Vedic Ritual o the Fire Altar*, Asian Humanities Press, Berkeley, pp. 95–126.

Selenius, C. (1975). Rationale of the Chakravala process of Jayadeva and Bhāskara II, *Historia Mathematica*, 2, 167–184.

Sen, S.N. (1962). Transmission of scientific ideas between India and foreign countries in ancient and medieval times, *Bulletin of the National Institute of Sciences of India*, 21, 8–30.

Sen, S.N. (1970). Influence of Indian science on other culture areas, *Indian Journal of History of Science*, 5, pp. 332–346.

Sen, S.N. and Bag, A.K. (1983). *The Sulbasutras*, Indian National Science Academy, New Delhi.

Senthil Babu, D. (2007). Memory and mathematics in the Tamil tinnai schools of South India in the eighteenth and nineteenth centuries, *International Journal for the History of Mathematics Education*, 2(1), pp. 15–37.

Shukla, K.S. (1963). (Ed. and Tr.). *Laghu Bhaskariya* , II.8, Department of . Mathematics and Astronomy, Lucknow University, Lucknow.

Shukla, K.S. (1976). *Aryabhatiya of Aryabhata, with the Commentary of Bhaskara I and Somesvara*, Indian National Science Academy, New Delhi.

Sihag, B.S. (2014). *Koutilya: The True Founder of Economics*, Vitasta Publications, New Delhi.

Smith, D.E. (1923/1925). *History of Mathematics*, 2 Volumes, Ginn & Co., Boston, MA; Reprinted by Dover, New York, 1958.

Sondheim, E. and Rogerson, A. (1981). *Numbers and Infinity*, Cambridge University Press, Cambridge.

Stewart, I. (1981). *Concepts of Modern Mathematics*, Penguin, London.

Srinivasiengar, C.N. (1967). *The History of Ancient Indian Mathematics*, World Press, Calcutta.

Sridharan, R. and Srinivas, M.D. (2011). Study of magic squares in India, In R. Sujatha *et al.*, *Mathematics Unlimited: Essays in Mathematics*, Taylor & France, London, pp. 383–391.

Staal, J.F. (1965). Euclid and Panini, *Philosophy East and West*, 15, pp. 99–116.

Staal, J.F. (1978). The ignorant brahmin of the Agnicayana, *Annals of the Bhandarkar Oriental Research Institute*, Diamond Jubilee Number, Pune, pp. 337–348.

Staal, J.F. (1988). *Universals, Studies in Indian Logic and Linguistics*, University of Chicago Press, Chicago.

Staal, J.F. (1995). The Sanskrit of science, *Journal of Indian Philosophy*, 23, pp. 73–127.

Staal, J.F. (1999). Greek and vedic geometry, *Journal of Indian Philosophy*, 27, pp. 105–127.

Staal, J.F. (2001). Squares and oblongs in the vedas, *Journal of Indian Philosophy*, 29, pp. 256–272.

Staal, F., Somayajipad, C.V. and Itti Ravi Nambudiri, M. (1983). *AGNI. The Vedic Ritual of the Fire Altar*, I–II, Asian Humanities Press.

Subbarayappa, B.V. and Sarma, K.V. (1985). *Indian Astronomy: A Source-Book*, Nehru Centre Publications, Bombay.

Subbarayappa, V. (1993). *Numerical System of the Indus Valley Civilisation. A Monograph on the Indus Script*, Indian Council of World Culture, Bangalore.

Tacchi Venturi (1913). *Matteo Ricci S.I., Le Lettre Dalla Cina* 1580–1610, Vol. 2, Macerata, 1613.

Tavernier, J.-B. (1889). *Travels in India*, Vol. 2, English Translation, Cambridge University Press, Cambridge.

Thibaut, G. (1875). On the Sulba-sutra, *Journal of the Asiatic Society of Bengal*, 44, pp. 7–75.

Van der Waerden, B.L. (1961). *Science Awakening*, Oxford University Press, New York.

Van der Waerden, B.L. (1976). Pells equation in Greek and Hindu mathematics, *Russian Math Surveys*, 31, pp. 210–222.

Van der Waerden, B.L. (1983). *Geometry and Algebra in Ancient Civilizations*, Springer-Verlag Berlin.

Warren, J. (1825). *Kalasankalita: A Collection of Memoirs on the Various Modes According to which the Nations of the Southern Parts of India Divide Time*, Madras.

Whish, C.M. (1835). On the Hindu quadrature of the circle and the infinite series of the proportion of the circumference to the diameter exhibited in the four Shastras, the *Tantrasamgraham, Yukti-Bhasa, Carana Padhati*, and *Sadratnamala. Tr. Royal Asiatic Society of Gr. Britain and Ireland*, 3, pp. 509–523.

Wicki, J. (1948). *Documenta Indica*, Vol. III, p. 307.

Witzel, M. and Farmer, S. (2000). Horseplay in Harappa: The Indus Valley Dicipherment Hoax, *Frontline*.

Name Index

Subject Index

A

Agnicayana (Vedic altar), 54, 64–65, 71, 73

Agnicayana alter, construction of an, 54, 56, 60

Algebra of Prthudakasvami: Solution of Equations, 225

Altar (Vedic) constructions: rituals or technology, 64

Ananta and, asmkhyata: Jain infinities, 126

Ancient Greece, India and China: Connections, 69, 72

Angula (basic unit of measure), 57, 59, 62

Antecedents of Pascal Triangle, 132

Apastamba *Sulbasutra*, 6, 59–60, 62, 72, 74, 78–79, 84, 96

Appearance of social arithmetic in Kautilya's *Arthashastra*, 99

Appearance of symbolism: Pingala, 99

Approach based on infinite series and integrals, 347

Approximations and accurate results in geometry, 237

Arc tan series, xvii, 331, 342, 375

Area bounded by circles, 239–240

Area of a *Vakrapaksa-Syena* (falcon altar), 75

Area of triangle, 88,171, 208, 274

B

Argument against a 'Greek' miracle, 67

Arithmetic of Calculation, 201

Arithmetic, 155, 182
operations and the eight procedures, 201–202

Arithmetic operations with positive and negative numbers and zero (*sunyaganita*), *see* sunyaganita

Aryabhata
Aryabhatiya, 157, 164
Aryabhatan numerals, 167–168
Computation of the first sine table, 400

Astronomy: the principal driving force of Indian mathematics, xviii, 156–158

B

Babylonian approach to calculating square root of 2, 94
Astronomy, 157

Bakhshali Manuscript, 23, 105, 145–147, 151, 153, 155, 158–160, 185, 287, 294, 428

Baudhyana*Sulbasutra*, 72, 75, 80, 92

Bhaskara I's three sine tables 45, 401

Bijaganita, 200

Bounded and unbounded infinities, 126

Printed in the United States
By Bookmasters